EARTHQUAKES

EARTHQUAKES

Editor

Hazan Burakgazi

scitus
academics

Earthquakes

Edited by **Hazan Burakgazi**

This Book is Printed on Acid-Free Paper. ♾

ISBN: 9781681178813

Library of Congress Control Number: 2017948908

SCITUS Academics LLC,

616, Corporate Way, Suite 2, 4766,

Valley Cottage, NY 10989

www.scitusacademics.com

Notice

Printed in United States of America.

Preface

Earthquakes are a naturally caustic effect of our earth's continually changing surface, and thousands of them happen day by day. Earthquakes, also called temblors, can be so tremendously destructive; it's hard to imagine they occur by the thousands every day around the world, usually in the form of small tremors. Some 80 percent of all the planet's earthquakes occur along the rim of the Pacific Ocean, called the "Ring of Fire" because of the preponderance of volcanic activity there as well. Most earthquakes occur at fault zones, where tectonic plates—giant rock slabs that make up the Earth's upper layer—collide or slide against each other. On average, a magnitude 8 quake strikes somewhere every year and some 10,000 people die in earthquakes annually. The onset of the 21st century has brought a new public awareness of natural hazards. Recent catastrophic events like the 2004 tsunami in Asia or the 2005 flooding of New Orleans have made it not only to the headlines in news publications around the globe but have also contributed to a more profound desire to accumulate knowledge about natural hazards in general among people all over the world. Earthquake research belongs to the most fascinating topics in the field of natural hazard research. Hardly any other hazard claims more lives, destroys more values and can lead to catastrophic after- effects (as can be seen when looking at the 2011 earthquake with resulting tsunami and nuclear disaster in Japan). In most cases scientists are not able to forecast when and where an earthquake may take place, but only the approximate region and the probability, not the precise date and the magnitude. Geoscientists therefore by and large concentrate on assessing and mapping regions that experienced earthquakes in the past. This work is devoted to different aspects of earthquake research as they can cause significant losses and deaths, it is really important to understand the process and the physics of this phenomenon. It does not focus on a unique problem in earthquake processes, but spans studies on historical earthquakes and seismology in diverse tectonic environments, to more applied studies on earthquake geology.

This book 'Earthquakes' is committed to a variety of aspects of earthquake research and analysis, from theoretical advances to practical applications. This book is devoted to various aspects of earthquake researches, especially to new achievements in seismicity that engages geosciences, assessment, and mitigation. Chapters contain advanced materials of detailed engineering investigations, which can help more clearly appreciate, predict, and manage different earthquake processes. The specific aim of the book is to encourage debate and future research to advance hazard assessments, spreading of earthquake engineering data and, eventually, the seismic provisions of building codes. It provides the state of the art on recent progress in earthquake engineering and management.

Table of Contents

CHAPTER 1

Earthquake Disaster Management Approach: The Case of Al-Ais, Medina Area in Saudi Arabia

Said Ali El-Quliti[1], Tawfiq Bin Saeed Al-Harbi[2], Mahdi Bin Salem Al-Yami[3], Ahmed Bin Matar Al-Ghamdi[4], Mohammed Bin Mattar Al-Shammari[5]

[1]*Department of Industrial Engineering, King Abdulaziz University, Jeddah, KSA*

[2]*Department of Geophysics at the National Center for Earthquakes and Volcanoes, Saudi Geological Survey, Jeddah, KSA*

[3]*Natural Studies and Natural Consulting Company, Jeddah, KSA*

[4]*Fire and Rescue Division of the General Directorate of Civil Defense in Jeddah, Jeddah, KSA*

[5]*Department of Fire and Rescue Services, Civil Aviation Authority Province of Hail, Jeddah, KSA*

ABSTRACT

The research tackles earthquakes as one of the most dangerous natural disasters. It defines disasters in general, which in turn include earthquakes, how to manage them, stages of their evolution, and their classification among other kinds of disasters such as hurricanes, floods, drought, desertification, etc. Afterwards, it tackles the impacts of disasters on man, buildings and infrastructure. It defines also the codes and laws existing in the Kingdom of Saudi Arabia to confront and treat the impacts of earthquakes in their different stages (before, during and after) and the authorities involved in managing them. Then, it moves to study the case of Al-Ais Earthquake, and identifies the points of deficiency in dealing with such a disaster. Moreover, the research tackles some global experiences in how to handle the situation, reaching to the presentation of a conceptual approach to confront disasters of earthquakes in their various stages.

Keywords: Earthquake, Disaster Management, Al-Ais, Saudi Arabia

1. INTRODUCTION

Allah made natural disasters, such as earthquakes, volcanoes and others act as a test for man and a reminder for him about the creator's capabilities compared to the limited liabilities of man. Because of natural disasters, the mainland came into existence with its continents and brought out their treasures, and because of them the life ends on earth. Man has tried frequently to explain and expect the various natural phenomena and disasters. In this respect, man took steps, as he still thinks and innovates ways that would alleviate their strike and their risks, and even get benefit of such disasters since man realized that it is impossible to prevent their occurrence. Therefore, man took pains to reduce the losses resulting therefrom, either concrete or abstract. Those attempts have contributed in several times to minimizing the size of human and concrete losses.

Natural disasters alone have caused, in the two recent decades of the last century, the death of about three million persons worldwide. In addition, they have negatively impacted on the life of not less than 800 million persons injured and vagrant, and concrete losses amounting to 629 billion dollars in nineteenths only [1] .

Earthquakes have been amounted globally in the period from 1975 to 2001 a number of 565 earthquakes, which represents 5.4% of the number of disasters in the world during this period. Since earthquakes represent the natural disasters that are most frequent and most resulting to losses, this study, which is about the development of mechanisms for managing the disasters resulting from earthquakes in Saudi Arabia and its neighboring countries, a number of which including Egypt, Jordan, Turkey and Iran could be exposed to frequent earthquakes, has been made.

1.1. Purpose of the Research

Lack of plans and curricula developed to treat and manage, especially repeated ones such as earthquakes in Saudi Arabia disasters, which exacerbate and increase the impact and damage of these disasters, lack of control and the occurrence of a lot of losses as a result of the impact of indirect resulting from the inefficiency of disaster management.

1.2. Goal of the Research

The research aims to develop a plan and a scientific approach to face the disasters resulting from earthquakes in Saudi Arabia, highlight the importance of the role of architectural plans in the decision-making process, and develop relevant strategies to limit the damages and losses resulting from the disasters of earthquakes. This is through studying and analyzing some of the local and international experiences, in order to reduce the material and human losses caused by earthquakes.

1.3. Methodology of the Search

The approach used in this research consists of three stages as follows:

- The first phase deals with the general definitions and concepts for the disaster, and disaster management and earthquakes, features and characteristics of these disasters.
- The second phase shall provide samples for disaster management in the Kingdom of Saudi Arabia and the developed world.
- The third phase identifies proposed approach to disaster management from earthquakes in three stages, before, during and after the disaster, through the assessment of past experiences.

1.4. Concepts and Definition of a Disaster

The concepts of the various elements of a disaster and the full definition of the problem, which is the subject matter of the study, are brought up as they play a key role in the disaster management. These definitions are reviewed as follows:

1.4.1. Disaster

It is an incident determined temporally and spatially, and making an entire community or a part of one be exposed to severe material risks and losses that could be caused to the individuals of such community, and which may affects the social structure by confusing its life and suspending the provision of the requirements essential for its continuity [2] . While a disaster is defined by R. Green as it is the sudden change having intense impacts, which takes place because of changes related in forces that lead to the collapse of the balance [3] .

The other definition is a system for confronting the emergency situations, and planning to deal with cases that cannot be avoided, when or before they occur, in order to control the results.

There is another concept defining the crises management as it is a special administrative process that would produce a strategic response to crisis situations through a set of pre-selected and specially-trained administrators in order to reduce losses to the minimum.

While the international definition provided by the International Civil Defense Organization stipulates that "it is a big incident resulting in huge losses of life and property, which may be natural disasters made by nature (floods-earthquakes-accidents- etc...), and may be technical disasters made by man, either voluntarily (intentionally) or involuntarily (negligence)". This definition also explains that a disaster, which confrontation requires the aid of the home or at the international level if its confrontation surpasses the capabilities of the homeland [4] .

1.4.2. Disaster Management

The disaster management can be defined as all of the means, procedures and activities carried out in the three stages of a disaster (before, during and after) (Figure 1), which aims to achieve the following:

- To prevent a disaster whenever possible.
- To face a disaster efficiently and effectively.
- To reduce losses to the minimum.
- To minimize the negative effects on the surrounding environment.
- To remove the psychological effects of the disaster on citizens.
- To analyze and get ready for incidents in the three stages of a disaster, and to improve and develop the self-capacities for confronting disasters.

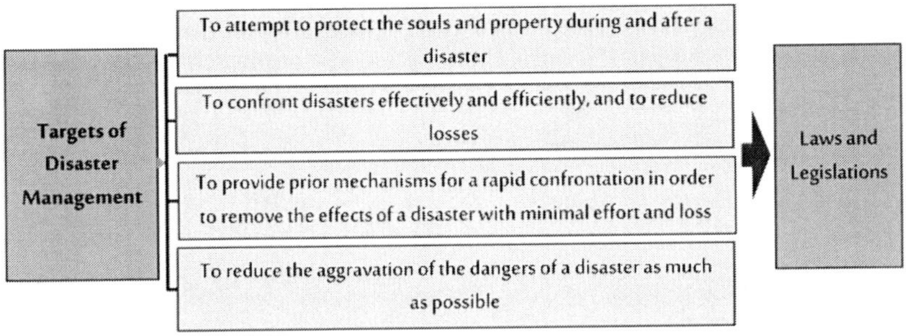

Figure 1. Targets of disaster management.

1.4.3. Disaster Alert

It alerts inhabitants by all available means to ensure that most of individuals are informed with the existence or removal of the various sources of disasters, in order to take the necessary measures to protect individuals and property.

1.4.4. Disaster Relief

It is defined as the provision of food, clothing, shelter, treatment, and all the requirements for those affected by disasters [5] .

2. CHARACTERISTICS, STAGES AND FEATURES OF DISASTERS

2.1. Characteristics of Disasters

Allah Barton has developed a table classifying the characteristics of disasters and has chosen to them five dimensions [6] , which are:

1) Scope of Impact:

It represents the scope or the circle containing those affected by a disaster, and whenever the geographic scope is less, the less the number of people affected and the size of the disaster will be reduced.

2) Surprise:

Whenever an incident is unexpected temporally and spatially, its impact will be greater and heavier, while the impact of preparations will be less, and vice versa.

3) Center of Disaster:

It constitutes the center of disaster or its start point, and has a great role in influencing, as the larger the regions have a population density and urban density, the greater the impacts and losses are, and vice versa.

4) Duration of Impact:

It illustrates the time period since the beginning of a disaster until the impact breakpoint. It also plays a key role in the size of the losses, as the more the time of the disaster increases, the more their devastating impact increases. A disaster may take seconds such as earthquakes, minutes such as storms, or hours such as floods or more.

5) Impacts resulted from Disaster:

It reflects the impacts of a disaster (that are resulted therefrom), as they may be more devastating than the disaster itself, such as the spread of epidemics and diseases, the suspension of services and others.

2.2. Features of Disaster

They are the criteria and indicators, by which the impact of disaster can be measured, which are:

1) Repetition Rate:

It expresses how often a disaster takes place during twenty years, and the more the repetition rate increases, the more its impact increases, and the more the importance of the exact planning appears to confront it.

2) Impact Extent:

It reflects the size of the losses and injuries of lives and property, and the costs required to return situation to its state before the disaster. There is a direct correlation among the scientific capability of expectation, the increase of efficiency of the disaster management, and the control of such disaster.

2.3. Disaster Management Stages

Disaster management stages can be divided into three stages, each of which has a great importance in reducing the impact resulted therefrom. Figure 2 illustrates disaster management stages, which are:

1) Pre-Disaster Stage:

It is concerned with the information collection, the definition of disaster, its characteristics and location, the probability of its occurrence, the method of preparation and training, and the development of plans and alternative plans to encounter it.

2) During-Disaster Stage (Period of Disaster Occurrence):

In this stage, it is dealt with the disaster according to the preparations and training made in the previous stage. It is also dealt with new developments, and proposed plans are applied.

3) Post-Disaster Stage:

It is the stage in which the disaster-confronting plan is completed, the impacts resulted there from are contained and treated, the performance during this disaster is evaluated, plans are developed, and new developments caused thereby, new technology developments and the experiences of others are added.

3. EARTHQUAKES AS A SOURCE OF DISASTERS IN SAUDI ARABIA AND ARAB COUNTRIES

3.1. Definition of Earthquake

An earthquake is a natural phenomenon occurring suddenly and rapidly as a result of the outburst of the energy concealed in rocks, and the spread of seismic waves in those rocks.

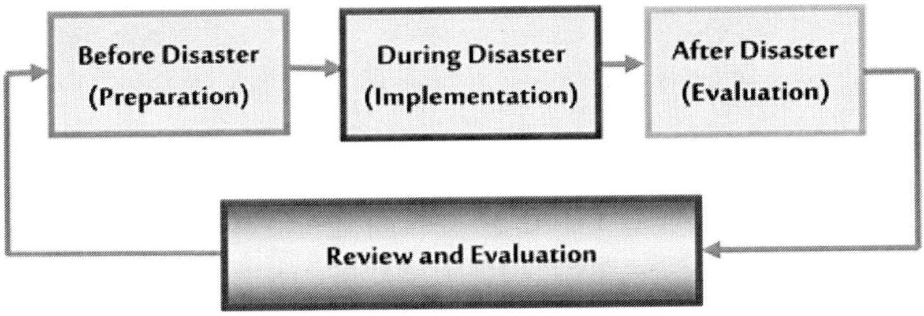

Figure 2. Disaster management stages applied to all disasters.

There is another definition identifying an earthquake as a tremor taking place suddenly due to underground disorders, either in the crust itself or in the formation underlying the crust, and such tremor lasts for few seconds.

In another definition, it is a sudden outburst of the energy hidden in the subsoil over time in the form of kinetic energy carried by seismic waves upon their occurrence, which are spread in all directions from the center of earthquake [7] .

3.2. Causes of Earthquake

Earthquake takes place due to the Elastic Rebound Theory, which is a result of the continuous accumulation and storage of energy along a weak line in the crust of the earth. Thus, intense pressure are generated, which are hard for rocks to resist, leading to the breakdown of rocks along the fault into two parts, each of which recovers its original shape through the elastic rebound as illustrated in Figure 3. Then, the energy erupts in the form of seismic waves and tremors. Earthquakes may take place because of man as a result of the construction of dams and artificial lakes, or nuclear explosions.

Classification of Earthquakes According to Their Causes

- Earthquakes resulted from the relative movement of tectonic plates.
- Earthquakes caused by the faults resulted from volcanoes.
- Earthquakes caused by the collapse of the ceiling of caves and giant caves.
- Earthquakes resulted from the human activity (explosions, filling and emptying dams, nuclear explosions).
- Industrial lakes-(injecting liquids into some oil exploration and extraction).

There are a large number of earthquakes that occur in the Arab region in general and Saudi Arabia in particular, because of the tectonic plates, as shown in Figure 4. However, there are also other causes of earthquakes, and in general the stock of earthquakes and their Pictures around the world show that the earthquake may occur in any area, as in Figure 5, and at the confluence of tectonic plates also.

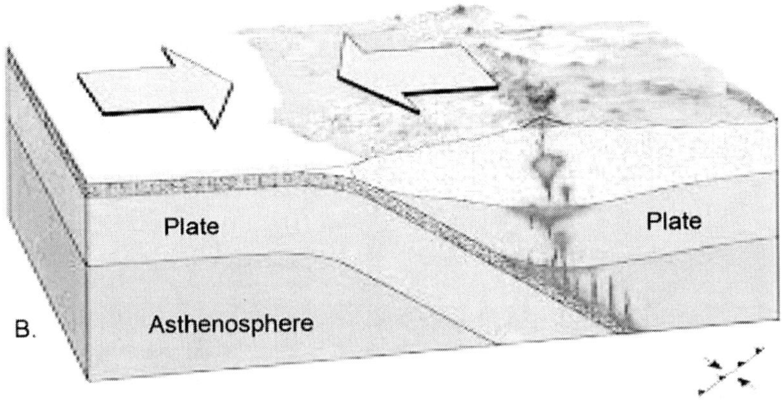

Figure 3. Various forms of movement.

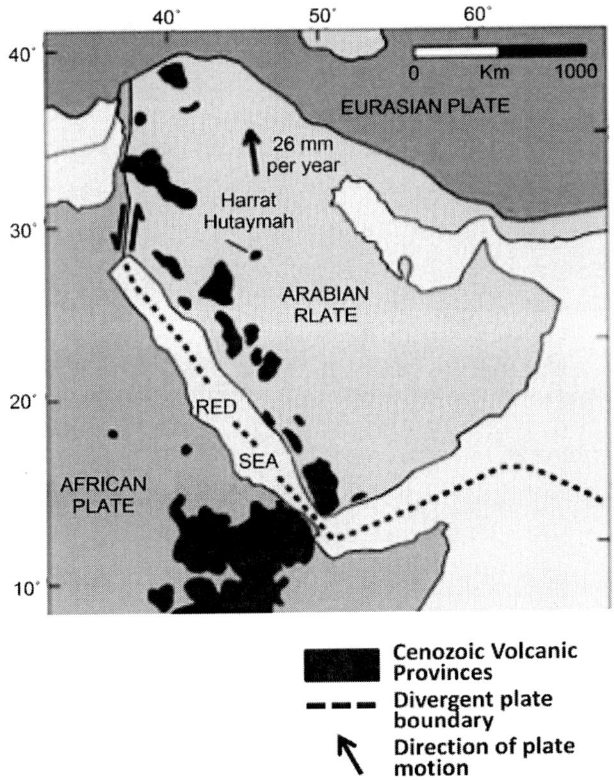

Figure 4. Earthquakes that occur in the Arab region and Saudi Arabia.

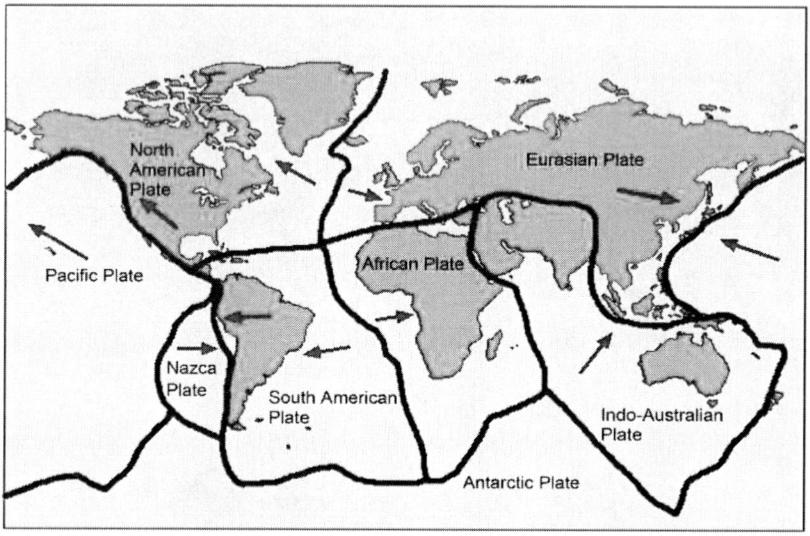

Figure 5. Tectonic plates around the world.

3.3. Seismic Standards

There are two major factors to classify earthquakes:

1) The Modified Mercalli Intensity Scale:

It is one of the scales depending on the knowledge of the damage caused by the earthquake. The Modified Mercalli Intensity Scale is divided into 12 classifications, Table 1. It is obvious that it is an inaccurate scale, and various elements may interfere into its impacts. Its grade is reviewed as follows [8] .

2) Richter Scale:

It explains the size of an earthquakes, and it is an open scale, the values of which are determined on the basis of the decimal logarithm of the great amplitude measured as a part per thousand of a millimeter at a distance of 100 km from the epicenter of the earthquake, and its values are amounted to 9 Richter as stated in Table 2 [9] .

3.4. Different Effects of Earthquakes

The earthquakes shall have effects on all districts and the components of the earth as follows:

3.4.1. Earthquake Effects on Human Lives

Earthquakes shall be deemed as one of the most natural disasters that claimed the lives of a large number of victims and inured on the earth as the number of victims in the past two decades have been three million killed people and 550 injured while in the Arab region, the number of victims in the 1992 Cairo earthquake has been only 561 killed and 1500 injured approximately and the number of killed by the earthquake of Boumerdes Algeria 2003 has been 2266 killed and 10,261 injured. The 1980 Chlef Algeria earthquake has led to the death of 5000 persons and the Agadir Morocco earthquake has led to the death of 12 thousand out of 50 thousand inhabitants in the city. In the worldwide the victims of one earthquake in china in 1976 has been 655 thousand killed and has been 50 thousand killed in the 1990 Iran earthquake in addition to the severe psychological effects and the horror incurred by millions of people during and after the earthquake for a long time [10] .

3.4.2. Economic Effects of Earthquakes

The resulting losses from the earthquakes shall be amounted by billions in addition to the destruction of large proportion of the infrastructure according to the severity and duration of the earthquake.

Table 1. Classification of earthquakes.

Intensity	Description
1	None feels he earthquake except a few number of people in certain conditions.
2	It is felt by a few people at rest, and light hanging things oscillate.
3	It is felt by people inside their houses, and parked cars shake.
4	Sleepers wake up, parked cars clash, and windows collide.
5	Dishes crash and pendulum wall clock stop.
6	It is felt by all people, who become panicked, and furniture moves.
7	House dwellers rush to outside, and it is felt by people in moving cars.
8	Poorly-constructed houses totter, and water change its passes in basins.
9	Well-constructed buildings falter, the earth is cracked, and underground water pipes smash.
10	Water comes out of riverbeds, and only a few buildings remain.
11	The underground water explodes, and the earth widely cracks.
12	Mass destruction happens, natural landmarks change, and things fly in the air.

Table 2. Richter scale grades impacts and horizontal velocity.

Earthquake Effects	Horizontal Velocity Cm/sec.	Grades by Richter Scale
Nobody feels the earthquake except some people under very special circumstances.	-	3
It is felt only by some people, and the things that are freely hanging tremble.	2 - 3	3.5
It is clearly felt in houses as it causes shakings such as the ones caused by the metro.	4 - 7	4
It is felt as if a truck collides with a building, and vehicles shake obviously.	7 - 15	4.5
Wall portraits fall, unbalanced things are turned over, and the pendulum of watches stops.	15 - 30	5
It is felt by everyone, heavy pieces of furniture move, and chimneys and weak buildings crack.	30 - 70	5.5
Cracking is remarkable in poor-concrete buildings and loadbearing walls.	70 - 150	6.0
Cracking is simple in earthquake-resistant installations. There is a big crack in buildings having loadbearing walls. Buildings separate from columns and beams, and some of them move out. Statues, minarets, chimneys and fences fall. Heavy pieces of furniture are turned over, and the level of well water changes.	150 - 300	6.5
Cracking is obvious in earthquake resistant installations, concrete installations. In addition, some buildings separate from their bases, and the soil splits clearly.	300 - 700	7
Most buildings having loadbearing walls are destroyed, and some concrete installations and their bases are ruined. The earth is split apart into big and serious cracks, and railways are bent. Sides of rivers and canals slide, and the sand soil and the silt move.	700 - 1500	7.5
Most buildings having loadbearing walls collapse, and concrete buildings are destroyed. Bridges fall, making in the earth breaks swallowing people and vehicles. Water and sewage networks collapse altogether.	1500 - 3000	8
The destruction is mass, as seismic waves can be seen on the ground obviously, and heavy objects fly in the air.	300 - 600	8.5

3.4.3. Earthquake Effects on the Buildings

The buildings shall represent the most effected element by earthquakes as it shall start with the appearance of superficial simple cracks and severe cracks though either structural building which is less affected by load-bearing walls. The damage shall reach up to the full depreciation of buildings in the region, located in the center of the earthquake.

4. ANALYTICAL STUDY OF THE EXPERIENCES OF DISASTERS MANAGEMENT (EARTHQUAKES)

4.1. Al-Ais Earthquake (May 2009)

Al-Ais (Figure 6) governorate shall be located in the northwestern part of Yanbu which space is amounted to 2863 km^2, representing 15.86% of the space of Yanbu. It is located in the northeast of Yanbu and 150 km from trading between Quraysh to Syria-ancient- and called in the way of Ma'rib Petraa. It is bounded by the city of Amlag with about 120 km and from the east with Medina at the distance of 240 km and from the north with approximately 200 km. such geographical location shall be distinguished for being located at the crossroads of ancient and modern groups and rising from the sea service with approximately 1200 m. the climate shall be hot in the summer and cold in the winter.

The number of population shall be 12,812 people. It shall be ranked at the second level in terms of the number of populations at the level of Yanbu and shall be surrounded by mountains from the most destinations. It shall be intertwined in the mountains with the steep slopes with a number of valleys that form an important tributary of the valley of (Al-Ais) and represent vast expanse of planted plains with wheat upon raining.

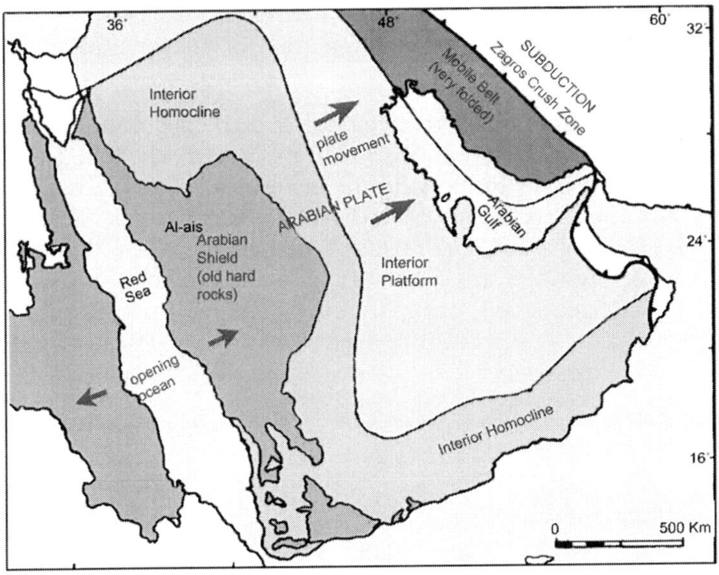

Figure 6. Map of Saudi Arabia—the Arabian Shield and Ais area.

Many of the mountains that form Al-Ais: Abu Elketta, Tora, Radwh, Alagrad, Hisham, Al Sala, Al Mokannaa Alracib, Al Habshi and these mountains are permeated by several valleys, including Toraa, Arafa, Al Ran, Hegag, Allahyan, Al Haver, Al Safeha, Baya, Al Ramas valley and otherwise.

A water tank for the area of Al Harra shall be formed and the water tank of the region has been hit. It has the length of 60 km approximately from the east to the west and approximately 25 km from the north to the south. It has been planted after the heavy rain in the winter and there are also known for their dormant volcanoes as in Halian [11] .

Al-Ais City is located within the active tectonic environment whereas it has been witnessed by large earthquakes in historic and useful periods. Recently, in May 2009, Al-Ais city has unveiled a series of earthquakes of 5.7 degrees on the Richter scale due to the volcanic activity in Heart Al Shaka (Lunayyir) as shown in Figure 7 and Figure 8 as more than 3000 earthquakes have been occurred during the month including 30 ones that reached to the degree of 3.7 on the Richter scale while the citizens have viewed only 15 shake thereof.

The authorities have begun the forced evictions on May 19, 2009 for the people around 20 km around Al Ais Volcano Center and the area is free of the populations till the date of June 27, 2009 for an indefinite period due to the continuity of the earthquake and at the rate of 4-5 daily shakes which power ranges between 4-3-5 degrees on the Richter scale. The nick name shall be deemed as located within the existing areas of the earthquake belt in the region [12] .

Figure 9 indicates the status of Al Ais province and the surrounding area thereto for the map of the Kingdom of Saudi Arabia and schedule 1 shall indicate the significant places in the volcanic eruptions (lanes) in the nick name province and the surrounding area thereto Nos. 3, 4 and the cracks shall be appeared in the ground as a result of the earthquakes in Lunayyir (Al Shaqa) explosion and other nearby places to Al-Ais Province. The figure represents the most important destinations of the volcanic eruptions (lanes) in Nick name and the surrounding area thereto [13] .

Figure 7. Cracks in the ground as a result of earthquakes in Lunayyir (Al-Sha- qa) eruption.

Figure 8. Cracks in the ground as a result of earthquakes in Alais area.

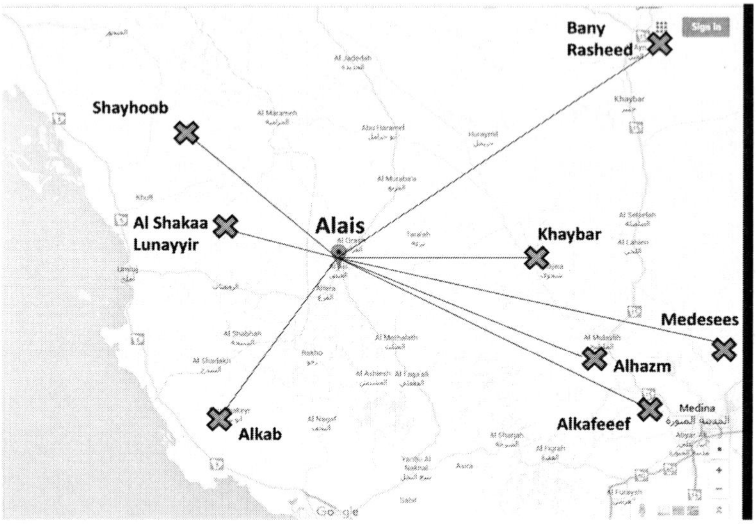

Figure 9. The most important volcanic eruption places (Harrat) in Alais and surrounding area.

4.2. Al-Ais Earthquake Disaster Management Process in 2009

Use of regulation in the status of the event as in Figure 10.

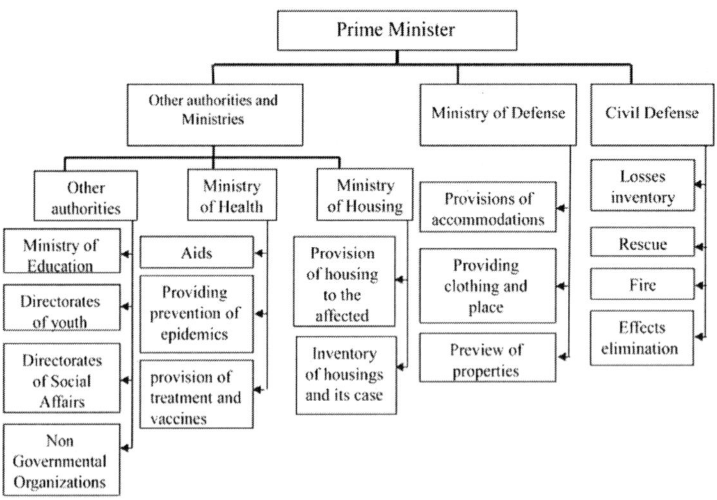

Figure 10. Management plan and authorities in Al-Ais earthquake in 2009.

Due to the lack of unwillingness and the surprise of the event, the event management has been less than the required level as if the popular participation has not the effective and entrusted role thereto as of not being qualified for the event management, however, once the event has occurred, several axes foe movement have been determined:

- Forming the operations room, operating around 24 hours for receiving the communications and to speedily evacuate the dilapidated buildings.
- Making a quick inventory of the areas, affected by the earthquake and assessment of the size of losses.
- The speed of implementing the emergent accommodation works for the effected from the earthquakes so as to accommodate all families along with the services of the facilities and subsistence as acceptable on the basis of being temporary phase.
- Considering the subsidies or compensations to be submitted to the effected.
- Developing a timetable for the transfer of the affected families by the emergent accommodation of the housing units.

The incident has been as a notification for starting the development of plans and preparations for any emergent accident and the competent authorities have taken the following procedures:

- The announcement of emergency cases in all the provinces apparatus, authorities and the facilities of different services.
- Raising the willingness degree in all the central hospitals and health units along with the provision of ambulances with sufficient numbers and readiness for immediately directing to the place of communication.
- Transferring the central rescue devices immediately to the incidents sites and focusing on the popular areas.
- Forming a supreme committee for housing and transferring the affected families which became homeless to the prefabricated and prepared housings.
- Forming engineering groups for previewing the dilapidated properties and evacuating what is subjected to collapse. The university professors and armed forces engineers as well as the Ministry of Construction have been taken advantage thereof.
- Developing a plan for the fast completion of the number of housings, built by the province and providing them with the facilities for the purpose of the speed of housing of all affected families, that live in the tents.

- Issuing decisions for the commissioning by business sector companies for removing the collapsed houses or reducing the floors to be removed or otherwise conducting the restoration works.
- Conducting the works of preventive health works by the Directorates of Health such as spraying the pesticides in the accommodation areas along with the vaccination works and health survey.
- Opening the youth centers by the Directorates of Youth and Sports at the provincial levels for setting up the tents in the form of aggregated camps.

5. PROPOSED APPROACH FOR EARTHQUAKE DISASTER MANAGEMENT

The disasters management, resulting from the earthquakes shall play an effective role in reducing the damages and losses, resulting from the earthquakes and Figure 11 shall indicate the proposed approach for managing the resulting disasters from the earthquakes.

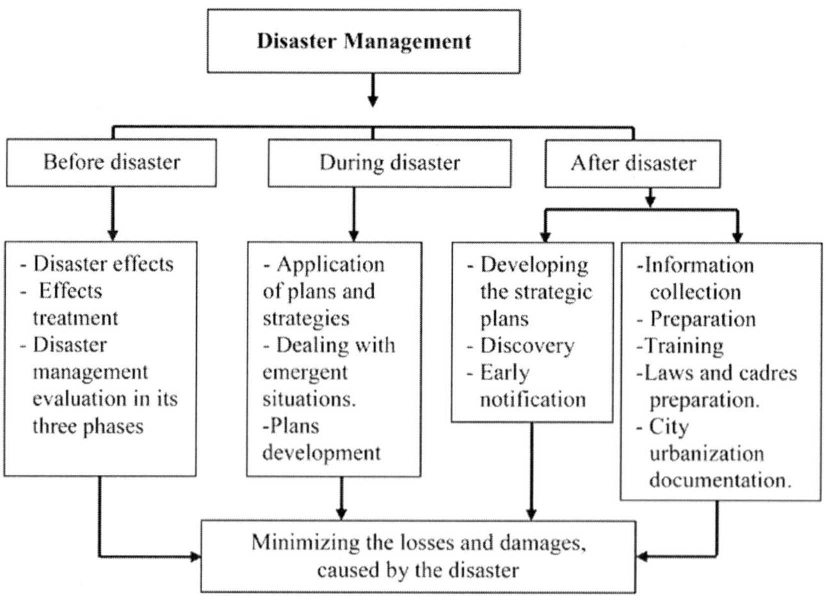

Figure 11. The proposed management approach for the earthquake disasters.

First: pre-disaster phase:

In this phase, all data and information of the resulting disaster for the earthquakes along with the method of notification and direction as well as development of operational plans upon the occurrence of disaster and training on these plans shall be gathered.

Development of the training plan on several axes which are:

1) Action seismic maps of the Kingdom of Saudi Arabia Private General of the Arab region and the development of codes and laws of the buildings.

2) Modeling the city through the GIS for indicating the points of danger and strength in the city along with the traffic axes as well as the method of accessing the most subject able areas for risks as soon as possible.

3) Raising and documenting the archeological buildings and urbanization in the city along with the networks and methods accurately and in a manner, guaranteeing identification of the city features before and after the disaster via the survey raising and using the computer applications as well as the GIS, photos and video clips.

4) Developing the plans for coping with the effects of earthquakes, rubble removal, accommodation and evacuation as well as buildings maintenance and the analysis of defects leading to the collapse of buildings and readiness with an alternative plan in which effective mobilization for the energies and resources as well as well-management with the least losses and as soon as possible shall be available.

5) Developing a training plan for the entrusted work for transfer so as the tasks shall be distributed in advance. Therefore, such matter will help to the non-occurrence of confusion in the relief and rescue operations between the personnel as well as the conflicting decisions between the decision makers through either making experimental testing statement for the case of the occurrence of earthquake so as to train them upon the relief and evacuation operations as well as the number of cam[s and lifting the rubble.

6) The central governments and the policy makers shall work on establishing an independent and permanent authority for managing the disasters in the different developing regions along with allocating the fund and mechanisms for the provision of appropriate and fast financial credits domestically and internationally through the civil bodies organizations and authorities along with making use of the experts and technicians who are trained upon the disaster management operations

and supplying them with the modern and appropriate equipment and devices.

Second: the phase of disaster occurrence:

1) Examination of facilities and receipt of communications for the effected buildings along with the formation of technical committees for preview.

2) Development of the evacuation priorities and accommodation of effected areas.

3) Lifting the rubble from the rundown buildings and the examination of infrastructure for status restoration as soon as possible.

4) Making the database for determining the size of losses and assessing the resulting damage for the buildings and infrastructure along with the number of effected persons, the size of necessary interim units and the restoration period.

5) Temporary facilities processing in accordance with the number of effected and the provision of necessary basic services.

6) Request of aids and development of automatic mechanism for the restoration with the local efforts or external aids.

7) Third: post-disaster phase:

8) The completion of relief and protection works from diseases and epidemics.

9) The transfer of resulting wastes from demolished buildings.

10) Evaluation of the resulting damage from the earthquakes in quantity and quality.

11) Restoration and rehabilitation of devastated areas and follow-up of various projects.

12) Assessment of set out codes and laws and developing them according to the incident effects assessment.

13) Limitation of the points of weaknesses and shortcomings in the prior plans for avoiding and treating them.

6. RESULTS

1) Earthquakes are the most natural disasters causing losses in the world and in the Arab nations.

2) The main reason for the occurrence of earthquakes is the movements of formative plats. There are other reasons for the occurrence of earthquakes which are unknown.
3) All the world regions are subjected to earthquakes of different scales.
4) The earthquakes cannot be fully predicted.
5) Identification of the disaster, characteristics and causes is important for the efficiency of its management.
6) The disaster management plays a significant role in reducing the various losses in a manner reaching about 70% in accordance with the real studies and experiences.

7. RECOMMENDATIONS

1) The construction of a centralized body for disaster management as a unique authority in the Arab League. It consists of the related governmental Ministries, and representatives of the main Arab relative organizations.
2) Directing the research to the enrichment of educational curricula and studies for disaster management.
3) The introduction of appropriate codes and laws for reducing the earthquakes risks in all fields.
4) Making contour models and enormous information for the earthquake regions in the Arab cities using GIS (pre-disaster preview).
5) To gain experiences from developed countries in earthquake disasters management. Training and simulation of disasters are among the interesting subjects.
6) Documenting and supporting historical buildings and classifying them on maps through GIS.
7) Participating in competent related international organizations.
8) Participating in teams of various relief operations around the world to increase friction, training and efficiency.

CITE THIS PAPER

El-Quliti, S.A., Al- Harbi, T.B.S., Al-Yami, M.B.S., Al-Ghamdi, A.B.M. and Al-Shammar, M.B.M. (2016) Earthquake Disaster Management Approach: The Case of Al-Ais, Medina Area in Saudi Arabia. Open Journal of Earthquake Research, 5, 219-235.

REFERENCES

1. The Ministry of Defense, Cairo (1991) Crisis Management and Negotiation, Strategic Reference. The Ministry of Defense, Cairo.
2. Fritz Charles, E. (1961) Disaster. In: Merton, R.K. and Nis-bet, R.A., Eds., Contemporary Social Problems, Harcourt, New York, 655.
3. De Greene, K.B. (1982) The Adaptive Organization: Anticipation and Management of Crisis. Wiley & Sons, New York, 183.
4. International Civil Defence Organization (ICDO) Website (2016). http://www.icdo.org/en/
5. Perry, R.W. and Nigg, J.M. (1985) Emergency Management Strategies for Communicating Hazard Information. Public Administration Review, 45, 12-11.
6. Nabarawy, A.M. (1995) Disasters, Floods. Unpublished Ph.D. Thesis, Ain Shams, Cairo, 8.
7. Nelson, S.A. (2015) Earthquakes and the Earth's Interior. Physical Geology, Tulane University. http://www.tulane.edu/~sanelson/eens1110/earthint.htm
8. Ailosh, M.N. (1996) The Basics of Earthquakes and Earthquake Engineering. Prentice-Hall International.
9. Earthquake Commission, The Egyptian Syndicate of Engineers (1992) Manual Guide. Earthquake Commission, The Egyptian Syndicate of Engineers, Cairo.
10. USGS Homepage (2016).https://earthquake.usgs.gov/earthquakes/world/world_deaths.php
11. Emirate of Madinah, Ministry of the Interior, Kingdom of Saudi Arabia (2015). http://www.imaratalmadinah.gov.sa/ar-SA/MedinaRegion/centers/Pages/ais.aspx
12. General Presidency for Youth Welfare Website (2016). http://www.saudinf.com/main/c6y.htm
13. Google Maps Website (2015). https://www.google.com.sa/maps/@24.266906,45.0983499,5z

CHAPTER 2

How to Identify Foreshocks in Seismic Sequences to Predict Strong Earthquakes

Giulio Riga[1], Paolo Balocchi[2]

[1]*Geologist, Independent Researcher, Lamezia Terme, Italy*

[2]*Geologist, Independent Researcher, Modena, Italy*

ABSTRACT

The time analysis of seismic events preceding several strong earthquakes occurred in recent decades throughout the world, has highlighted some foreshocks' characteristics, which are helpful for their discrimination compared to other types of events. These features can be identified within the seismic sequence and used as strong events' precursors. Through the energy release pattern analysis, which precedes any strong earthquakes, in this study we describe some graphical procedures suitable for distinguishing a foreshock from any other type of earthquake. We have broadly divided foreshocks into two classes, depending on their position within the energy release pattern, by describing some relationships between the foreshock's magnitude and the following earthquake's. The results obtained show how the energy release pattern of some major earthquakes has distinctive features and repeatability which it is possible to obtain information from in order to perform sufficiently reliable short-term forecasts.

Keywords: Foreshock, Mainshock, Aftershock, Earthquake, Microsequence, Hierarchization

1. INTRODUCTION

From the observation of several seismic sequences, we infer that some earthquakes are preceded by events of increasing magnitude related to the main shock, which are called foreshocks [1] .

Several studies conducted on seismic sequences show how the seismicity increased significantly before the mainshock. For example, 10 days before the 2009 L'Aquila mainshock occurred, the foreshock sequence was concentrated in the hanging-wall domain of the normal Paganica fault [2] .

The foreshocks are one of the few well-documented precursors to large earthquakes: Therefore, understanding their very nature is crucial for earthquake prediction and hazard mitigation. Several studies use foreshock sequences with (combined with) a statistical and probabilistic approach [2] to predict earthquakes (retrospectively) [3] .

We believe that foreshocks are the manifestation of an ongoing energy release process, with different spatial and temporal scales, which leads to the mainshock. During its manifestation time, the seismic sequence analysis may be an appropriate tool to highlight the various stages of preparing a strong event. Under this approach we have noticed that an area seismicity before major earthquakes shows a certain organization [2] , according to a repetitive developmental pattern of "Progressive Earthquakes"-type, [1] consisting of a succession of one or more foreshocks of various order and one or more mainshocks.

The analysis performed suggests that the space-time occurrence of foreshocks, ranges from a few hours up to years before the mainshock, and from a few kilometers to thousands of kilometers from mainshock. It is in fact believed that foreshocks may also occur in areas relatively far from the mainshocks and be part of an external energy release process, which influences the area by triggering the main event [4] .

For a better understanding of foreshocks temporal organization during the energy release phase, we propose some simple graphical and numeric procedures that can be used for the preliminary forecast of a major event.

2. FORESHOCKS' CHARACTERISTICS

Foreshock sequences are the most obvious precursor to large earthquakes: therefore, understanding their origin and relation to mainshocks is crucial for earthquake prediction and hazard mitigation. Previous studies conducted on immediate foreshocks in California suggest that these events may be part of a mainshock rupture nucleation process, because estimated Coulomb stress changes from foreshocks are too small to produce stress triggering and observed foreshock areas scale with mainshock magnitude, are consistent with nucleation instead of earthquake-to-earthquake triggering [2] [5] [6]

Foreshocks occur close to the mainshock and most likely are part of the nucleation process [7] . They may be both physical and tectonic. Most short- and impending-term foreshocks that occur close to major shocks are considered as physical. In general, foreshocks occurring around the epicentral area can be considered as tectonic precursors, as they result from occasional plates' shifting or blocks shifting along the fault plane that will break [8] . They most frequently occur in areas featuring moderately heterogeneous rocks [9] and in the presence of a moderate pre-stress and rupture stress [10] .

Foreshocks' magnitude values and position during the energy realease phase allow the implementation of useful models for predicting a strong earthquake. However, it is necessary to know the features that distinguish them from other earthquakes. Through the graphical analysis of the seismic sequence, it is possible to highlight the foreshocks' characteristics and their grouping in various orders, based on magnitude and period.

Defined as the "foreshocks' height", magnitude represents the distance, in terms of energy, between a minimum and the subsequent maximum, while period is the time or the number of shocks between a foreshock and the next.

Depending on the period or the number of shocks between a foreshock and the next within the seismic sequence, it is possible to identify foreshocks of different magnitude orders. First order foreshocks characterize short-medium term time windows, while those of second order usually occur immediately before the mainshock and exactly in the same area where the mainshock occurs.

The time elapsing between the last second order foreshock and the mainshock may vary but it typically consists of few days, while those of first order may happen in different times and areas.

Based on foreshocks magnitude values and temporal position, which characterize the energy release phase, we obtain the following "Progressive Earthquakes"-type developmental patterns [1] :

a) From one or more first order or short to medium-term foreshocks and mainshock (Figure 1(a));

b) From one second order or impending-term foreshock and mainshock (Figure 1(b));

c) From one or more first order or short to medium-term foreshocks, one second order or impending-term foreshock and mainshock (Figure 1(c));

d) From one or more various orders foreshocks and from multiple mainshocks (Figure 1(d)).

By observing the graphs shown in Figure 1, we understand how each earthquake is followed by a energy accumulation (red segment) and by an energy re lease phase (green segment) triggered by small magnitude shocks. Major events, complementing the "Progressive Earthquakes" pattern, in turn trigger a long- term energy accumulation phase that consists of two or more energetic aftershocks.

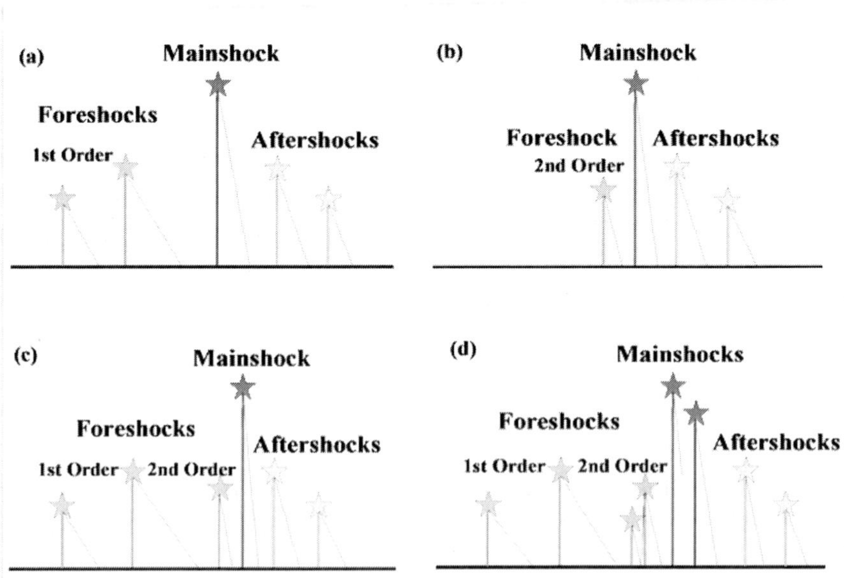

Figure 1. Schematic representation of Progressive Earthquakes patterns [1] . The green and red lines respectively show the energy release and accumulation phases. The green star indicates the foreshock, the yellow star the aftershock, while the red star shows the mainshock.

Figure 2 and Figure 3 show the magnitude values trend and different order foreshocks position in relation to the monthly long-term seismic sequence's mainshock in Japan and to the short-term one preceding the earthquake occurred in L'Aquila on 6 April 2009.

By observing the graph of the Japanese seismic sequence we can notice two "Progressive Earthquakes"-type patterns consisting of two first-order foreshocks and one mainshock, in which the temporal distance between the second foreshock and the mainshock is greater compared to that between the first and the second foreshock, while the magnitude values decrease.

In the graph showing the seismic sequence of L'Aquila, we see that the energy release pattern consists of three first order foreshocks, two second order foreshocks and one mainshock. The three first-order foreshocks show an "additive/constant" model (green line) in which the foreshocks' magnitude values show a proportionality relationships as the number of seismic events increases, while the second order foreshocks and the mainshock, show a multiplicative/amplified model (red line) where the magnitude values between the foreshocks and mainshock amplify as the number of seismic events raises. This amplified effect is due to the short-term occurrence between foreshocks and the mainshock, which usually lasts just a few days.

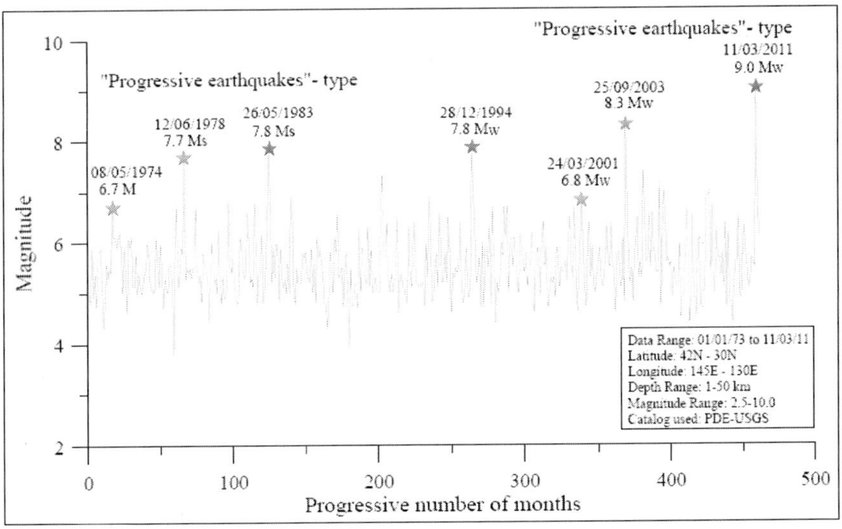

Figure 2. Earthquake in Japan on 3 March 2011. Green stars show mid-term foreshocks (first order) while red stars indicate the mainshocks. The vertical axis shows the magnitude values, while the horizontal axis displays the progressive number of months recorded in the period analyzed.

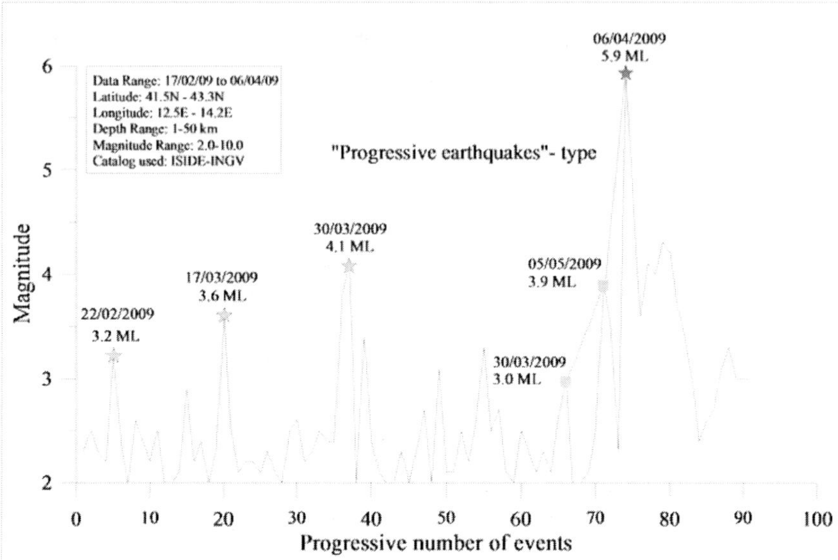

Figure 3. Earthquake in L'Aquila on 6 April 2009. Green stars show short-term foreshocks (first order), green squares indicate impending-term foreshocks (second order), while red show the mainshock. The vertical axis shows the magnitude values, while the horizontal axis displays the progressive number of events recorded in the period analyzed.

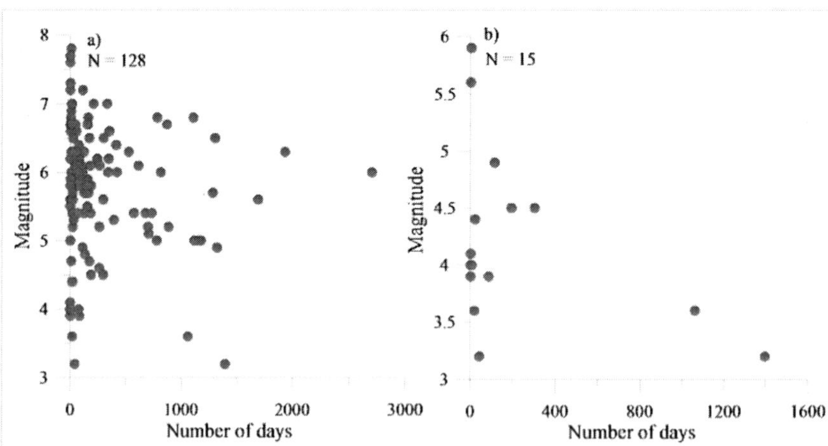

Figure 4. Relationship between foreshock magnitude (y-axis) and the number of days elapsed between foreshock and mainshock (x-axis): (a) related to the whole world; (b) related to Italy.

Figure 4(a) and Figure 4(b) show the relationship between the foreshock magnitude and the number of days elapsed since the mainshock, processed using earthquakes in the "Progressive Earthquakes"-type energy release pattern concerning 128 earthquakes occurred in various areas of the world and in Italy.

Figure 5(a) shows the relationship between the distance between the foreshock epicenter and the mainshock's and the number of days elapsed between foreshock and mainshock, while Figure 5(b) shows the relationship between the foreshock magnitude and the distance between the foreshock and mainshock epicenters.

The graphs and Table 1 display the following significant trends related to foreshocks frequency rate and characteristics:

1) Many earthquakes occur less than thirty days after the last foreshock, but quite a number of foreshocks occurs twenty-four hours before the mainshock;
2) The frequency rate of foreshocks occurring within a radius of 50 km and over thirty days, decreases as distance and time increase;
3) Closer events in time and space, show higher magnitude;
4) Magnitude values decrease as distance and number of days increase.

The results obtained show that in some seismic sequences, the "Progressive Earthquakes"-type energy release phase tend to develop near the epicenter of the main shock.

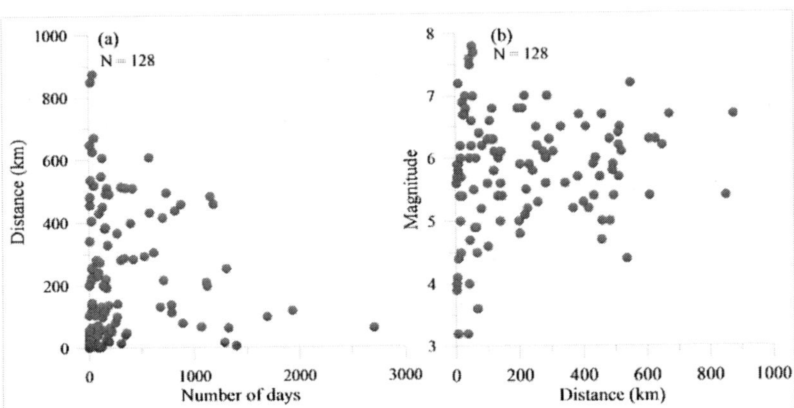

Figure 5. (a) Relationship between the number of days elapsed between foreshock and mainshock (x-axis) and the distance between the foreshock and mainshock epicenters (y-axis); (b) the relationship between the distance between the foreshock and mainshock epicenters (x-axis) and the magnitude of the foreshock (y-axis).

Table 1. Correlations between foreshocks and earthquakes.

No	Earthquake	Foreshock		Mainshock		Time (days)	Distance (km)
		Date	M	Date	M		
1	Colfiorito - Italy	26/09/1997	5.6 ML	26/09/1997	5.8 ML	1	1.24
2	L'Aquila - Italy	05/04/2009	3.9 Mw	06/04/2009	6.1 Mw	1	1.90
3	Emilia - Italy	19/05/2012	4.0 Mw	20/05/2012	5.8 Mw	1	2.14
4	Central Italy	26/10/2016	5.9 Mw	30/10/2016	6.5 Mw	4	8.70
5	Vanuatu	09/07/1980	6.9 Mw	17/07/1980	8.0 Mw	8	20.8
6	Vanuatu	28/11/1985	7.2 Ms	28/11/1985	7.6 Ms	1	8.53
7	Vanuatu	16/11/1985	6.8 Ms	21/12/1985	7.6 Ms	5	31.2
8	Afghanistan	05/03/1990	6.0 Ms	25/03/1990	6.3 Ms	10	15.1
9	Japan	09/03/2011	7.3 Mw	11/03/2011	9.0 Mw	2	44.3
10	Japan	18/07/1992	5.8 Mw	18/07/1992	6.9 Mw	1	6.96
11	Nicaragua	10/08/1992	5.7 Mw	02/09/1992	7.7 Mw	23	3.95
12	Gulf of Alaska	17/11/1987	7.0 ML	30/11/1987	7.9 Mw	13	29.5
13	Filippine	17/05/1992	7.1 Mw	17/05/1992	7.3 Mw	1	13.1
14	Colombia	17/10/1992	6.7 Ms	18/10/1992	7.3 Ms	1	26.3
15	Denali-Alaska	23/10/2002	6.7 Mw	03/11/2002	7.9 Mw	11	23.3
16	Grecia	01/02/2014	5.0 Mb	03/02/2014	6.1 Mw	2	14.0

3. FORESHOCKS-IDENTIFYING METHOD

3.1. Use of Trendlines

Today it is believed that a foreshock is physically indistinguishable from any other earthquake, until a subsequent mainshock classifies it as such [11] [12] [13] .

The methodology proposed to identify foreshocks is based on identifying the dynamic trendline that characterize the development of the energy accumulation and release phases [1] .

As we can infer from Figure 6, to highlight the energy accumulation and release phases multiple points of relative maximum both in the case of increasing (points 1, 2, 3) and decreasing (points 4, 5, 6) magnitude values, have been joined with a straight line.

As their numerical value varies over time, these straight lines are called dynamic trendlines. Trendlines clearly show the direction the magnitude values are moving in both in the impending and long-term and until the trend remains unchanged.

A seismic events's trend is considered unchanged as long as there are no clear signs of reversal, such as the trendline break. In fact, the crossing of the energy accumulation trendline (transition point) marks the beginning of an energy release phase that can be of "Flash earthquakes-" or "Progressive Earstquakes-" type [1] . The first is characterised by one or more mainshocks, the second by an initial foreshock followed by a first magnitude values' fall, which means that the magnitudo values are returning close or under the energy accumulation phase's trendline. Subsequently, the magnitude values begin to raise again up to form a second foreshock from which it is possible to obtain the direction of the energy release phase's trendline.

Besides showing the ongoing trend, it seems that the trendlines momentarily prevent the magnitude values from raising and, in some cases, they allow knowing in advance certain levels of magnitude that will be achieved in the future. In fact, the Figure 6 shows also how the knowledge of the magnitude levels achieved in points 1 and 2 allows calculating in advance the magnitude value of point 3 in the energy accumulation phase or of point 6 in the energy release phase after points 4 and 5 have formed.

The trendline drawn from points 6 (first order foreshock) and 7 (aftershock) allows identifying the second order foreshocks 8 and 9.

An earthquake in the "Progressive Earthquakes"-type energy release pattern cannot be identified as foreshock or mainshock until a subsequent energetic aftershock is formed. In the absence of an energetic aftershock, the foreshock or mainshock is identified as provisional.

Figure 7 reports the seismic sequence of the Denali earthquake (Alaska) occurred on 3 November 2002.

The earthquake was preceded by three first order foreshocks (green stars) and by a second order foreshock (green square). As can be seen, the trendline drawn from points 1 and 2 shows a hypothetical medium—long term trend and allows estimating point 3's magnitude (next minimum), while the intersection of the straight line passing through points 3 and 4 with the vertical line passing through the mainshock occurence point (point 5), provides the dynamic mainshock's magnitude value. Alternatively, where the seismic sequence includes multiple first order foreshocks, the dynamic mainshock's magnitude value is provided by the linear interpolation between them (dashed green line).

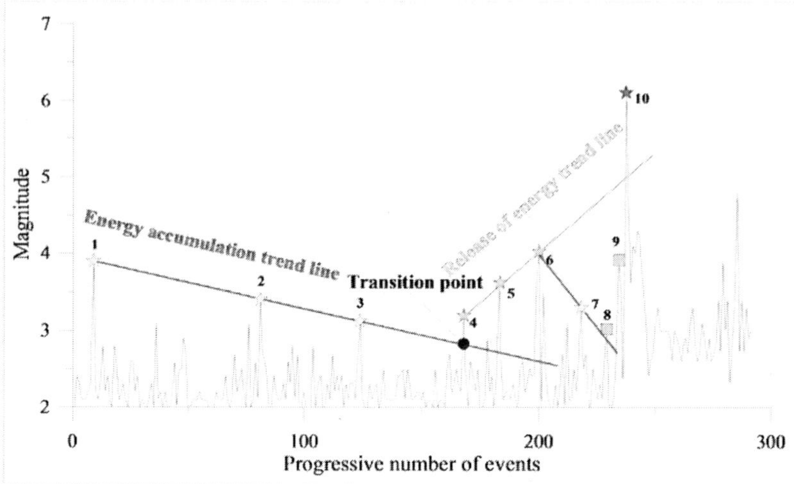

Figure 6. Short-term sequence of L'Aquila earthquake on 6 April 2009 (red star). The yellow stars indicate the strongest aftershocks in the energy accumulation phase, the green stars indicate the first order foreshocks, while the green square shows the second order foreshocks. The solid black circle indicates in the transition point from the energy accumulation to energy release phase.

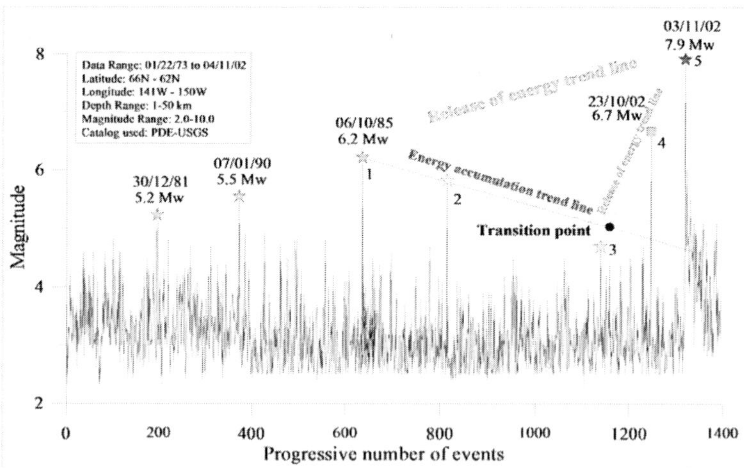

Figure 7. Long-term sequence of Denali earthquake (Alaska) on 3 November 2002 (red star). The yellow stars indicate the strongest aftershocks in the energy accumulation phase, the green stars indicate the first order foreshocks, while the green square shows the second order foreshocks. The solid black circle indicates in the transition point from energy accumulation to energy release phase.

3.2. Seismic Sequence Hierarchization

The use of a hierarchization process in the seismic sequence analysis, allows locating foreshocks and mainshock during the energy release phase. More accurate data on the foreshocks formation can be obtained from the seismic grid of the energy accumulation and release phases (Figure 8) representing the cyclic pattern of reference that includes cycles of different duration.

Obtained through the hierarchization process, the graph [4] , highlights how during the energy release phase, only the energy release seismic branches of different order (1r, 2r, 3r, 4r) convey to the foreshocks (nodes F1, F2, F3) or mainshock (node M), while in the energy accumulation phase, both the energy accumulation (3a) and release (2r, 1r) seismic branches convey to the aftershock (nodes A1, A2, A3) of a previous event,

The extension of the first higher order seismic branch of the branched structure (branch 3a) up to the point of calculation allows identifying the transition point (full black circle) and the various orders foreshocks of the energy release phase. We can note how the various orders foreshocks and the mainshock are positioned above the dashed red line that represents the descending trendline drawn from the higher order seismic branch.

In some seismic sequences, it is apparent how a longer period after the last first order foreshock corresponds to a greater magnitude of the subsequent foreshock or mainshock.

3.3. Index Simplified Force (ISF)

ISF is an simple and very sensitive short-term oscillator, which can be used to monitor the strength and the developmental state of a seismic sequence and to detect the foreshocks as well as the energy accumulation phase's trend.

ISF uses two types of seismic data: the magnitude values and the number of events recorded during the day. The greater is the variation of the magnitude values and/or of the number of events, the greater the strength of the sequence. Therefore, whenever you have a daily increase of the magnitude value due to the occurrence of a foreshock or mainshock or a number of seismic events or of both, a peak is formed on the graph, followed by a gradual reduction in the amplitude of the ISF oscillation over time.

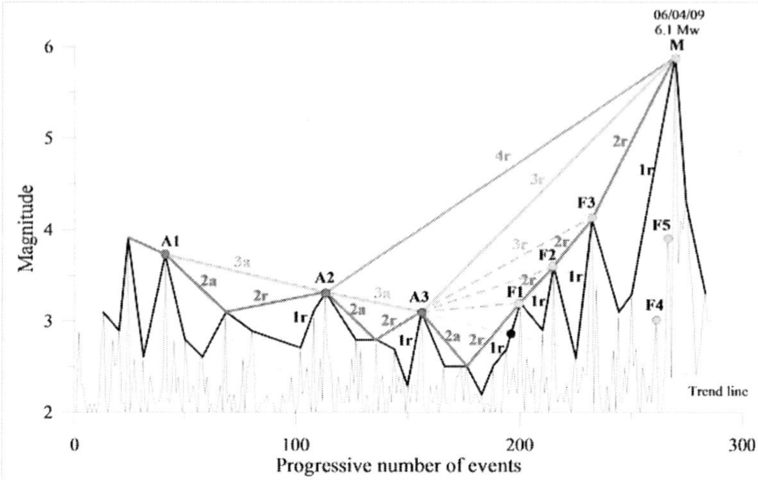

Figure 8. Seismic sequence of the L'Aquila earthquake on 6 April 2009.

Due to the effect of the values fluctuations above the zero line, ISF has the appearance of a seismogram where, values above zero indicate the energy release phases, while negative values indicate the energy accumulation phases. ISF's high values are associated with an increase in the number of earthquakes and/or high magnitude values. Positive and increasing index values show an energy release phase that is being strengthened, while positive and decreasing values show an energy accumulation phase. Values close to zero show that none of the two phases is ongoing and both anticipate a trend reversal.

Figure 9 shows an example of a seismic sequence structured for ISF calculation. The graph is built taking into account the following conditions:

a) If the magnitude value increases compared with the previous day, it means that an energy release phase is ongoing (green segment) ;

b) If the value decreases, it means that there is an energy accumulation phase ongoing (red segment);

c) If the value is equal, we have the same phase as the day before.

From the graph we can infer how, as the events progress, the magnitude values are not random but follow a trend according to increasing trends up to the process reversal. We can notice also growing trend phases, characterized by increasing maximum and minimum, and decreasing trend phases characterized by decreasing maximum and minimum.

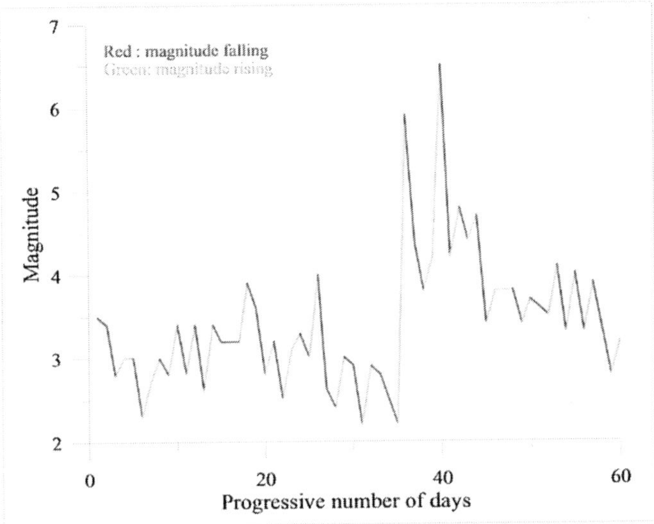

Figure 9. Seismic sequence structured for IFS calculation.

The analysis of many seismic sequences revealed the existence of three main types of trends that can be identified in advance: (a) the main trend (lasting a few years); (b) the intermediate trend (lasting several months); (c) the minor trend (lasting a few weeks). It is therefore evident that there is not only one type of trend but different trends (one inside another) according to the observation time horizon.

To calculate the ISF, we simply identify the value of the daily maximum magnitude and multiply it by the number of daily shocks, attributing a positive or negative value depending on whether the value of the maximum magnitude of the day considered is greater (green segment) or smaller compared to that of the previous day (red segment). In the case where the maximum magnitude value of the day considered is equal to the previous day's, the previous day ISF sign is attached to ISF.

$$ISF = \pm M_{max} \cdot N_{events} \quad (1)$$

Figure 10 and Figure 11 show the ISF obtained from seismic sequences related to central Italy in 2016 and Japan in 2011.

The ISF graph related to central Italy (Figure 10) from 16 to 23 August 2016 shows low values, then a first peak due to the occurrence on 24 August (green star) of the first order foreshock whose magnitude was 6.0 Mw.

Later, after a brief energy accumulation phase highlighted by a progressive decrease of ISF values, we observe a second peak in correspondence of the second order foreshock (green star) on 26 October whose magnitude was 5.9 Mw. The positive peak is followed by a negative minimum which corresponds to the triggering point of the energy release phase (red triangle) that ends on 30 October with the mainshock whose magnitude is 6.5 Mw, which is indicated on the graph by a positive ISF peak. The epicenter of the second order foreshock on 26 October was located approximately 8.70 km from the main shock epicenter. The energy accumulation phase that is triggered after the mainshock is represented by a progressive and slow decrease in the oscillations amplitude of the ISF values.

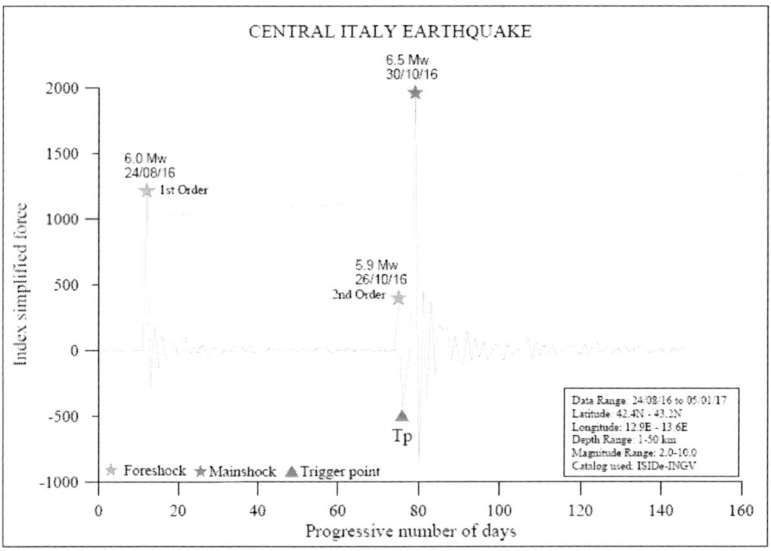

Figure 10. Seismic sequence force index in central Italy (2016).

Table 2. Succession of earthquakes in the "Progressive Earthquakes"-type pattern (Central Italy earthquake) between 24 August and 30 October 2016.

No	Earthquake Date	Magnitude	Identification	Order	Time (days)	Distance (km)
1	24/08/2016	6.0 MW	Foreshock	1st Order	-	-
2	26/10/2016	5.9 Mw	Foreshock	2st Order	64	24.98
3	30/10/2016	6.5 Mw	Mainshock	--	4	8.70

Table 3. Succession of earthquakes in "Progressive Earthquakes"-type pattern (Central Italy earthquake) between 16 and 30 October 2016.

No	Earthquake Date	Magnitude	Identification	Order	Time (days)	Distance (km)
1	16/10/2016	4.0 Mw	Foreshock	1st Order	-	-
2	26/10/2016	5.4 Mw	Foreshock	1st Order	10	15.2
3	26/10/2016	5.9 Mw	Foreshock	1st Order	<1	3.23
1	29/10/2016	4.1 Mw	Foreshock	2st Order	3	11.24
2	30/10/2016	6.5 Mw	Mainshock	--	1	8.70

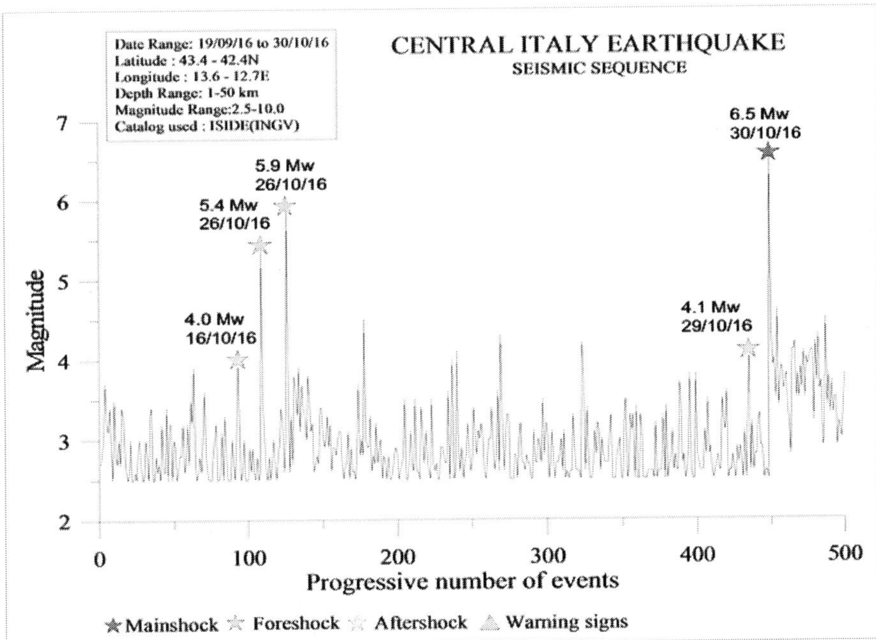

Figure 11. Seismic sequence and succession of earthquakes in "Progressive Earthquakes"-type pattern (Central Italy earthquake) between 16 and 30 October 2016.

Table 2 displays the short-term "Progressive Earthquakes"-type earthquakes in the energy release phase in central Italy, the occurrence time and the distance between them, while Table 3 and Figure 11 shows the impending-term pattern of the strongest earthquakes recorded in Central

Italy between 16 and 30 October 2016, the occurrence time and the distance between them.

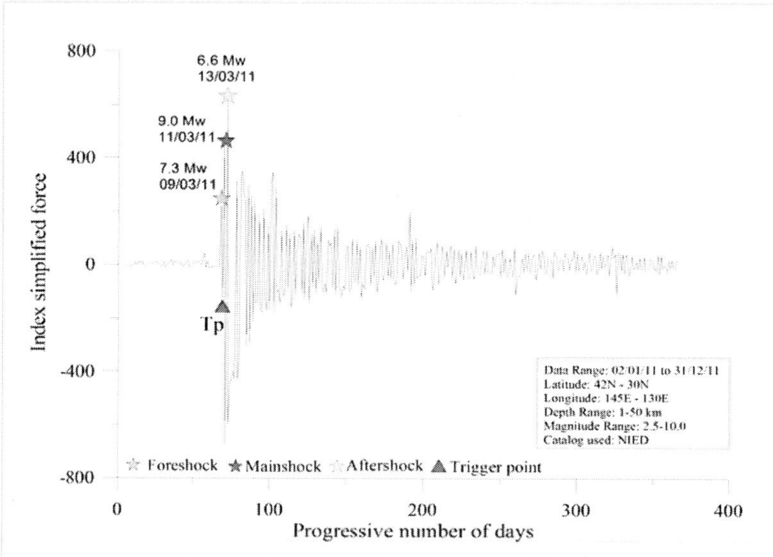

Figure 12. Seismic sequence force indicator of Japan earthquakes in 2011.

Table 4. Succession "Progressive Earthquakes"-type pattern earthquakes (Japan earthquake).

No	Earthquake Date	Magnitude	Identification	Order	Time (days)	Distance (km)
1	09/03/2011	7.3 MW	Foreshock	1st Order	-	-
2	11/03/2011	9.0 Mw	Mainshock	---	2	44.3

Figure 12 shows the ISF seismic sequence in Japan during the time window from 2 January to 31 December 2011. Up to 8 March 2011, the graph shows an initial portion of cyclical fluctuations of small amplitude, followed, on 9 March, by a first peak due to the occurrence of the foreshock with a magnitude of 7.3 Mw (green star) associated with an increase in the number of recorded events and shortly afterwards by a further ISF increase due to main shock with a magnitude of 9.0 Mw (red star) occurred on the same day. A third peak of greater amplitude is formed following the aftershock occurrence on 13 March whose magnitude was 6.6 Mw (yellow star) associated with a strong increase in the number of daily events recorded. After the 13 March earthquake, the ISF index value decrease over time and, at the end of December, they reach values similar to pre-foreshock's on 9 March 2011.

Table 4 reports the "Progressive Earthquakes"-type pattern earthquakes in the energy release phase of the Japan earthquake.

3.4. Mainshock Magnitude Calculation

3.4.1. Graphical Method

A quick method for estimating the mainshock magnitude is shown in Figure 13 that reports the seismic sequence in the earthquake occured in the Gulf of Alaska on 30 November 1987.

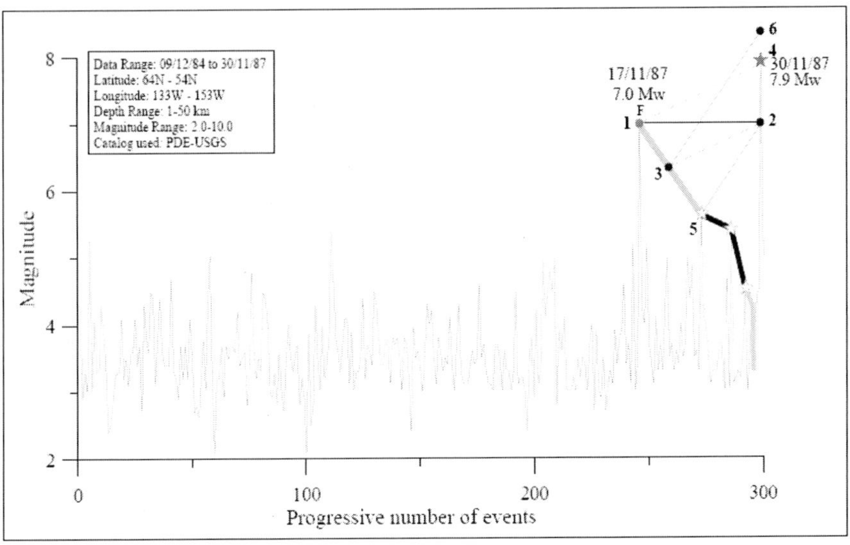

Figure 13. Short-term sequence of the earthquake occurred in the Gulf of Alaska, on 30 November 1987. Graphical procedure to determine the fluctuation range of mainshock magnitude value.

As can be seen, the mainshock was preceded by a foreshock with a magnitude of 7.0 Mw recorded on 17 November 1987. To calculate the magnitude of the mainshock, the first step is to draw the branched structure [4] with its point of origin located in the foreshock and then locate on higher order seismic branch the point corresponding to 50% (point 3). From foreshock, we draw a horizontal dashed line up to meet the vertical line passing through the mainshock's occurrence time (point 2) and thereafter, we join the point 2 with the intermediate point of the seismic branch (point 3). Finally, from the foreshock we draw its parallel line (solid black lines). The intersection of this half line with the vertical line passing through the time selected for the calculation, provides the value of the minimum expected magnitude. To calculate the maximum value, we join the points 2 and 5 of the seismic branch, then from the intermediate point 3 we draw its parallel line (solid black lines). The intersection of this parallel with the vertical line passing through the time selected for the calculation, provides the value of the maximum expected magnitude.

3.4.2. Numerical Methods

The empirical relations developed can be used to assess the magnitude of various order foreshock or "Progressive Earthquakes"-type pattern mainshock by know- ing the magnitude of the previous foreshock.

The relationships were obtained from the study of M > 3.8 foreshocks recorded by the Italian seismological network NIED network and by the USGS network between 1970 and 2016.

The average Mainshock Magnitude (MM) value is estimated through the following empirical relationship, obtained from the graphs displayed in Figure 14:

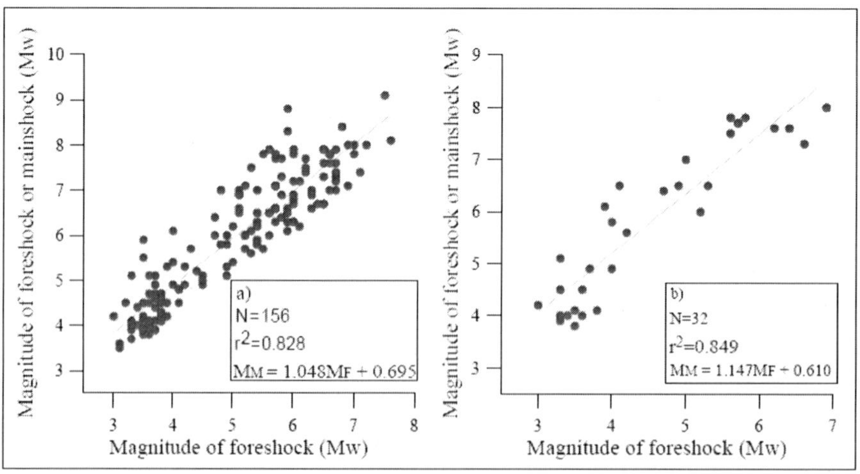

Figure 14. Relationship between foreshocks and mainshock: (a) valid for all areas of the world; (b) valid for Italy.

Relationship valid for all areas of the world with a foreshock magnitude (MF) in Mw:

$$M_M = 1.048M_F + 0.695 \tag{2}$$

Relationship valid for the Italian territory with a foreshock magnitude (MF) in Mw.

$$M_M = 1.147M_F + 0.610 \tag{3}$$

In the second procedure, the expected earthquake magnitude value is obtained with a formula empirically derived from a statistical analysis of the magnitude values of first and second order foreshocks that preceded the mainshocks in different areas of the world.

The equation to calculate the magnitude is as follows:

Relationship valid for all areas of the world with a foreshock magnitude (MF) in Mw:

$$M_M = 3.407 \cdot \ln(M_F) + 1.318 \tag{4}$$

4. CONCLUSIONS

The study of various order foreshocks the "Progressive Earthquake"-type energy release phase consists of, has provided a different methods of comparison between foreshocks and big earthquakes, leading to a better understanding of the phases regulating the preparation process of strong earthquakes.

The results obtained show that in some seismic sequences, the "Progressive Earthquake"-type energy release phase tends to develop near the main shock epicenter.

In particular, the analysis of the distances in time and space between foreshocks and mainshock has highlighted how some foreshocks occur shortly before the mainshock, within a radius of 50 kilometers from its epicenter. We also pointed out how the foreshocks frequency rate decreases as the distance from the mainshock epicenter and time increase, while the magnitude values decrease as the distance and the number of days increase.

The graphical analysis of the seismic sequence has shown how the foreshocks lie above the trendline that joins decreasing magnitude maximum values that form during the energy accumulation or above the seismic branch extension whose order is greater compared to the branched structure.

An in-depth examination of the branched structure showed how during the energy release phase only the branches seismic energy release of different order convey to foreshocks or mainshock, while in the energy accumulation phase the seismic branches of both energy accumulation and release phases convey to aftershock.

ISF (Index Simplified Force) can be used to obtain information about the strength of the seismic sequence in a given moment, to locate the foreshocks and to monitor the energy accumulation and release phase progress.

Based on the data resulting from the observations of several seismic sequences, we obtained some graphical and numeric procedures to determine the mainshock's magnitude value using the trendline, the branched structure's higher order seismic branch or the foreshock magnitude value.

We believe that the described method shows the foreshock identification limit during seismic sequences. Actually, as is not always possible to identify the event that precedes a strong earthquake, one prefers to classify it as provisional foreshock. The subsequent evolution of the seismic sequence will provide more information so that an appropriate foreshocks classification can be implemented.

The aforementioned procedures show how, starting from the analysis of the "Progressive Earthquakes"-type energy release phase, it is possible to obtain information to locate the foreshocks and make sufficiently reliable short-term forecasts. Seismic sequence's evolution monitoring is essential to identify correctly the foreshock that precedes strong events.

CITE THIS PAPER

Riga, G. and Balocchi, P. (2017) How to Identify Foreshocks in Seismic Sequences to Predict Strong Earthquakes. Open Journal of Earth- quake Research, 6, 55-71.

REFERENCES

1. Riga, G. and Balocchi, P. (2016) Seismic Sequence Structure and Earthquakes Triggering Patterns. Open Journal of Earthquake Research, 5, 20-34.

2. Papadopoulos, G.A., Charalampakis, M., Fokaefs, A. and Minadakis, G. (2010) Strong Foreshock Signal Preceding the L'Aquila (Italy) Earthquake (Mw 6.3) of 6 April 2009. Natural Hazards and Earth System Sciences, 10, 19-24.
3. McGuire, J.J., Boettcher, M.S. and Jordan, T.H. (2005) Foreshock Sequences and Short-Term Earthquake Predictability on East Pacific Rise Transform Faults. Nature, 3377, 1-5.
4. Riga, G. and Balocchi, P. (2016) Short-Term Earthquake Forecast with the Seismic Sequence Hierarchization Method. Open Journal of Earthquake Research, 5, 79-96.
5. Dodge, D.A., Beroza, G.C. and Ellsworth, W.L. (1996) Detailed Observations of California Foreshock Sequences: Implications for the Earthquake Initiation Process. Journal of Geophysical Research, 101, 22371-22392.
6. Chen, X. and Shearer, P.M. (2013) California Foreshock Sequences Suggest Aseismic Triggering Process. Geophysical Research Letters, 40, 2602-2607.
7. Das, S. and Scholz, C. (1981) Theory of Time-Dependent Rupture in the Earth. Journal of Geophysical Research, 86, 6039-6051.
8. Ishibashi, K. (1988) Two Categories of Earthquake Precursors, Physical and Tectonic, and Their Roles in Intermediate-Term Earthquake Prediction. Pure and Applied Geophysics, 126, 687-700.
9. Jiao, M., Tang, C., Zhang, G., Shi, Y. and Hou, W. (2003) A Numerical Test on Influence of Mesoscopic Heterogeneity on Macroscopic Behavior of Rock Failure and Seismic Sequance Types. Chinese Journal of Geophysics, 46, 943-953.
10. Chen, Y.T. and Knopoff, L. (1987) Simulation of Earthquake Sequances. Geophysical Journal of the Royal Astronomical Society, 91, 693-709.
11. Christophersen, A. and Smith, E.G.C. (2008) Foreshock Rates from Aftershock Abundance. Bulletin of the Seismological Society of America, 96, 2133-2148.
12. Felzer, K.R. (2004) A Common Origin for Aftershocks, Foreshocks, and Multiplets. Bulletin of the Seismological Society of America, 96, 88-98.
13. Reasenberg, P.A. (1999) Foreshock Occurrence Rates before Large Earthquakes Worldwide. Pure and Applied Geophysics, 96, 355-379.

CHAPTER 3

Earthquake Prediction

Khaled Ghaedi* and Zainah Ibrahim

[1] *Department of Civil Engineering, University of Malaya, Kuala Lumpur, Malaysia*

ABSTRACT

Among the countless natural disasters, earthquakes are capable to inflict vast devastation to a large number of buildings and constructions at the blink of an eye. Lack of knowledge and awareness on earthquake as well as its comeback is conspicuous and results in disaster; leading to bitter memories. Therefore, earthquake forecast has been a polemical study theme that has defied even the most intelligent of minds. In this chapter, an attempt was made to do an extensive overview in the area of the earthquake prediction as well as classifying them into the main strategies comprising short-, immediate-, and long-term prediction. An example of each strategy was carried out by mentioning their corresponding approaches/algorithms, such as ΔCFS, CN, MSc, M8, ANN, FFBPANN, KNN, GRNN, RBF, and LMBP; depending on the importance of each strategy. Based on these, it was concluded that, after the Tohoku-Oki earthquake with M9.0, the current orientation of the Headquarters for earthquake Research Promotion of MEXT in Japan declare that, their mission would be long-term statistical forecast of seismicity. Even, it is claimed that they do not emphasize on short-term forecasting. Besides, intermediate-term estimations are not capable to be used for prevention of all damages and protect all human life, but they may be utilized to undertake certain affordable activities to decrease damage, losses, and modify postdisaster relief. And, despite the long-term prediction is more concerned by researchers, there is no certain satisfactory level to content them. De facto, the made covenant of 1970 that investigators will be capable to forecast/predict ground excitations within a decade, still remains unmet.

Keywords: earthquake prediction, artificial intelligence, short-term prediction, intermediate-term prediction, long-term prediction, neural network

1. INTRODUCTION

Seismologists try to help human beings across the globe by taking different strategies and techniques to be aware of the most dangerous disaster, earthquake (EQ). The first usage of seismology dated back to 100 years ago as a recorder of EQ. Earthquake is the result of the faults mutation and their failure process. For failure of the faults, a massive energy has to be collected to the level where friction is broken down. Recently, seismologists have realized the source of energy. Activities of the flexible plates located at the earth surface contribute in coolness of the globe by heat transfer. Physically, the degree which is recognized an occurrence is frequently evaluated by how truly it can be predicted. Consequently, the request is not how EQ can be detected; it is preferably how accurate they can be predicted. No scientific estimation is feasible without precise description of the predicted phenomenon and the instructions, which specify evidently before it; whether the anticipation is affirmed or not [1]. Perchance, the simplest adaptive formulation in report of future incidences was carried out by the eminent Panel on Earthquake Prediction [2] in 1976. Based on it, for prediction of an EQ the geographical zone, the time period in which it would occur with sufficient meticulousness and the predictable magnitude range must be taken into consideration so that the final failure or success of the EQ prediction can willingly be judged. Furthermore, scientists should correspondingly allot a level of confidence for each prediction.

In this matter, researchers have studied and examined various techniques and approaches, such as numerical and mathematical models [3–11] to predict EQs in a desired range. A good example of finite meshing of a complex earthquake was evaluated by Landry and Barbot [8], in which a new numerical approach was presented as a numerical solution to the elastostatic equations with embedded discontinuities. The method was performed in a new earth modeling code, Gamra. In particular, recommendation of recent studies done encourage the use of several seismicity indicators or features comprising geophysical information corresponding to EQ incidence in order to predict EQ [12–14]. The relationship of such indicators using the binary class showed that some indicators displayed information fetch up nine to zero [15]. However, some investigators [16] tried to work out one step forward since all of those indicators were used within a baseline arrangement and only used the standard values, neglecting the fact that, changes in arrangements or configurations might lead to opposed results or in some conditions to better consequences.

Review of the abovementioned studies as representatives of investigations in the field of EQ prediction shows serious and heavy works, but according to studies done, to predict a potent motion, the estimation of pertaining parameters is vital for the aim of seismic analysis, seismic design, and seismic retrofitting. This study attempts to present new current methods for prediction of EQ in seismic regions. However, the EQ prediction, as mentioned, is a very difficult task which has been broadly addressed by means of various techniques, but it still sounds to achieve precise results, there is a long way to go. Therefore, to have an overview of utilized approaches, a category of these has been carried out. Herein, an effort was made to figure out those strategies used such as short-term (SHT hereinafter), intermediate-term (IMT hereinafter) and long-term (LT hereinafter) prediction in order to familiarize readers in this filed. Moreover, this chapter will go briefly through artificial intelligence branches that are being comprehensively implemented by scientists across the globe. **Figure 1** demonstrates a general process of the chapter.

2. EARTHQUAKE PREDICTION STRATEGIES

Prediction of earthquakes is routinely categorized into three divisions: SHT, IMT, and LT prediction. They may differ based on the used methods, purpose, and accuracy. The SHT prediction requests precursors. Even though some encouraging precursors are informed, the dominant pieces of evidence in Japan and other places are excessively pessimistic. The cynicism basically roots as a fact is that, the SHT precursor is commonly nonseismic and the developed tools in the seismology field are not planned to find them. Regrettably, SHT predictions that ran to practical activities have rarely been accomplished and doubtful views are prevalent in seismological society. Both IMT and LT predictions are in nature predictions of the actuarial possibility of earthquakes. LT prediction contracts with the possibility of EQ incidence on a time-scale between 10 and 100 years and mainly is based on the geological investigations of faults and epochal seismicity records. IMT prediction, unlike to LT prediction is mainly in a time-scale of 1–10 years, utilizes more of current instrumental seismology and geodesy data. Despite the negative opinions such as that of Geller et al. [17], substantial development has been created in the study of precursory model changes of seismicity like [18–20] and in case of the IMT estimation of strong EQs around the globe is already proven in the statistically stage [21]. More lately, even the attempts to simplify the main time to the SHT range are being carried out [1, 22]. Therefore, this chapter attempts to present aforesaid strategies in the EQ prediction field to update and familiarize readers with current progress of the area.

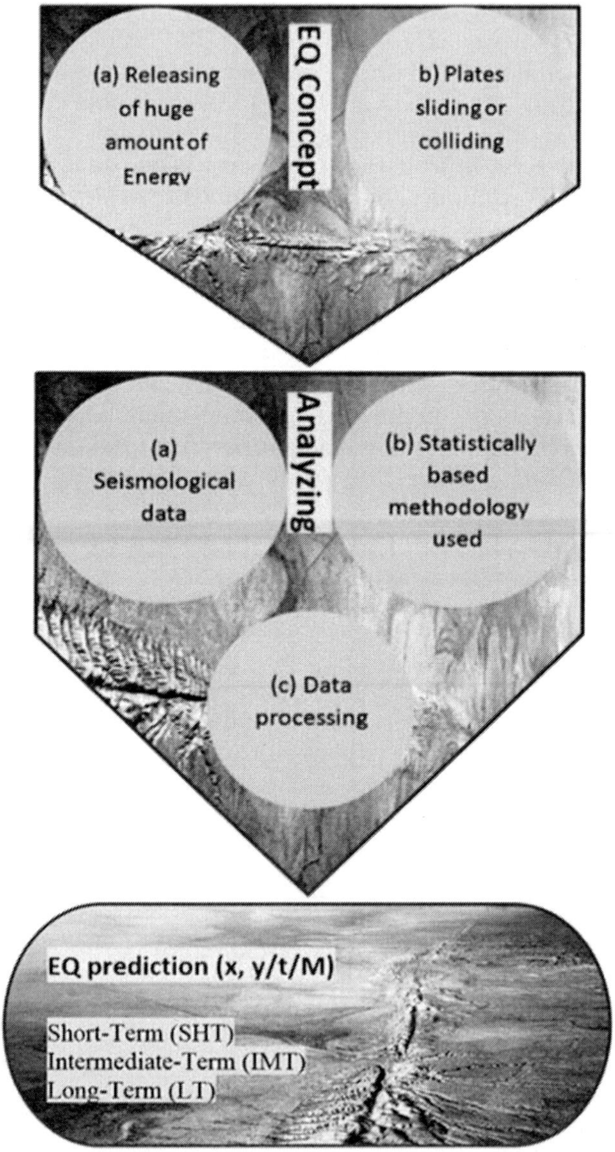

FIGURE 1. The general process of EQ prediction used by researchers is investigated in this chapter to clarify the current progress in the relevant field.

2.1. Short-Term (SHT) Prediction

Historically, a broad variety of methods have been practiced to predict EQs. Nowadays, in addition to attempts at IMT and LT prediction, SHT forecasting is on progress too. This would postwarnings at the very early signs of a momentous EQ or tsunami. EQ prediction has to determine the epicenter, EQ size and domain, and time with high precision. Amongst the short-, mid, and long-term predictions, SHT prediction may be significant, but as mentioned before, unfortunately, it has not achieved yet. Some precursors are thoroughly required for SHT detection. Several types of EQ precursors like Uyeda et al. [23] were determined. Seismological events such as fore-shocks can be considered as precursors. Nevertheless, most of precursors cannot be considered as seismological. In order to predict EQ, the national scheme (launched in Japan, 1965) has no success even a unique achievement. The explanation of this failure was that it failed to understand those precursors as explained by Uyeda [24]. In 1995, the Kobe EQ with M7.3 occurred with no prediction (**Figure 2**), in the seventh 5-year plan. With no prediction, the national project was violently criticized. After long deliberations at different points and stages, they resulted to disuse SHT forecast since precursors were highly knotty to cope with (see Swinbanks [25]) and struggles must have focused on the fundamental study that really was seismology [26]. Taking this exercise, the mission not only outlived the criticism, but investing was augmented too. After 1995, the no SHT prediction strategy was stepped up even to "make decision" that, precursors are not existent and their investigation is not science. A few years later, namely in 2011, the Tohoku-Oki EQ with M9.0 hit Japan (see **Figure 3**) and produced a destructive tsunami, which caused huge explosions, smelt at Fukushima Nuclear Plant No.1 [27] and over 20,000 people were killed.

Such an EQ was described by the severity model, however, none of them even visualize that an EQ with M9.0 occur. With this broad disappointment, researchers now even converse about eliminating a well-disciplined working group (Seismological Society of Japan) in the field of EQ forecast/prediction. The current authorized attitude of the Headquarters for EQ Research Promotion of MEXT (Ministry of Education, Culture, Sports, Science, and Technology) of Japan declares that, their mission would be long-term statistical forecast of seismicity. Even, it is claimed that they do not emphasize on SHT forecast. These are weird positions for accountable authorities, once the people instantly need all conceivable information on next ground motions, for example, the Ryukyu trench as shown in **Figure 4** with high potential, particularly after the ruinous Tohoku EQ, 2011 with M9.0. This is inoperative. Recently, even the national project title has been altered to "Promotion of EQ and Volcano Observation Research Project to Contribute to Disaster Mitigation", eliminating "prediction" term.

Prediction of such ground excitations are not any more a simple monetary source, but the ship has shifted the helm to "catastrophe prevention".

FIGURE 2. A damaged massive bridge and railways by Kobe EQ, 1995, January 17, M7.3.

FIGURE 3. Uprising the Sekii River and burning the port of Sendai during the Tohoku-Oki tsunami, 2011, March 3 and 11, M9.0.

Herein, the SHT prediction has been described as an untrusted strategy compared to other prediction techniques. Therefore, an attempt is made to present an example of this strategy in a brief manner just in order to acquaintance.

In this regard, the Chiayi zone (southwestern Taiwan) is selected with a significant deformation rate which has been hit by several devastating EQs, e.g., Chiayi EQ, 1792 [28]. This EQ can be due to the Meishan fault rupture. Moreover, the Meishan EQ sequence (during 1904 and 1906) caused wide fatalities and numerous buildings collapsed. Nevertheless, a SHT probabilistic seismic hazards assessment was applied, as shown in **Figure 5**, to the Chiayi zone that involved the Meishan EQ sequence. **Figure 6** and **Figure 7** show the Touliu, 1904, Meishan, 1906, and Yangshuigang EQ, 1906 which is used as the source events in order calculate ΔCFS. Here, the rate and state friction model [29] is considered for assessing SHT seismicity-rate evolution. The Coulomb failure stress changes, ΔCFS, by EQ are computed for the model application. Based on the constant apparent friction model, the general state for ΔCFS can be written as follows:

$$\Delta \mathrm{CFS} = \Delta\tau + \mu'\Delta\sigma_n \quad (1)$$

in which Δτ is the changes of shear stress calculated along the receiver fault slip. μ′μ′ is the coefficient of apparent friction; and ΔσnΔσn is the changes of normal stress which is perpendicular to receiver fault. Former studies, e.g., [30] have discovered that a positive change of stress boosts following events. In contrast, a negative change of stress debars the future seismicity events. Additionally, the focal mechanisms or receiver fault mechanism should be used as an information key for calculation of the ΔCFS. According to those previous investigations [30, 31], a temporally fixed fault is assumed and procedure is made. The ΔCFS should be estimated on each grid (e. g., 0.01° × 0.01°) and solved partial changing of receiver faults by means of an appropriate program such as Coulomb 3.3.

FIGURE 4. An EQ probability along the Ryukyu trench, Japan.

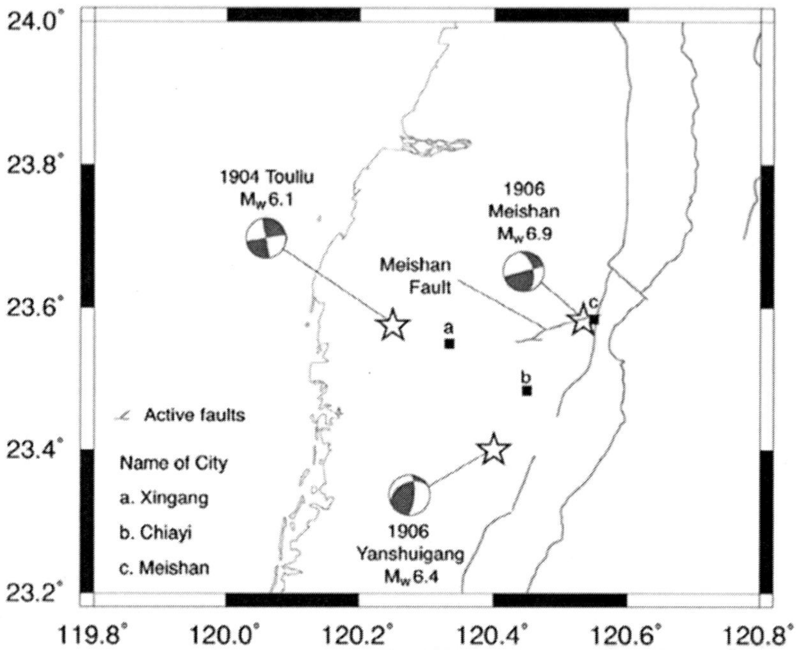

FIGURE 5. Propagation of three EQs with M ≥ 6.0 for the Meishan EQ sequence between 1904 and 1906. Three Xingang, Chiayi, and Meishan cities considered for seismic hazards assessments and are depicted as black squares. Stars display the EQs epicenters. The location of the illustrated region is given in **Figure 7**.

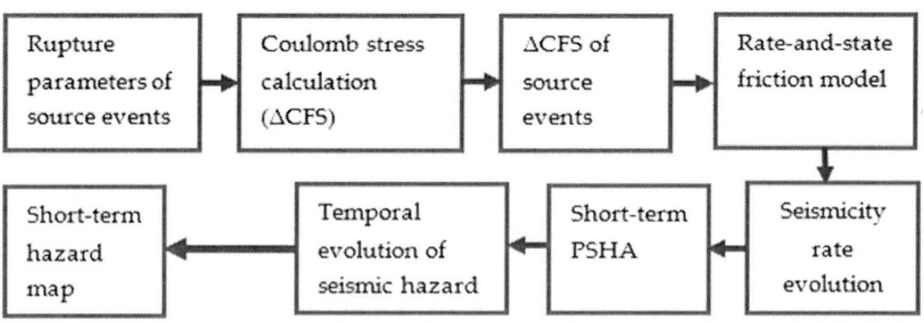

FIGURE 6.The flowchart for an approach of SHT prediction.

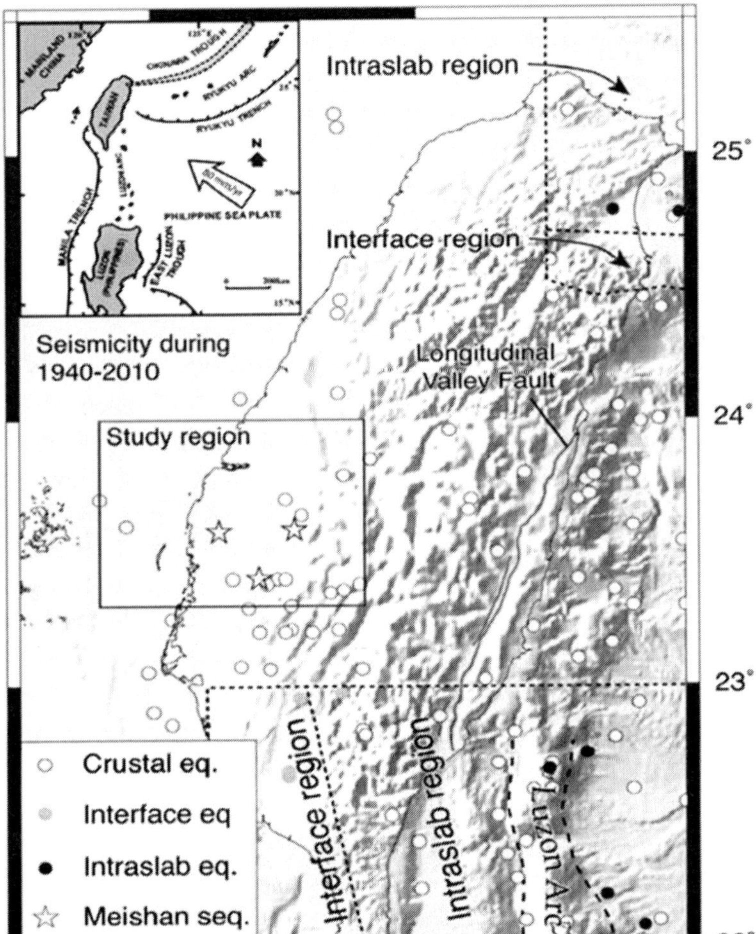

FIGURE 7. The repartition of declustered EQs with *M* ≥ 5.0 during 1940 and 2010. Note that, some interface actions in northeastern Taiwan occurred at the longitude 122° east are not revealed in this figure.

The ΔCFSs divulged by the Meishan EQ sequence have been assessed in the Chiayi zone (**Figure 8**). The results showed notable intensifications in the neighborhood of each event. The M6.9 Meishan EQ (**Figure 8(b)**) produced a stress, that increased with a vaster range and a severer magnitude. In contrary, the M6.1 Touliu EQ (**Figure 8(a)**) generated less considerable stress disturbance. Such an inconsistency can be ascribed to the difference of magnitude between EQs. In short, the outcomes demonstrated that, a large EQ close to the epicenter zone of the Yanshuigang EQ could be predicted. It is noteworthy mentioning that, some areas with the highest increases of Coulomb stress did not take place subsequent to devastating EQs (**Figure 8**). Such consequences can be attributed to the low rate of cumulative stress (away from next coming rupture, or to relevant tectonics not found with seismogenic).

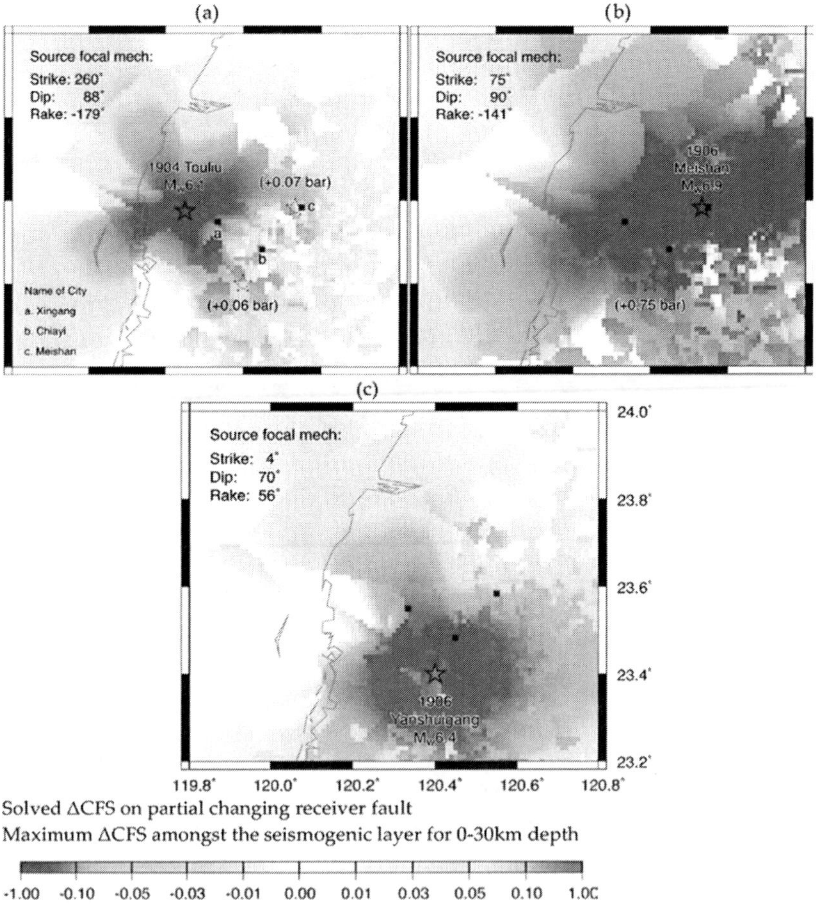

FIGURE 8. The Coulomb stress changes divulged by (a) the Touliu, 1904, (b) the Meishan, 1906, and (c) the Yanshuigang EQ, 1906. Changes of stress nearby the epicenters of subsequent EQs that presented as dash stars are illustrated. (a) The Touliu EQ, 1904. (b) The Meishan EQ, 1906. (c) The Yaunshuigang EQ, 1906. Solved ΔCFS on partial changing receiver fault, maximum ΔCFS among the seismogenic layer for 0–30 km depth.

2.2. Intermediate-Term (IMT) Prediction

In the last few decades the serious EQ model, that is according to considerations pertaining to accelerating seismic transformation and dynamic concepts of the critical point, has been suggested by several seismologists as an advantageous tool in order to predict IMT EQ. IMT EQ prediction is divided into various algorithms such as CN, MSc (Mendocino Scenario), M8, and M8S. The common methodology to the different algorithms creates usage of generic concepts of pattern identification that allow dealing with several sets of EQ precursors, and permits for a regular seismicity monitoring as well as for an extensive testing of the predictions. In fact, predictions may be helpful if their precisions are recognized, but not certainly high. It is the standard exercise to all forecasting difficulties comprising national defense. For instance, the MSc algorithm reduces the estimation area to a fine range (2–3 times the EQ source zone size) and for some cases approximately a precise location (inside the source zone dimension). The M8 algorithm forecasts EQ with several years (intermediate-term) middle-range (5–10 times the EQ source zone size) precision [32]. Though, the long-running global testing IMT and middle-range EQ forecasts support the theory that, the algorithm of M8 can be efficient to globally decrease the effect of strong EQs ($M \geq 8.0$) [32, 33], there is no connexion to implement events and improve EQ readiness in a reaction to them. This is not ended so far; the EQ prediction might have been used to fulfill measures and amend EQ preparedness beforehand; unfortunately, this was not achieved, in part owing to the limited distribution of predictions and the absence of applying current approaches in order to use IMT detections to take action and make decisions [32]. Overall, IMT estimations are not capable to be used for prevention of all damages and protect all human life, but they may be utilized to undertake certain affordable activities to decrease damage, losses and modify postdisaster relief. Davis et al. [34] proposed examples and methodologies on how activities may be occupied in reflex to predictions. Davis [35] explained how to employ economic parameters to ignored equations for optimization of activities that may be affected a forecast. Such activities are fruitful to improve the normal LT seismic risk reduction approaches, such as building codes or standard disaster preparation methods. As an example, these processes were recognized [34] and models were prepared for the Tohoku EQ on how prudent, cost effective, and reasonable conclusions can be made to decrease destructive EQ influences. Information provided through M8–MSc algorithms may reduce the worldwide effects from the strongest EQs. Retrospectively, there were many precursor-like anomalies earlier the Wenchuan EQ, 2008 (see [36] for review), but no anomaly was so decisive until the SHT or IMT alarm was used. Indeed, so long as seismologists deal with the SHT to IMT EQ forecast, which has many uncertainties and is frequently linked to emergency management counteractions, e. g., evacuation activities, this public stress becomes even larger.

A simple explanation is prepared herein to express some potentials on how the EQ forecasting may have been used by means of the IMT strategy, e. g., EQ prediction algorithms using CN, M8, and MSc, real-time forecasts using the M8-MSc algorithms and other applications of M8 algorithm. In brief, the CN, MSc, and M8 algorithms, as the main algorithms for IMT prediction of EQs, are present as below.

2.2.1. CN Algorithm

CN algorithm is structured accordance with a pattern identification scheme to permit an analysis of the times of increased probability (TIP) for the event of strong EQs. It represents the possible occurrence (within a specified time window and region) with a magnitude larger than an immovable threshold, based on a measurable investigation of the seismic study. The seismicity patterns are obtained using a set of experiential time functions (assessed on the order of the incidences occurred in the analyzed zone) and description of the seismic activity level), seismic reticence and space-time collecting of events. Therefore, CN algorithm makes usage of the information arranged by minor and moderate EQs, having more or less good stats within the surrounded region, to forecast the severer EQs, which are infrequent events.

Since the period of TIPs varies from few months to few years, CN estimations are categorized with a time uncertainty over the years and with a space uncertainty of several kilometers, the so-called medium-range predictions, in relation to an entire single monitored area. Accordance with CN, once a TIP is notified, the strong EQ could take place in any location of the alerted region; accordingly, defined regions ought to be small. Nonetheless, the algorithm is on the basis of precursors, which can be hosted in a region with linear dimensions extremely larger than the dimension of the anticipated source. At the end, taking into consideration the accuracy of CN algorithm in global tests and the low proportion of occurrence of the strong EQs, it may be possible to predict the contingent possibility for a TIP about 40%. So, as revealed by Peresan et al. [37] a notified TIP has roughly 60% probability to be an incorrect alarm, whilst if TIP is not designated, at 96% of probability no potent EQ will happen.

2.2.2. MSC Algorithm

Algorithm MSc was introduced and named by retroactivity examination of the local seismic catalog before the Eureka EQ, 1980 with *M*7.2, in California nearby Cape Mendocino. By having a TIP recognized for a given territory

(U) at the time (T), the algorithm is considered to detect U, in a smaller area (V) in which the predicted EQ can be anticipated. An application of this algorithm needs a rationally complete catalog of EQs with ranged magnitudes ($M \geq 4.0$), that is lesser than the least threshold normally utilized via M8 algorithm. The nature of MSc algorithm is briefly described as follows:

a. Coarse-grained U (territory) is changed to fine squares with $s \times s$ dimension. Let say i and j is the center coordinates of the squares. In each coordinate (i, j), the number of EQs n_{ij} (k, the subsequence quantity of a specific time window) including aftershocks is computed for consecutive (short time windows), months long (u), onset from initial time to onward (=6 years), to permit the EQs which is associated with the TIP's recognition. The considered time-space should separate into small cases (i, j, k) of the dimensions (s × s × u).

b. "Quiet" cases are separate from each small square with i, j coordinate; they are differentiated by n_{ij} (k) that is under the n_{ij} Q percentile/centile.

c. The clusters of more quiet cases or Q linked in place or in a time are known. The area (V) is the regional pattern of these clusters.

The adjusted values of parameters have been standardized for the Eureka EQ and are as follows:

$$u = 2 \text{ months}$$

$$Q = 10\%$$

$$q = 4$$

$$s = 3D/16$$

where, D is the circle diameter used in M8 algorithm. Note that, the phenomenon utilized in the MSc algorithm may reflect the shorter-term (second) phase of premonition increase of seismic motion near the initial source of main-shock.

2.2.3. M8 Algorithm

The M8 algorithm was introduced and named by retroactivity examination of the seismicity prior to the greatest EQs worldwide (M8+). It is according to an ordinary physical system of forecast, which is briefly presented as subsequent writing. M8 is a written program in Fortran 77 which agrees in order to predict the EQ in the IMT by means of the M8 algorithm. Also, the algorithm is used to evaluate a time series of integral numbers according to transient seismicity within a region.

2.2.3.1. Prediction is designed at eqs with magnitude m ≥ 0

Various values with a step of 0.5 for M_0 are considered. Overlying circles with diameter of D (M_0) monitor the seismic region. In each circle, EQ sequence is deliberated with after-shocks eliminated {ti, hi, mi, bi(e)} where i = 1, 2,....

Here, ti is the source time and ti ≤ ti + 1; hi and mi stand for focal depth and magnitude and bi(e) represents the number of after-shocks within the first (e) days. The sequence is standardized by the less magnitude (C), in which C is the standard value for the annual average number of EQs in the sequence. The used magnitude scale must reflect the EQ source size.

2.2.3.2. Calculation of numerous running norms in the time windows

In this case, they illustrate diverse measures of severity in EQ flow, its deviance from the trend of long-term and grouping of EQs. These averages contain:

the number of main-shocks, $N(t)$, the deviance of $N(t)$ from the trend of long-term, $L(t)$, and the linear concentration of the main-shocks, $Z(t)$, considered as the average diameter ratio of the source, i. e., l, to the r (average distance) between them. The EQ sequence, {i}, is deliberated within the time window. The functions of N, L, and Z are calculated for C = 20 and 10. Hence, the EQ sequence gives a strong averaged statement by seven functions; namely N, L, and Z (twice each) and B (the maximum number of after-shocks). **Figure 9** shows the criterion of the M8 algorithm in the seismic extended standardized phase space.

2.2.3.3. A tip or an alarm (increased probability time)

It is announced for 5 years once at least six out of seven aforesaid functions including B become too large within a limited time window; namely (t–u, t). To fix prediction, this statement is necessary for two successive periods, t and t + 0.5 years.

To sum up, prediction algorithms of an EQ, through above mentioned algorithms, are in accordance with a comprehensive general scheme demonstrated in **Figure 10**. A seismically energetic region is considered with an example of areas, generally, (CIs) circles of investigation (a), in which the areas have their own seismic events "history" of various magnitude (b). Each "history" of seismic activity is described based on the specified exactly definable moving numbers (c), which synthesis is expose to pattern diagnosis of "precursor" representing whether the next periods are a TIP or not, for the expected target EQ occurrence (d).

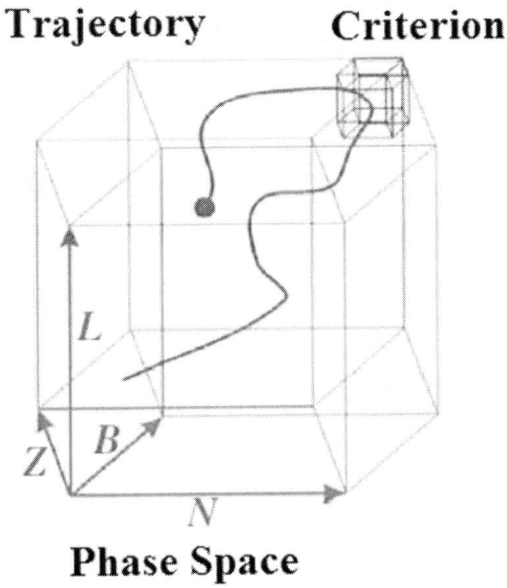

FIGURE 9. The parameters of N, L, Z and be for the *M*8 algorithm in the seismic extended standardized phase space.

2.3. Long-Term (LT) Prediction

Despite the serious effort done and the several models developed [38], no prosperous technique has been detected yet. Because of the random actions of EQs, it may not be possible to determine the exact location, magnitude, and time of the next fatal EQ. Most recently, Tohoku, Japan was hit by an EQ with M9.1, on March 11, 2011. This excitation caused approximately 20,000 deaths, more than $300 billion (USD) in detriment and it proved impossible to predict the LT disasters (it ranks amongst the most devastating natural hazards ever recorded). As regards, it occurred in a well-prepared and well-organized country where great efforts have been made to reduce seismic hazard, it was a stupendous experience. Long-term procedures are occasionally defined as timeless historical risk valuation, whereas SHT preEQ processes mean a detected process happening minutes to months earlier than an EQ. Considering long-term seismicity properties is much more problematic than similar investigations of SHT and IMT changes of EQ event rates [39]. A reason for this can obviously be lack of good documents and analogous long time catalogs of EQs. Existing historical catalogs are not strongly homogeneous both in space and time. In fact, primary investigations of active faults as well as their LT seismicity spreading disclosed that, majority of active regions instrumental seismicity catalogs do not include the seismic sequence. Even, in active areas where the epochal seismicity covers more than 1000 years, e.g., the Fault of Dead Sea, paleoseismology prepared substantial results on the return period of fatal EQs with fault properties and corresponding magnitude of past seismic actions that complete old scripts [40, 41]. Investigations of the LT performance of seismogenic faults and correlated paleoseismic statistics are today often mandatory in seismic hazard characteristics of applications for building services and as participation to national and international requirements for seismic safety [42]. More significantly, current works query the credit of seismic hazard maps mentioning the weakly constrained EQ limitations resulting from overlooking paleoseismic data for the risk analysis [43]. Prospective EQ predictions make scientific theories of EQ occurrence refutable, transparent, and testable. A main step along this direction was undertaken by the Regional Earthquake Likelihood Models (RELM) working group, which demanded LT (5 year) predictions for California city in a particular format to ease syllogistic testing [44, 45].

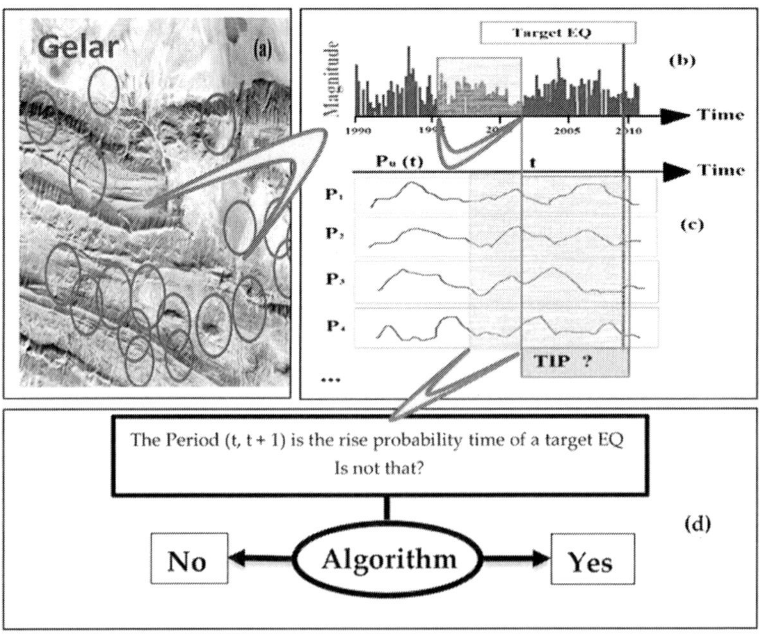

FIGURE 10. A general design of an EQ detection tool.

The aim of this section is to present the current progress in LT EQ prediction and its effect on the seismic risk assessment in zones with long epochal EQ excitation records. Herein, the LT forecast has been described as a strategy which has a long path to complete the mission that is EQ prediction with a secure and accurate manner. Therefore, an effort is made to present an example of this prediction, in the particular application of neural networks, in a short manner, as presented in the subsequent writings (Section 2.3.1), to acquaint. Despite using Bat-ANN algorithm (a combination of Bat Algorithm, BA, with Artificial Neural Network, ANN, algorithm) in one of the most recent investigations [46] to predict EQs in Pakistan regions, but in conclusion the research is confessing that, due to more diversity provided by the method and stochastic approach, used algorithm has "further chances" to detect global goals (still waveringly). In a nutshell, an EQ location of a given magnitude range and time prediction can be arranged into the listed categories as illustrated in **Table 1** based on spatial accuracy and its temporal [47].

TABLE 1. Precision classification of EQ prediction.

Temporal, in years		Spatial, in source zone size	
Short-term (SHT)	0.01–0.1	Narrow	2–3
Intermediate-term (IMT)	1	Middle-range	5–10
Long-term (LT)	10	Long-range	Up to 100

2.3.1. Artificial neural networks (ann)

A number of EQ forecast approaches including artificial intelligence (AI) have used. One of these techniques is the ANN which has presented a good ability for detecting solutions in different fields. Variations of different algorithmic have been designed and proposed to expand the ANN accuracy. The ANN algorithm is a challenge to model things with sophisticated software or professionalized hardware, the several layers of processing elements via neurons. The human brain as the most fundamental element is a particular kind of neuron that furnishes one with the capacities to think, remember, and apply past skills to every practice. The benefit of this frame is that, the ANN delivers a black-box method and the operator does not require to distinguish much about the process nature being simulated. Having this in mind, ANNs have been promising methods in order to predict and detect locally imminent EQs according to reliable seismic information. For these reasons, the ANN has recently been broadly applied to different areas to prevail the problem of exclusive and nonlinear relationships.

Herein, the section tries to briefly investigate the methodical possibilities of EQ forecast by means of Neural Networks (NNs). Moreover, this short investigation provides an accurate layout of different precursors; namely PGA, and main-shock detection, however, other precursors such as radon detection, liquefaction, and aftershocks should also be taken into consideration. In addition, it is also discussed how these precursors are used by NN in order to EQ detection and prediction. A corresponding network analysis is stated for each seismic precursor, beside the type used NN.

2.3.1.1. Prediction of peak ground acceleration (pga)

A PGA is an acceleration measure of EQ due to extensive ground motion and it is induced by a powerful energy released from an EQ, leading to earth deformation like landslides, surface ruptures, and liquefaction. In this regard, Derras and Bekkouche [48] presented a comprehensive method to assessing the maximum PGA by means of a feed-forward back propagation neural network (FFBPANN). The result was compared with those ground motion prediction equations (GMPEs) demonstrated by Ambraseys and Douglas [49]. The GMPEs were utilized as a substitute method for the prediction of the PGA where accelerogram monitoring stations do not exist. Such an approach needs a wide data of PGA values as well as the site coefficients. An FFBPANN was projected with an overall selected 1000 epochs set and a tangent hyperbolic sigmoid activation function containing five input parameters; namely the depth of focus where an EQ was activated, the Japanese Meteorological Agency magnitude (MJMA)(MJMA), the epicentral distance, the resonant frequency, fxfx, and the sedimentary layers' thickness, ZxZx. Prominently, shear wave velocity with x = 800 m/s is constant for both fxfx and ZxZx parameters. **Figure 11** presents a visual view of the site seismic parameters used for a PGA estimation [48].

The configuration of ANN used in **Figure 11** contains of 1850 testing data taken out from Kiban Kyoshin network (KiK-net) data and 326 training. An evaluation between an FFBPANN and selected GMPE model [49] indicated that, the FFBPANN performance is far compared to those GMPEs. The determination coefficient (R2)(R2) predicted by an NN was 0.94 in comparison of those two GMPE methods, which were 0.82 and 0.76. In the same site, the normalized root mean square error (NMRSE) for the NN models was noticeably smaller, 0.11%, in comparison with the selected GMPE approaches, which depicted a corresponding NMRSE of 0.17 and 0.25%. It is approving that, a PGA estimation by means of an NN is much better than the GMPE approaches from both accuracy and performance

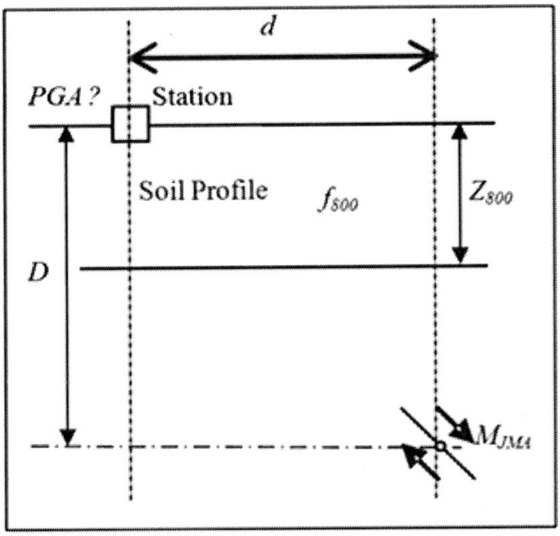

FIGURE 11. Site seismic parameters in assessment of a PGA.

point of view. Specifically, the outcomes showed that, the parameter of epicentral distance heavily affects the result of a PGA value, procuring the best mean square error (MSE)and *R* values of 0.075/0.076 and 0.51/0.48 for the testing phases and training, respectively. In contrary, the site parameters and focal depth have the minimum effect on the PGA value outcome. In addition to these, a mixture of altogether five parameters was seen to return the optimum results, keeping the MSE value with 0.0205 and 0.0203 and an *R*-score (correlational coefficient) with 0.84 and 0.85, for the testing phases and training. The *R*-score shows the independence degree betwixt an output and an input. An *R* coefficient closer to ± 1 displays a powerful correlation, though, $R = 0$ is considered as an inaccurate prediction.

For prediction of PGA values in a specific site a different direction was discovered [50] using an ANN algorithm and microtremor measurements. On the other hand, microtremor measurements are an experiential method to collect all vibrations of ground in a very scant time interval, thereby taking an obvious superiority over a traditional record database. Applying predefined data for training and network validating utilizing a BP syllogism, the ANN method was developed. Particularly, three types of input neurons were assessed: the focal depth, epicentral distance, and EQ magnitude. The results revealed that, using input parameters obtained 0.972 for an *R*-score, which was interestingly higher superior to two other input parameters (on an average 0.6–0.9) and one input parameter (on an average under 0.6). A comparative study of this issue has given clear consequences that microtremor measurements are literally efficient once time is limited, but do not have performance ability. On the contrary, an ANN is noteworthy in performance, nonetheless as a weak point extremely relies on predefined data. The models of three neural networks are shown in **Figure 12** and a comparison between microtremor measurements and a used ANN for the PGA approximation is shown in **Figure 13**, respectively [50].

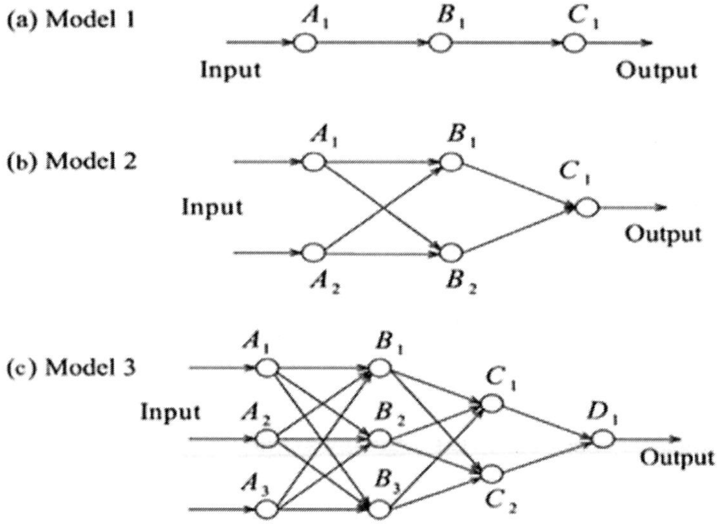

FIGURE 12. Neural networks models.

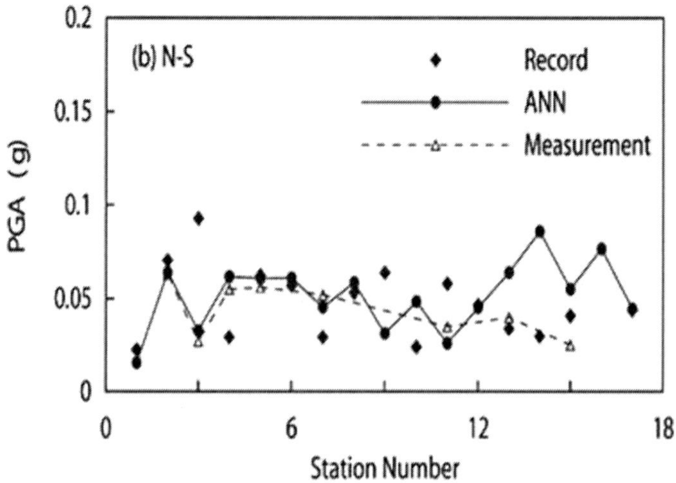

FIGURE 13. PGA approximation.

Additionally, several novel methods have been examined to estimate PGA values, giving different accuracy rates and performance levels. Günaydın and Günaydın [51] presented a comparative method to PGA forecast by assessing three ANN types; namely a generalized regression neural network (GRNN), feed-forward back propagation (FFBP) and a radial basis function (RBF). All these ANN types were trained by means of a back propagation (BP) procedure through solely one hidden layer and were also assessed using four input request parameters; namely the EQ moment magnitude, focal depth, site conditions, and hypocentral distance. These request parameters, through 15 accelerometers for 95 records, were taken from three wave directions including east-west, north-south, and up-down). In summary, for forecasting horizontal and vertical values of PGA the FFBPANN and GRNNANN were the ideal options.

2.3.1.2. Prediction of large-magnitude of main-shocks

Panakkat and Adeli [13] performed an analysis using three different types of NNs; namely a Levenberg-Marquardt back propagation neural network (LMBP)recurrent neural network (RNN), and a radial basis function (RBF) neural network, to eventually forecast a strongest EQ one month earlier. The input request parameters were alike to the neurons described in [52]. Comparative results amongst three different networks indicated that, for ranging from M5.0 to M6.0, the RNN showed 0.20–0.51 for an *R*-score, while LMBP and RBF networks acquired ranging from 0.01 to 0.14 and 0.12 to 0.37, respectively, for an *R*-score. Consequently, it was apparent that, an RNN is able to determine large EQs (M6.0+) more accurate and faster than both RBF and LMBP networks. **Figure 14** shows an RNN layout. Despite the fact, the presented EQ forecast cannot be completed with a satisfactory degree of certainty.

Meantime, Reyes et al. [12] lately developed an approach for determining EQs in Chile by employing a BPANN in a three layer using the *b*-value, Omori-Utsu's and Bath's law. The network result is twofold [53]. First, it shows the probability of a magnitude beyond a defined data threshold. Next, the network will produce a possible magnitude that might take place within an interval of 5 days. A comparison of different NNs like a KNN [54], an ANN, K-means clustering [55], and an SVM [56] indicated that, except SVM, a KNN, an ANN, and K-means clustering give better forecast precisions. For validation of the used network, 500 eons were utilized in four unlike zones in Chile. A comparison among the NNs demonstrated that, the overall measured performance according to sensitivity and specificity values was highly location dependent. Specifically, the sensitivity of 35.7, 42.9, and 50% was recorded for the ANN, KNN, and K-means clustering, respectively. Inconclusive results were given for the SVM.

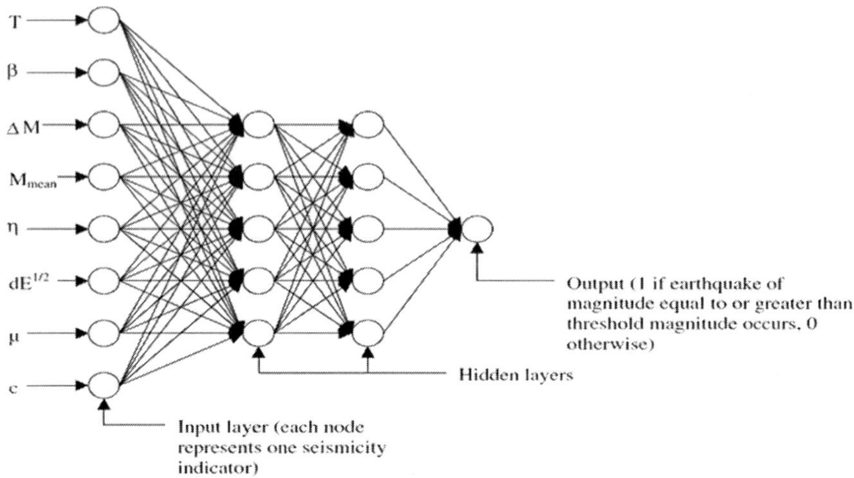

FIGURE 14. Structure of the back propagation NN to predict an EQ occurrence [**13**].

3. CONCLUSIONS

A brief description in anatomy of an EQ causes has been made and the existing problem in relation to predict and detect this fatal phenomenon has been stated. Therefore, the prediction strategies have been classified into three main scenarios; namely short-term (SHT) prediction, intermediate-term (IMT) prediction and long-term (LT) prediction. For each strategy, an attempt has been made to present either an example or a general trend of that strategy to familiarize readers more in the corresponding field as well as presenting an overall background of the researches done by several investigators worldwide. Based on these explanations the below conclusion is drawn:

In the field of EQ prediction, numerous researchers have vastly done serious investigations by means of diverse techniques to improve seismic reliability, recognize warning of EQ in a specific zone, and eventually decrease the negative effects of this natural hazard on human life. The improvement of science and equipment leads experts to develop the strategies of earthquake prediction more accurately. Based on the practical, numerical, and equational studies, it can be detected that majority of SHT predictions cannot be fruitful. Indeterminate and random surface displacement is normally seen in the LT predictions. Practical applications need the knowledge of ground shaking especially in analyses of liquefaction. In addition to these, AI branches/algorithms such as ANN, KNN, SVM, RBF LMBP, and RNN, however, have been proven with more accuracy and intensively used in EQ estimation field because of obtaining more satisfactory results, but regrettably, no general beneficial approach to precisely predict EQs has been detected yet. As a matter of fact, it may not be possible to determine the exact time once a devastating EQ will take place. It is because, as soon as sufficient strain has made, a fault possibly will become innately unstable and all teeny background EQ may or may not keep rupturing and become a large EQ. In particular, the major novelty of these efforts is the growth of a system capable of forecasting EQ occurrence for phases of time from statistical point of view, in order to permit the administrations to organize cautionary policies.

4. ACKNOWLEDGEMENTS

The authors gratefully acknowledge the supports given by University of Malaya Research Grant (UMRG - Project No. RP004A/13AET) and Fundamental Research Grant Scheme, Ministry of Education, Malaysia (FRGS - Project No. FP028/2013A).

REFERENCES

1. V. G. Kossobokov, "Testing earthquake prediction methods: The west pacific short-term forecast of earthquakes with magnitude MwHRV z 5 . 8 r," *Tectonophysics*, vol. 413, pp. 25–31, 2006.
2. C.R. Allen, W. Edwards, W.J. Hall, L. Knopoff, C.B. Raleigh, C.H. Savit, M.N. Toksoz, and R.H. Turner, "Predicting earthquakes: A scientific and technical evaluation with implications for society," *Panel on Earthquake Prediction of the Committee on Seismology, Assembly of Mathematical and Physical Sciences, National Research Council, U.S. National Academy of Sciences, Washington, DC.*, 1976.
3. A. Borghi, A. Aoudia, R.E.M. Riva, and R. Barzaghi, "GPS monitoring and earthquake prediction: A success story towards a useful integration," *Tectonophysics*, vol. 465, no. 1–4, pp. 177–189, 2009.
4. E. Florido, F. Martínez-álvarez, A. Morales-esteban, J. Reyes, and J.L. Aznarte-mellado, "Detecting precursory patterns to enhance earthquake prediction in Chile," *Comput. Geosci.*, vol. 76, pp. 112–120, 2015.
5. T. Hongho, S. Chongyang, W. Jin, W. Jian, and X. Songhai, "Simulation of coseismic effects of the Ms7. 0 Lushan earthquake," *Geod. Geodyn.*, vol. 4, no. 3, pp. 12–18, 2013.
6. L. Huang, Y. Du, J. Zheng, and Y. Zhang, "Detection of precursory wave using a novel sensor and its application to earthquake prediction," *Procedia Soc. Behav. Sci.*, vol. 3, pp. 613–618, 2012.
7. M. A. Jafari, "Statistical prediction of the next great earthquake around Tehran, Iran," *J. Geodyn.*, vol. 49, pp. 14–18, 2010.
8. W. Landry and S. Barbot, "Gamra: Simple meshing for complex earthquakes," *Comput. Geosci.*, vol. 90, pp. 49–63, 2016.

9. R. Kumar, H. Nankali, B. Kundu, P. Patel, and V. K. Gahalaut, "Finite fault slip models for the 11 August 2012 Varzaghan-Ahar, NW Iran earthquakes (Mw 6 . 4 and 6 . 3) from near-field GPS measurements of coseismic offsets," *J. Asian Earth Sci.*, vol. 115, no. August, pp. 268–272, 2016.
10. Z. Shoubiao, "A finite-element simulation of 2008 Wenchuan earthquake and implications for numerical earthquake prediction," *Geod. Geodyn.*, vol. 2, no. 2, pp. 39–42, 2011.
11. A. Szakács, "Earthquake prediction using extinct monogenetic volcanoes: A possible new research strategy," *J. Volcanol. Geoth. Res.*, vol. 201, no. 1–4, pp. 404–411, 2011.
12. J. Reyes, A. Morales-Esteban, and F. Martínez-Álvarez, "Neural networks to predict earthquakes in Chile," *Appl. Soft Comput.*, vol. 13, no. 2, pp. 1314–1328, 2013.
13. A. Panakkat and H. Adeli, "Neural network models for earthquake magnitude prediction using multiple seismicity indicators," *Int. J. Neural Syst.*, vol. 17, no. 1, pp. 13–33, 2007.
14. A. Morales-esteban, F. Martínez-álvarez, and J. Reyes, "Earthquake prediction in seismogenic areas of the Iberian Peninsula based on computational intelligence," *Tectonophysics*, vol. 593, pp. 121–134, 2013.
15. F. Martínez-Álvarez, J. Reyes, A. Morales-Esteban, and C. Rubio-Escudero, "Determining the best set of seismicity indicators to predict earthquakes. Two case studies: Chile and the Iberian Peninsula," *Knowl. Based Syst.*, vol. 50, pp. 198–210, 2013.
16. G. Asencio-cortés, F. Martínez-álvarez, A. Morales-esteban, and J. Reyes, "A sensitivity study of seismicity indicators in supervised learning to improve earthquake prediction," *Knowl. Based Syst.*, vol. 101, pp. 15–30, 2016.
17. R. Geller, D. Jackson, Y. Kagan, and F. Mulargia, "Earthquakes cannot be predicted," *Am. Assoc. Adv. Sci.*, vol. 275, no. 5306, pp. 1616–1617, 1997.
18. Q. Huang, "Search for reliable precursors: a case study of the seismic quiescence of 2000 western Tottori prefecture earthquake," *J. Geophys. Res.*, vol. 111, p. B04301, 2006.
19. M. Wyss and A. Martirosyan, "Seismic quiescence before the M7, 1988, Spitak earthquake, Armenia," *Int. J. Geophys.*, vol. 134, pp. 329–340, 1998.
20. Q. Huang, G. A. Sobolev, and T. Nagao, "Characteristics of the seismic quiescence and activation patterns before the M = 7.2 Kobe earthquake, January 17, 1995, 2001.," *Tectonophys*, vol. 237, pp. 99–116, 2001.
21. V. G. Kossobokov, L. L. Romashkova, V. I. Keilis-Borok, and J. H. Healy, "Testing earthquake prediction algorithms: statistically significant real-time prediction of the largest earthquakes in the circum-pacific, 1992–1997," *Phys. Earth Planet. Inter.*, vol. 111, no. 3–4, pp. 187–196, 1999.
22. S. Uyeda, T. Nagao, and M. Kamogawa, "Short-term earthquake prediction: Current status of seismo-electromagnetics," *Tectonophysics*, vol. 470, no. 3–4, pp. 205–213, 2009.
23. S. Uyeda, T. Nagao, and M. Kamogawa, "Earthquake precursors and prediction," Solid Earth Geophysics, Part 5, *Springer*, pp. 168–178, 2011.
24. S. Uyeda, *On earthquake prediction in Japan*, vol. Ser. B 89, pp. 391–400, 2013.
25. D. Swinbanks, "Quake panel admits prediction is 'difficult'," Nat. 388, p. 4, 1997.
26. S. Uyeda, "Current affairs in earthquake prediction in Japan," *J.Asian Earth Sci.* vol. 114, pp. 431–434, 2015.
27. S. Tanaka, "Accident at the Fukushima Dai-ichi nuclear power stations of TEPCO – outline & lessons learned," *Proc. Japan Acad.*, vol. Ser. B 88, pp. 471–484, 2012.
28. C. H. Chan, Y. M. Wu, C. T. Cheng, P. S. Lin, and Y. C. Wu, "Scenario for a short-term probabilistic seismic hazard assessment (PSHA) in Chiayi, Taiwan," *Terrest. Atmosph. Ocean. Sci.*, vol. 24, no. 4 PART II, pp. 671–683, 2013.
29. J. Dieterich, "A constitutive law for rate of earthquake production and its application to earthquake clustering," *J. Geophys. Res.*, vol. 99, pp. 2601–2618, 1994.
30. C. H. Chan, Y. M. Wu, and T. L. Lin, "A short term seismic hazard assessment in Christchurch, New Zealand, after the M 7.1, 4 September 2010 Darfield earthquake:

An application of a smoothing Kernel and rate- and-state friction model," *Terr. Atmos. Ocean. Sci.*, vol. 23, pp. 231–236, 2012.
31. C. H. Chan, Y. M. Wu, and J. P. Wang, "Earthquake forecasting using the rate-and-state friction model and a smoothing Kernel: Application to Taiwan," *Nat. Hazards Earth Syst. Sci.*, vol. 12, pp. 3045–3057, 2012.
32. C. Davis, V. Keilis-borok, V. Kossobokov, and A. Soloviev, "Advance prediction of the March 11, 2011 Great East Japan Earthquake: A missed opportunity for disaster preparedness," *Int. J. Disaster Risk Reduct.*, vol. 1, pp. 17–32, 2012.
33. V. G. Kossobokov, "Earthquake prediction: 20 years of global experiment," *Nat. Hazards*, vol. 69, no. 2, pp. 1155–1177., 2012.
34. C. Davis, V. Keilis-Borok, G. Molchan, P. Shebalin, P. Lahr, and C. Plumb, "Earthquake prediction and disaster preparedness: interactive analysis," *ASCE Nat. Hazards Rev.*, vol. 11, no. 4, pp. 173–84, 2010.
35. C. A. Davis, "Loss functions for temporal and spatial optimizing of earthquake prediction and disaster preparedness," *Pure Appl. Geophys.*, vol. 169, no. 11, pp. 1989–2010, 2012.
36. T. F. Ma and Z. L. Wu, "Precursor-like anomalies prior to the 2008 Wenchuan earthquake: a critical-but- constructive review," *Int. J. Geophys.*, vol. 2012, Article ID 583097, p. 13 pages, 2012.
37. A. Peresan, G. Costa, and G. F. Panza, "A proposal of regionalization for the application of the CN earthquake prediction algorithm to the Italian territory," *Annali di Geofisica.*, vol. 42, pp. 281–306, 1999.
38. K. F. Tiampo and R. Shcherbakov, "Seismicity-based earthquake forecasting techniques: ten years of progress," *Tectonophysics*, vol. 522–523, pp. 89–121, 2012.
39. Y. Y. Kagan and D. D. Jackson, "Long-term earthquake clustering," *Geophys. J. Int.*, vol. 104, no. 1, pp. 117–134, 1991.
40. M. Meghraoui, "Evidence for 830 years of seismic quiescence from palaeoseismology planet, archaeoseismology and historical seismicity along the Dead Sea Fault in Syria," *Earth Sci. Lett.*, vol. 210, pp. 35–52, 2003.
41. P. Galli, F. Galadini, and D. Pantosti, "Twenty years of paleoseismology in Italy," *Earth-Sci. Rev.*, vol. 88, no. (1–2), pp. 89–117, 2008.
42. "International Atomic Energy Agency, Safety Standard Series, Evaluation of Seismic Hazards for Nuclear Power Plants: Safety Guide No NS-G-3.3," *IAEA, Vienna*, p. 31, 2002.
43. S. Stein, R. J. Geller, and M. Liu, "Why earthquake hazard maps often fail and what to do about it," *Tectonophysics*, vol. 562–563, pp. 1–25, 2012.
44. D. Schorlemmer, J. D. Zechar, M. J. Werner, E. H. Field, D. D. Jackson, and T. H. Jordan, "First results of the regional earthquake likelihood models experiment," *Pure Appl. Geophys.*, vol. 167, pp. 8–9, 2010.
45. D. Schorlemmer and M. Gerstenberger, "RELM testing center," *Seismol. Res. Lett.*, vol. 78, no. 1, pp. 30–36, 2007.
46. S. Saba, F. Ahsan, and S. Mohsin, "BAT-ANN based earthquake prediction for Pakistan region," *Soft Comput.*, 2016.
47. J. F. Shroder and M. Wyss (Eds.), "Hazards and Disasters Series; Earthquake Hazard, Risk, and Disasters," *Acad. Press, Elsevier*, vol. 53, no. 9, pp. 1689–1699, 2014.
48. B. Derras and A. Bekkouche, "Use of the artificial neural network for peak ground acceleration estimation," *Lebanese Sci. J.*, vol. 12, p. 101, 2011.
49. N. N. Ambraseys and J. Douglas, "Near-field horizontal and vertical earthquake ground motions," *Soil dyn. earthq. eng.*, vol. 23, no. 1, pp. 1–18, 2003.
50. T. Kerh and D. Chu, "Neural networks approach and microtremor measurements in estimating peak ground acceleration due to strong motion," *Adv. Eng. Softw.*, vol. 33, no. 11, pp. 733–742, 2002.
51. K. Günaydın and A. Günaydın, "Peak ground acceleration prediction by artificial neural networks for northwestern turkey," *Math. Probl.* Eng., vol. 2008, Article ID 919420, p. 20 pages, 2008.

52. H. Adeli and A. Panakkat, "A probabilistic neural network for earthquake magnitude prediction," *Neural Netw.*, vol. 22, no. 7, pp. 1018–1024, 2009.
53. A. Sriram, S. Rahanamayan, and F. Bourennani, "Artificial neural networks for earthquake anomaly detection," *J. Adv. Comput. Intel. Intel. Informat.*, vol. 18, no. 5, 2014.
54. T. M. Cover and P. E. Hart, "Nearest neighbor pattern classification," *IEEE Tran. Inform. Theory.*, vol. 13, no. 1, pp. 21–27, 1967.
55. J. MacQueen, "Some methods for classification and analysis of multivariate observations," in *Proceedings of the 5th Berkeley Symposium on Mathematical Statististics and Probability*, 1968, pp. 281–297.
56. V. V. C. Cortes, "Support-vector networks," *Mach. Learn.*, vol. 20, no. 3, pp. 273–297, 1995.

CHAPTER 4

How to Predict Earthquakes with Microsequences and Reversed Phase Repetitive Patterns

Giulio Riga[1], Paolo Balocchi[2]

[1]*Geologist, Independent Researcher, Lamezia Terme, Italy*

[2]*Geologist, Independent Researcher, Modena, Italy*

ABSTRACT

A strong earthquake is always preceded by groupings of shocks whose identification and understanding constitute a sound method for improving short-term earthquake forecasts. Thanks to a graphical method, we have identified and classified some microsequences and reversed phase repetitive patterns that precede the hazardous events. The seismic microsequences include a series of information useful to know in advance the beginning of energy release and accumulation phases that usually precede and follow a moderate-to-high magnitude earthquake. Their identification and correct interpretation allow us to determine various warning signals. In particular, through the analysis of their shape and position in the seismic sequence we can claim that the strongest earthquakes occur shortly after the formation of some peculiar micro-sequences. The checks carried out on large data sets related to earthquakes occurred in the past have shown that the analysis procedures developed do not depend on the size of the area analyzed while predicting a high percentage of moderate-to-high magnitude earthquakes.

Keywords: Microsequences, Foreshocks, Pattern, Triggers, Earthquake, Mainshock

1. INTRODUCTION

In any earthquake prediction study, it is essential to identify the precursory phenomena within the seismic data to be analyzed [1] . To this end, it is necessary to analyze a solid scientific database to identify and classify the premonitory models that precede big seismic events, whose nature and shape, variable from one event to another, are probably controlled by the tectonic environment (geometries of the fault, strain rate) and the fault plan heterogeneity [2] . The seismic sequence of a given area includes all the information needed to understand its evolution in space and time [3] ; through an in-depth analysis of its inner structure it is possible to draw conclusions related to strong events prediction. In fact, the long-term sequence includes smaller-scale, medium-term sequences, which, in turn, contain short- and imminent-term sequences that influence the sequence evolution in relation to development speed and magnitude (space-time evolution).

The analysis performed on several seismic sequences has identified in short and imminent-term time windows, particular groupings of shocks (microsequences and repetitive patterns), which anticipate a phase inversion [4] . Usually, micro-sequences and repetitive patterns are formed before the beginning/end of the energy release energy, since the distribution of shocks over time is not random but follows rules that allow us to understand the development level achieved by the sequence and to make more reliable predictions.

The analysis of the seismic sequence structure is performed using: a) a spatial coverage of variable datasets ranging from regional scale to international scale (or global, which covers an area ranging from few tens of km^2 to thousands of km^2); b) a range of magnitude values of 2.0 - 10 M; (c) a range of depths of 1 - 50 km.

2. METHODOLOGY

2.1. Microsequences TT-7S and DB-3SE

We all are aware of the relative importance of the small and large earthquakes to change static stresses and trigger earthquakes. The small shocks observed before strong earthquakes, suggest that they have specific properties that can be used as of earthquakes precursors. Even though big earthquakes are unquestionably more important compared to smaller ones in relation to the energy release, small shocks collectively have the same influence [5] .

Microsequences TT-7S (Triple Maximum—Seven Shocks) and DB-3SE (Double Minimum—Three small shocks) are graphic patterns that develop inside the seismic sequence and allow us to locate the point of closure/beginning of an energy release phase.

In microsequence TT-7S (Figure 1), we identify a peculiar trend, which involves the magnitude values fluctuations that is completed during the energy release phase: it develops a first maximum (point 2) followed by a minimum value (point 3). Hereinafter, the magnitude values begin to rise again (point 4) reaching higher values than the previous maximum, and a second minimum (point 5), which generally has greater value compared to the previous one. The next step is characterized by a third peak (point 6) lower than the second maximum, followed by the last minimum (point 7) that is placed below the "transition line" [6] joining the minimum points 3 and 5, respectively. This completes the microsequence TT-7S [7] .

Overall, the microsequenceTT-7S appears as a strong fluctuation of the magnitude values that occur and are completed during the energy release phase.

In Figure 2 the green circles indicate the fourth shock in the microsequence whose magnitude M(4) must comply with the following condition:

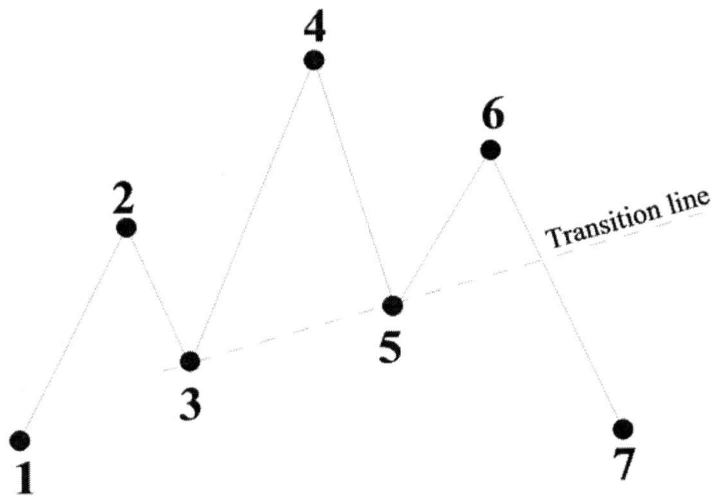

Figure 1. Microsequence TT-7S pattern.

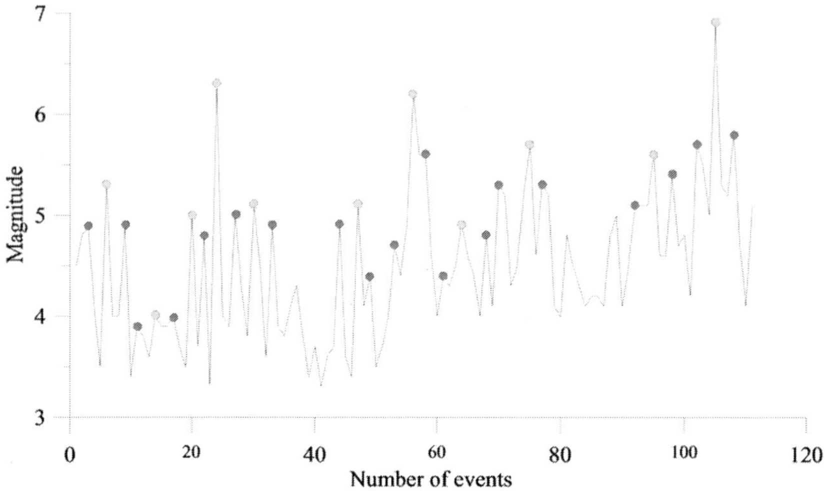

Figure 2. Microsequences TT-7S fluctuations.

$$M(2) \leq M(4) \geq M(6) \tag{1}$$

While the red circles indicate the points 2 and 6, respectively.

The microsequence BD-3SE [4] show more reliable features for imminent-term predictions. It is characterized by three small shocks fluctuations within two shocks of greater magnitude (Figure 3), where their completion establishes both the reversal of the energy accumulation phase and a critical condition, in which the successive small shocks trigger either average magnitude or catastrophic events. The microsequence's characteristic lies in the presence of a double minimum (points 1 and 3), and can be either symmetric, if the two minimums have the same magnitude value (Figure 3(a)) or asymmetrical, if the two minimums values are different (Figures 3(b)-(e)).

The microsequence BD-3SE is identifiable on charts concerning both the daily/monthly magnitude values trends or the progressive number of shocks, and generally they are completed after a very short period. Conversely, the release phase is active for one to eight days/months or eight events (in relation to the sequence's development time) after the trigger point, which is represented by the second minimum of the microsequence DB-3SE (red triangle). After the trigger point, it is possible that the magnitude values progressively increase up to the final target, thus generating, a progressive earthquake- or flash-earthquake type energy release phase [4] .

The procedure for identifying the microsequence DB-3SE is the following:

1) Identify the absolute maximum and relative values P(n), which represent the extreme points (Figure 4);

2) Identify the extreme points $P_1(n)$ (second shock in the microsequence DB-3SE) whose magnitude $M_1(n)$ must comply with the following conditions:

$$M_1(n-1) \geq M_1(n) < M_1(n+1) \tag{2}$$

$$M_1(n-1) > M_1(n) \leq M_1(n+1) \tag{3}$$

where $M_1(n-1)$, $M_1(n+1)$ are the magnitude values of the extreme point that precedes and follows the point $P_1(n)$, respectively (Figure 5).

3) Draw the line joining the second shock in the microsequence DB-3SE (blue line) (Figure 6);

4) Draw the line joining the fourth shock in the microsequence TT-7S (black line).

During a seismic sequence development it is possible to identify the beginning of an energy release phase by assuming the following condition:

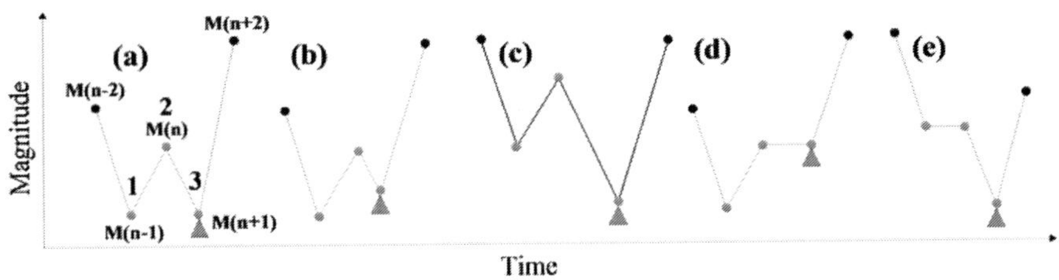

Figure 3. Schematic representation of microsequence DB-3SE. The red circles indicate the microsequence, while the red triangle indicates the trigger point of the energy release phase. The black circles represent the microsequence's starting and completion shocks.

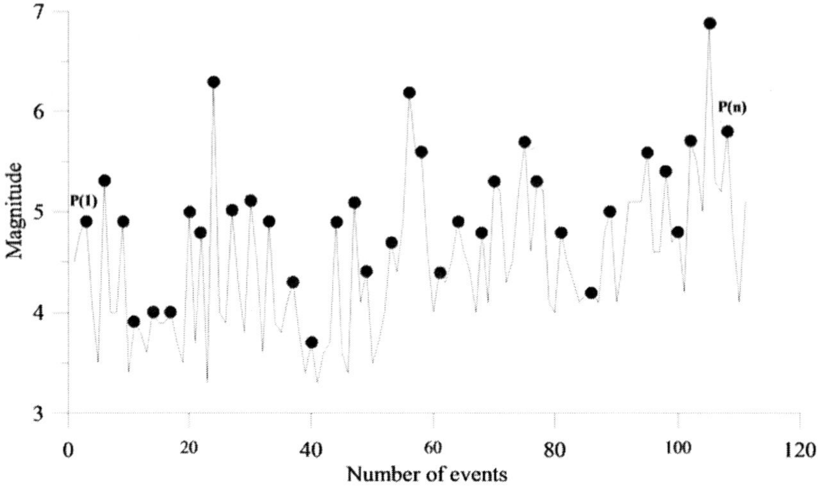

Figure 4. Identification of absolute and relative maximum values (black circles).

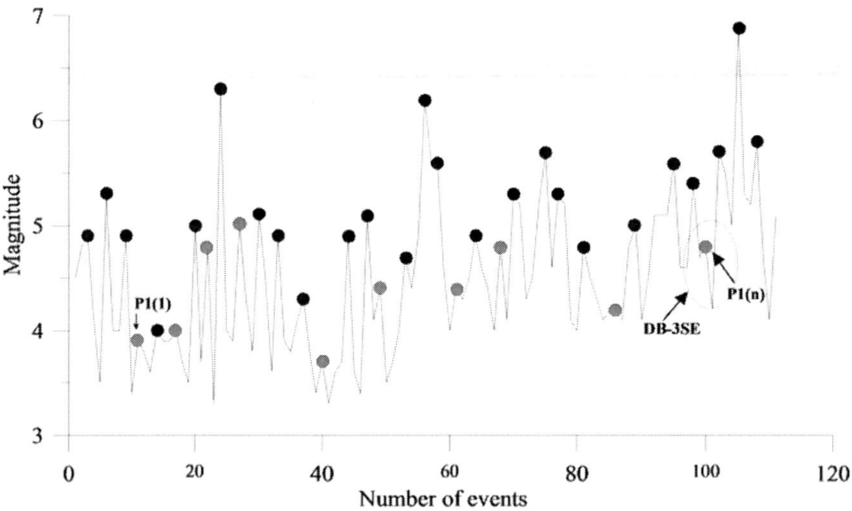

Figure 5. Procedure apt to identify the microsequences DB-3SE. In the red ellipse an example of microsequence DB-3SE is shown. The red circle indicates the second shock in the microsequence. The black circles indicate the extreme points (absolute and relative maximum values)—Earthquake occurred in China on 13 April 2010.

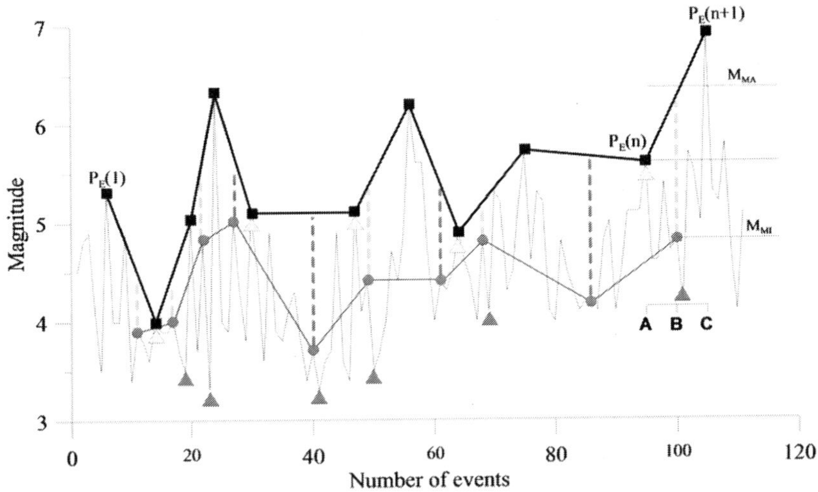

Figure 6. Definition of energy accumulation and release phases. The yellow triangles represent the first-order warning signals, while the red triangles represent the second and third order ones; graphical identification of minimum (M_{MI}) and maximum magnitude (M_{MA}).

If $M_E(n) \geq M_E(n-1)$ is the beginning of an energy release phase (yellow triangles). Where $M_E(n)$ is the magnitude of the fourth shock $P_E(n)$ in the microsequence TT-7S that must be confirmed by the following condition as well:

$$AB < BC$$

where AB is the temporal distance or number of shocks between the point $P_E(n)$ (yellow triangle) and the second shock in the microsequence DB-3SE (red circle), while the segment BC represents the temporal distance or the number of shocks between the microsequence DB-3SE and the last recorded event during the current phase.

The AB > BC condition indicates an ongoing energy accumulation phase, while the AB = BC condition is regarded as the "watershed" between the energy accumulation and release phases.

In general, the magnitude value of the second shock in the micorsequence DB-3SE corresponds to the minimum expected magnitude (M_{MI}) during the energy release phase, while an indicative value of the maximum expected magnitude (M_{MA}) shall be calculated by adding the difference between the magnitude of the latter point and the minimum expected magnitude value (M_{MI}) to the magnitude value of the fourth shock in the microsequence TT-7S (point $P_E(n)$).

In order to obtain a more accurate method, however, in some cases it is necessary to use other types of signals to confirm the ongoing phase. For example, a clear signal of phase inversion is given by the fluctuations of the red line (Figure 7) that joins the extreme points of the seismic sequence to the second shock in the microsequence DB-3SE. In the lower part of the graph, the amplitudes of the imminent-short term energy accumulation and release phases are indicated by red and green horizontal bars.

From the graph it is possible to infer the following information:

1) The microsequence DB-3SE indicates the end of the energy accumulation phase and the beginning of the energy release phase (point A).
2) Point B represents an imminent-short term foreshock and the first maximum in microsequence TT-7S.
3) The point C represents the mainshock or a stronger shock and the second maximum in microsequence TT-7S.
4) The point D represents the first shock of the energy accumulation (aftershock) and also the third maximum in microsequence TT-7S.

The seismic sequence development takes place according to an evolution that follows this pattern: after the microsequence DB-3SE formation (point A), a single shock of greater magnitude than the second shock in microsequence DB-3SE or more shocks whose magnitude increases

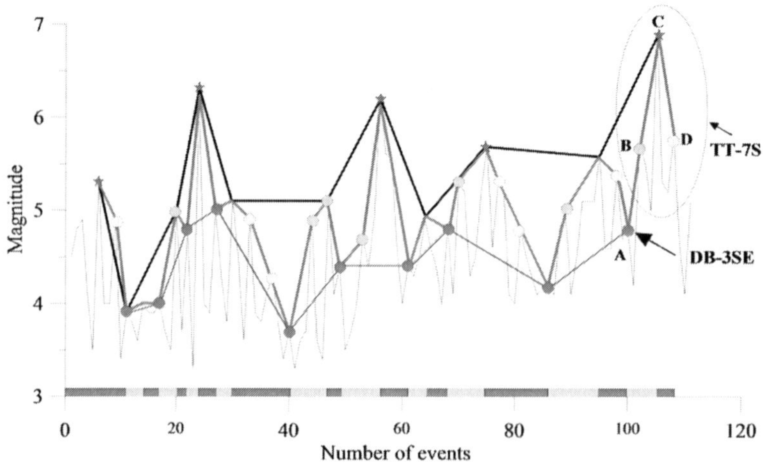

Figure 7. Alternating accumulation and release of energy phases.

over time (points B and C) may occur. In point C the closure of the energy release p occurs if the third maximum in the microstructure TT-7S is formed (point D), which may coincide with a second mainshock or with an imminent-term aftershock (yellow circle), while the imminent-term energy accumulation phase ends with the formation of a microsequence DB-3SE.

2.2. Warning Signs

A traditional use of the microsequence DB-3SE involves the triggering of a first alert in correspondence of the occurrence of the third shock it consists of. However, this method does not allow to know the extension of the energy release phase.

A more reliable method is to associate alerts in increasing order with the microsequence DB-3SE and with the shocks that follow (Figure 8). For example, the third shock in microsequence DB-3SE that precedes an extreme point $P_E(n)$ of magnitude $M_E(n) \leq M_E(n-1)$ is associated with a first-order alert (green triangle), while the extreme point $P_E(n)$ is associated with second order alert (yellow color).

The imminent-term, higher order signals (red triangles) are associated with microsequences DB-3SE that forms after the second order signal.

The graph clearly shows how the warning signs occur following a repetitive 1-2-3 or 1-2-3-4 order that allows to follow the sequence development during the energy release phase and to establish the levels of danger that grow over time.

It is possible to use a different system to define the various order warning signs, using the time/number distance

of shocks between the extreme point $P_E(n)$ of magnitude $M_E(n) \leq M_E(n-1)$ and the second shock in micro-

sequence DB-3SE. The extreme point $P_E(n)$ constitutes a first order warning sign, the third shock in micro- sequence DB-3SE is a second-order warning sign, while the point C, placed at a distance from the point B equal to AB, represents a third order alert (Figure 9).

2.3. Repetitive Patterns

A strong earthquake is preceded by shocks of different magnitude organized in repetitive patterns that anticipate the energy release phase, whose identification and understanding constitute a sound method to make short and medium term predictions.

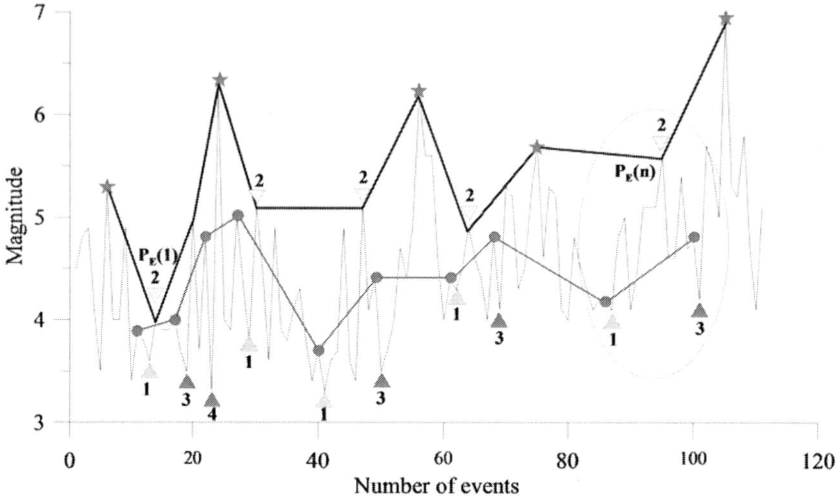

Figure 8. Various orders warning signals. The red star indicate the strongest earthquakes.

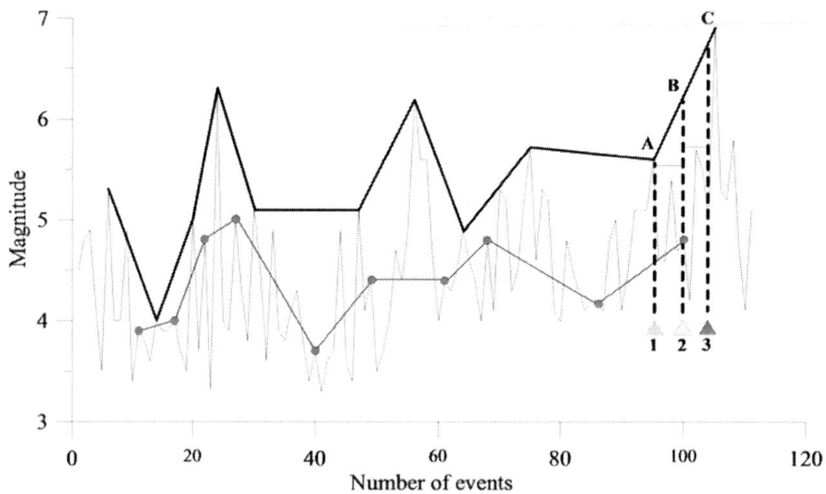

Figure 9. Various order warning signs.

The method for identifying the repetitive patterns consists in drawing the line that joins the fourth shock in microsequence TT-7S (magenta-colored upper line) and the line joining the low extreme points (magenta-colored bottom line) whose magnitude values must comply with the following conditions (Figure 10):

$$M_I(n-1) \geq M_I(n) < M_I(n+1) \tag{4}$$

$$M_I(n-1) > M_I(n) \leq M_I(n+1) \tag{5}$$

where M_I (n − 1) and M_I (n + 1) are the magnitude values of the low extreme points that precede and follow the point $P_I(n)$ of magnitudo $M_I(n)$.

The two lines constitute the fluctuation band of the extreme points values.

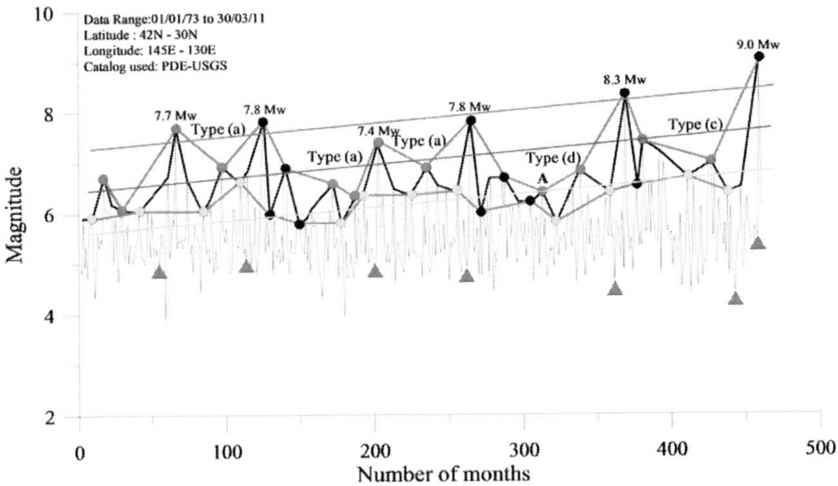

Figure 10. Procedure to identify the repetitive patterns. We observe some structures of repetitive patterns based on the type shown in Figure 11. The red triangles indicate the third-order warning signs generated by the microsequence DB-3SE.

The distinctive feature of the band is that the spaces between the top and the bottom line will vary as a function of the ongoing phase. Basically, the fluctuation band provides information on the ongoing phase in the seismic sequence: if the sequence is currently in an energy release phase then the band undergoes a progressive widening, while if it is in an accumulation phase, the band undergoes a shrinkage. In particular, during the activation step of the energy release phase the band shrinks further to contain the small shock.

The flattening of the fluctuation band usually means that a trigger point of the energy release phase is about to form.

If the magnitude values begin to exceed the interpolation line of the extreme points values (solid blue line), it means that the motion will continue usually upwards (with increasing magnitude values), while a downwards crossing will indicate the presence of an energy accumulation phase.

By default, after a strong earthquake the magnitude values always tend to return below the straight interpolation line.

The band upper line and the straight interpolation line movements provide some information on the magnitude values of the strongest earthquakes.

The method adopted involves the drawing of the parallel line to the interpolation straight line (green line) from the minimum point of the oscillation band upper line (point A) and the projection of a second parallel (line red) on it, whose amplitude is equal to that between the interpolation straight line and the lower parallel line. The magnitude values supplied by the second parallel line over time may be assumed as the strongest earthquakes' expected target.

Figure 11 shows the five repetitive patterns consisting of four shocks (two maximum and two minimum) generated by the oscillation band of the extreme points.

Models a-b-c will form at the end of a energy accumulation cycle, while models d-e during the energy release cycle (they are the most important, because the shock S3 is a foreshock).

In addition, the pattern are followed by warning signals generated by the microsequence DB-3SE that anticipate the strongest shocks.

3. RESULTS

In Figure 12 shows the results obtained from the application of the methodology explained above to the seismic sequence of the earthquake occurred in 1989 in Loma Prieta.

The main shock occurred on 18 October 1989 (6.9 Mw) was preceded by a progressive earthquake-type energy release phase [6] characterized by a short-term foreshock of magnitude 5.4 ML recorded on 08 August 1989 (point F1) and preceded by a fifth-order imminent-term warning sign generated by the microsequence DB-3SE.

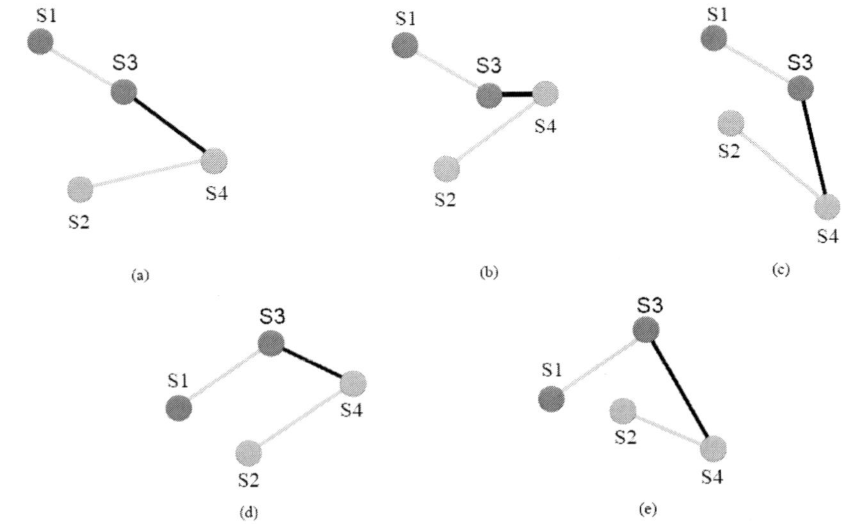

Figure 11. The repetitive phase inversion patterns.

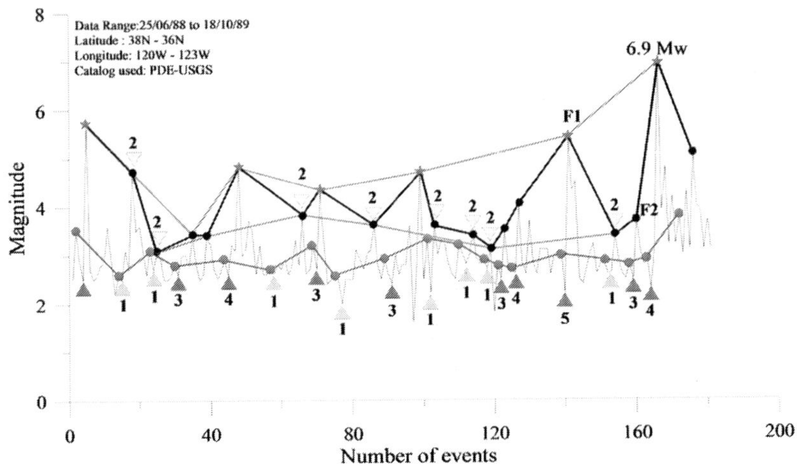

Figure 12. Warning signs before the earthquake in Loma Prieta, 1989.

The subsequent evolution of the seismic recordings over time shows the development of an impending-term accumulation phase consisting of shocks of magnitude less than 3.5 M, some of which are organized in microsequences DB-3SE that have generated various-order warning signs.

In particular, we note the formation of a first-order signal (point 1) generated by the third shock in microsequence DB-3SE, which has formed after the first foreshock (point F1), followed by the second-order signal generated by the extreme point (point 2) and by subsequent points (points 3 and 4). The latter preceded respectively the second foreshock (point F2) and the mainshock with magnitude of 6.9 Mw, generated by the third shock in microsequence DB-3SE. In addition, prior to the energy release phase, the oscillation band shrinks, while after the first foreshock it expands.

Even in this case, the oscillation band formed by the extreme points lines (black line) and by the second shocks in microsequence DB-3SE (blue line) can be used to identify phase inversion patterns (Figure 13). In fact, this oscillation band is an excellent oscillator to understand when the seismic sequence reaches a critical stage where small shocks can trigger shocks of greater magnitude. Essentially, when the band narrows, we reach the terminal step of the energy accumulation phase: this may suggests that in the short term the magnitude values could begin to grow. The band provides also significant warning signs (you should always bear in mind that this is a predictive analysis, i.e. it tries to predict future events and therefore not yet occurred) generated by the intersection between the descending sections extensions of the line joining the extreme points and the ascending sections extensions of the line joining the second shock in the microsequence DB-3SE (red triangle).

The magnitude minimum value (M_{MI}) expected in the energy release phase and associated with the warning sign, corresponds to the magnitude value of the extreme point that precedes the signal (green circle).

In order to further reduce the errors related to the interpretation of the signals which inevitably the procedure can generate, we recommend the adoption of one or more filters as the simplified Aroon oscillator [4] , with the purpose of identifying the energy accumulation areas in relation to the data provided by the oscillation band.

Starting from the previous example, in the graph shown in Figure 14 various order warning signs are reported (green, yellow and red triangles), which were identified through the distance between the extreme point (black circle) and the second shock in microsequence DB-3SE (red circle).

The first order warning sign coincides with the extreme point. The main shock has been triggered by an asymmetric microsequence DB-3SE (second order signal), while the third-order signal, placed at a distance BC = AB (point C) has formed a few seconds before the mainshock.

The graph displayed in Figure 15 shows some details of the microsequences TT-7S that have formed during the energy release phase and after the most hazardous seismic events.

In particular, it is possible to observe two imminent-term microsequences TT-7S in which the second maximum coincides with the most energetic shocks (foreshock and mainshock), while the third maximum, which closes the energy release phase, is represented by an aftershock.

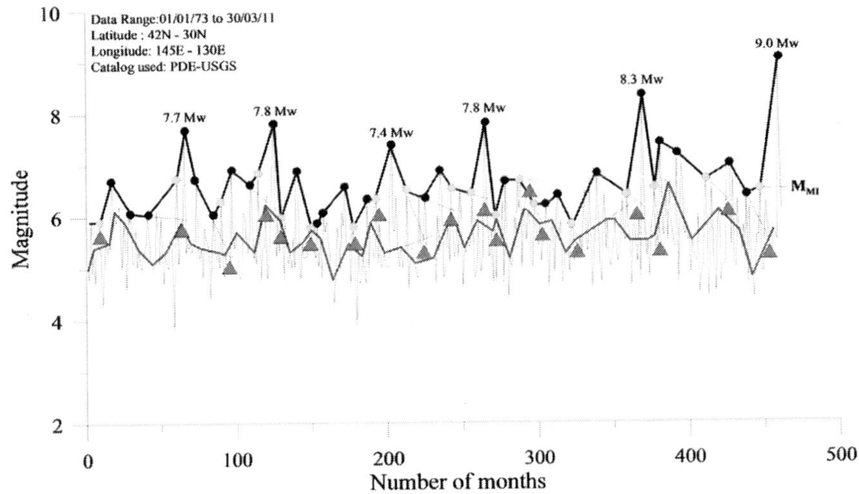

Figure 13. Warning signs (red triangles generated by the oscillation band of the extreme points-microsequence DB-3SE.

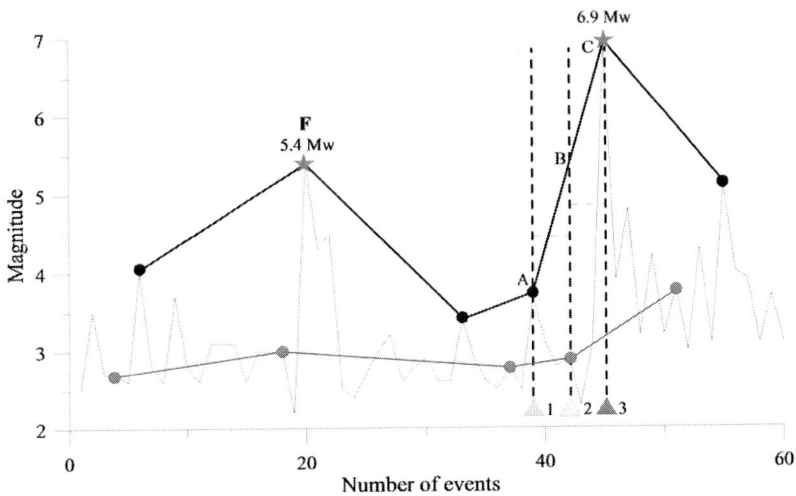

Figure 14. Warning signs before the earthquake in Loma Prieta, 1989. The red star indicates the mainshock, the letter "F" the foreshock.

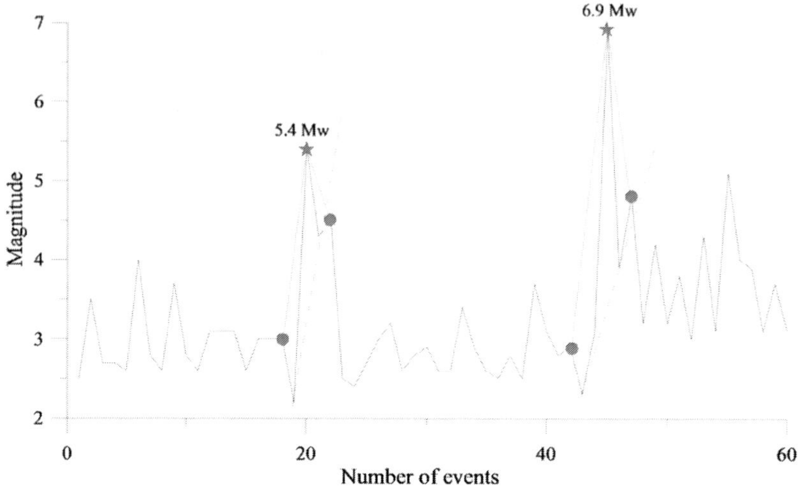

Figure 15. Microsequence TT-7S in energy release phase. The red symbols indicate the three maximum values in the microsequence.

The transition from the energy release to the energy accumulation phase is also confirmed by the magnitude values decrease below the "transition line" (dashed blue line) [6] .

4. CONCLUSIONS

A seismic sequence that occurs in a given area consists of microsequences and repetitive patterns formed by a grouping of seismic events with particular features. Their analysis allows studying the short-term evolution of a seismic sequence by identifying the warning signals that precede a strong earthquake.

In particular, the microsequence TT-7S allows us to locate the closure point of an energy release phase and identify a second order warning sign.

The microsequence DB-3SE, identifies the beginning of the energy release phase and explains how from low energy earthquakes moderate to high magnitude events can result, whose occurrence time can be known with sufficient advance.

The band allows us to understand when the seismic sequence is located within an energy release or accumulation phase; therefore if it reaches a critical stage where small shocks can trigger shocks of greater magnitude.

The aforementioned method, based on the microsequence analysis and on repetitive patterns can be easily applied to the study concerning earthquakes forecasts both at regional and global level. The method is based on a simple graphical method that can be used through the data of different seismicity catalogues.

CITE THIS PAPER

Giulio Riga, Paolo Balocchi (2016) How to Predict Earthquakes with Microsequences and Reversed Phase Repetitive Patterns. *Open Journal of Earthquake Research*,**05**,153-164.

REFERENCES

1. Wakita, H., Nakamura, Y. and Sano, Y. (2013) Case 3: Short-Term and Intermediate-Term Geochemical Precursors. In: Wyss, M., Ed., Evaluation of Proposed Earthquake Precursors, American Geophysical Union, 15-20.http://onlinelibrary.wiley.com/doi/10.1029/SP032p0015/summary
2. Kanamori, H. (1981) The Nature of Seismicity Patterns before Large Earthquakes. In: Simpson, D.W. and Richards, P.G., Eds., Earthquake Prediction, American Geophysical Union, Washington, 1-19.
3. Mantovani, E., Babbucci, D., Tamburelli, C. and Viti, M. (2009) A Review on the Driving Mechanism of the Tyrrhenian-Apennines System: Implications for the Present Seismotectonic Setting in the Central-Northern Apennines. Tectonophysics, 476, 22-40.
4. Riga, G. and Balocchi, P. (2016) Short-Term Earthquake Forecast with the Seismic Sequence Hierarchization Method. Open Journal of Earthquake Research, 5, 79-96. http://www.scirp.org/Journal/PaperInformation.aspx?PaperID=66332
5. Helmstetter, A., Kagan, Y.Y. and Jackson, D.D. (2005) Importance of Small Earthquakes for Stress Transfers and Earthquake Triggering. Journal of Geophysical Research: Solid Earth.
6. Riga, G. and Balocchi, P. (2016) Seismic Sequence Structure and Earthquakes Triggering Patterns. Open Journal of Earthquake Research, 5, 20-34. http://www.scirp.org/Journal/PaperInformation.aspx?PaperID=63538
7. Bulkowski, T. N. (2002) Trading Classic Chart Patterns. John Wiley & Sons, Inc.http://fx-arabia.com/vb/uploaded/178_01275676445.pdf

CHAPTER 5

Earthquakes and Structural Damages

Burak Yön[1*], Erkut Sayın[2] and Onur Onat[1]

[1] *Department of Civil Engineering, Tunceli University, Tunceli, Turkey*

[2] *Department of Civil Engineering, Fırat University, Elazığ, Turkey*

ABSTRACT

Earthquakes are the most destructive natural hazards throughout human history. Hundreds of thousand people lost their lives and loss of billions of dollars' properties occurred in these disasters. Occurred medium or high-intensity magnitude earthquakes in last twenty years showed that these loses continue. For reinforced concrete (R/C) buildings, inappropriate design such as soft and weak stories, strong beam–weak column, short column, hammering, unconfined gable wall and in-plane/out-of-plane movement of the walls causes damages. These are the main reasons. In addition to this, low quality of structural materials, poor workmanship, lack of engineering services, and construction with insufficient detailing of the structural elements are the another reasons of damages. Main reasons of masonry building damages in terms of design faults can be shown as heavy earthen roofs, inappropriate detailing of wall to wall connection and wall to roof connection, absence of bond beams, large openings. However, construction of buildings by using local materials with poor workmanship on the base of traditional rules is the other reason of failures for these buildings. In this book chapter, earthquakes and reasons of damages arose from earthquakes for reinforced concrete and masonry structures were presented. In addition to this, appropriate solutions are suggested.

Keywords: earthquakes, reinforced concrete buildings, masonry buildings, structural damages

1. INTRODUCTION

Earthquakes are one of the most destructive natural hazards that cause huge amount of loss of life and property. Nearly 10,000 people were killed every year because of these hazards. Moreover, annual economic loss is in the billions of dollars. In the last quarter century, severe earthquakes on the world like 1995 Kobe, Japan, 1998 Afghanistan, 1999 Kocaeli, Turkey, 2001 Gujarat, India, 2003 Bam, Iran, 2004 Indian Ocean, 2008 Wenchuan, China, 2009 L'Aqulia, Italy, 2010 Haiti, 2010 Chile, and 2011 Van earthquakes experienced construction industry to take severe measures to prevent collapse and to decrease damages of the structure; for example, after 1995 Kobe, Japan earthquake, it was reported that more than 6434 people lost their lives; nearly 4600 of them were from Kobe. In 1999 Kocaeli earthquakes, more than 17,000 people were killed and more than 40,000 people were injured and 300,000 people became homeless. In the year 2008, an earthquake hit the Sichuan China. Measured magnitude of earthquake from surface is 8.0. It was reported that nearly 70,000 people were dead, 95% of this death toll is in Sichuan province. In addition, more than 370,000 injured and 18,000 people missing. In Italy, 308 people were killed and more than 1500 people were injured after L'Aquila earthquake in 2009. However, total economic loss was 16 billion dollars during this earthquake. Many historical structures were collapsed and heavily damaged. The last earthquake tragedy for Turkey, very close to present time, is Erciş (Van) and Edremit (Van) earthquakes. These earthquakes struck Erciş (Van) district and Edremit (Van) district on 2011. After these earthquakes, 604 people were killed and 4852 people were injured, among of them 1301 people were seriously injured. A total of 2307 multistorey structures were collapsed. In addition, nearly 8% of the total province population became homeless.

The purpose of this chapter was to present earthquake characteristics and structural defects, damages, and possible solutions. The scope of the book chapter is depicted in the **Figure 1**.

Section 1 gives information about last destructive earthquakes.

Section 2 shows structure of the earth, plate tectonics, seismic waves, faults, and effects of earthquakes.

Section 3 gives failure reasons of reinforced concrete (R/C) buildings.

Section 4 presents failure reasons of masonry building damages.

Section 5 evaluates lessons learned from earthquake damages and failure experience from this type of natural hazards. Also in this part, suggestions were presented in order to prevent earthquake damages.

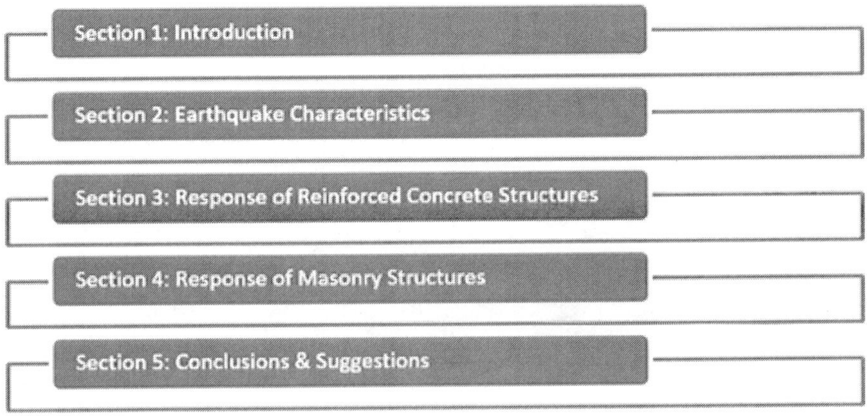

FIGURE 1. Scope of the book chapter.

2. EARTHQUAKE CHARACTERISTICS

2.1. Structure Of The Earth

The earth consists of layers which have different properties. The outer layer of the earth is called as "crust." The thickness of this layer is between 35 and 70 km for continents, and this thickness varies between 5 and 10 km thickness for ocean floor. Mantle layer, existed under the crust, is divided as lower mantle and outer mantle. This layer is approximately 2900 km thickness. Convection current occurred in the mantle causes plate tectonics in the crust. Core is the innermost layer and divides into two parts as fluid outer core and solid inner core. The outer layer is 2300 km, and inner layer is 1200 km thickness [1]. The internal structure of the earth is shown in the **Figure 2**.

2.2. Plate Tectonics

Plate tectonic deals with movement and strain of earth crust. According to the state of the art of Plate tectonic, the earthquake occurs in some parts of the plate and these parts act relative to each other. Pressure shift arose from these action and cooling stages in mantle causes stresses in the earth crust. When the increased stresses reached to bearing capacity of the crust on faults, this event causes sliding (breakthrough). Sliding movement spreads outward starting from hypocenter. Strain energy, which cumulated for a long time, discharges with sliding and causes earthquake shaking. Propagation of wave from hypocenter that results surface sliding is perceived as earthquake [3]. **Figures 3** and **4** show the tectonic plates and worldwide earthquake distribution, respectively.

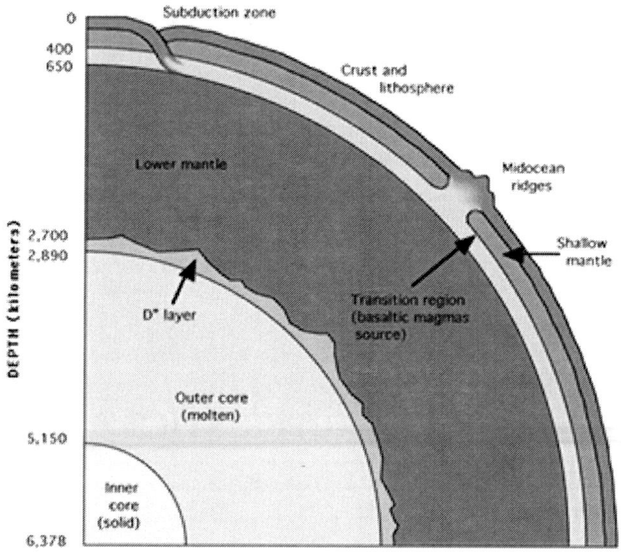

FIGURE 2. The internal structure of the earth [**2**].

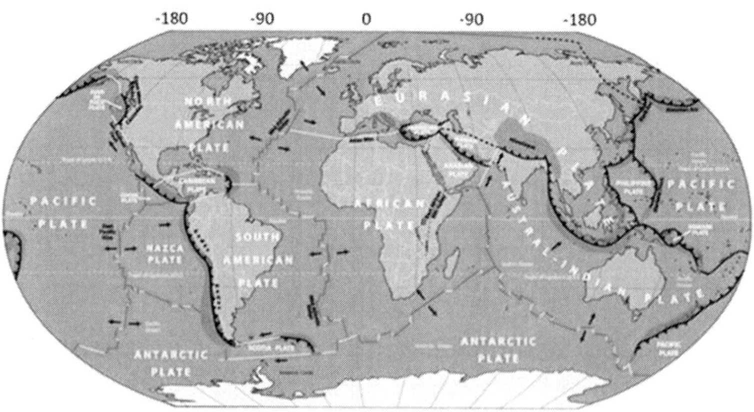

FIGURE 3. Tectonic plates [**4**].

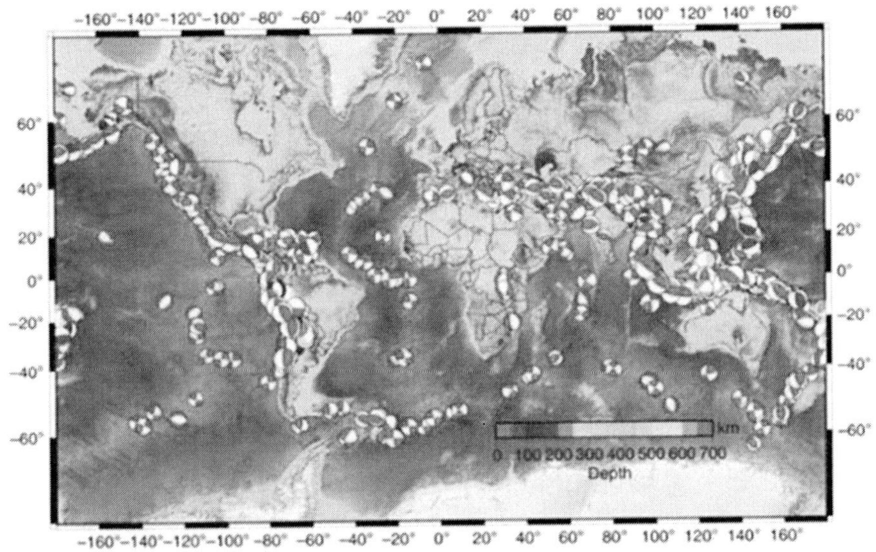

FIGURE 4. Geographical distribution of the 1700 earthquakes on the worldwide [*5*].

2.3. Faults

When two plates move with respect to each other, elastic strain energy is accumulated in results of tectonic processes. Thus, these two plates are released through the rupture of the interface zone. The shapeless blocks show immediate reaction towards equilibrium. As a result of this reaction, a seismic motion is produced. This process is called as "elastic rebound" theory. The resulting fracture in the crust of earthquake is defined as "fault". **Figures 5–7** show fault mechanisms [6].

FIGURE 5. Normal fault graphical presentation and mechanism [*7*].

FIGURE 6. Reverse fault graphical presentation and mechanism [7].

FIGURE 7. Strip slip fault graphical presentation and mechanism [7].

2.4. Seismic Waves

2.4.1. Body Wave

Seismic activity that results in earthquake generates two types of seismic waves: body and surface. Body waves move through the interior layers of earth's. Body waves include primary waves (known as P-waves) and secondary waves (also called as S-waves). P-waves generate sequential push (or compression) and pull (or tension) in soil as shown in below **Figure 8a**. P waves have relatively little damage potential. On the contrary, S-wave propagates horizontal and vertical motion. S-waves produce shear stresses in the soil along their paths [6] as shown in **Figure 8b**.

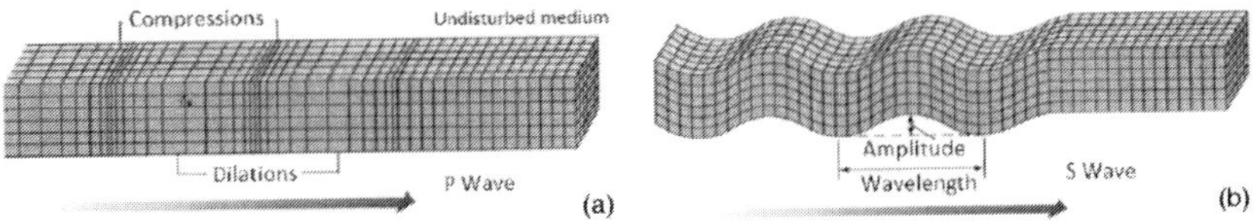

FIGURE 8. (a) Primary (P) wave and (b) Secondary (S) wave [*8*].

2.4.2. Surface Waves

Surface waves include Love (L) and Rayleigh (R) waves that propagate through the outer layers of the crust. These waves are generated by body waves move through parallel to the ground surface and various underpass the layer boundaries. These waves cause large displacements. These types of waves take various forms at a further distance away from the earthquake source. Surface waves are occurred during shallow earthquakes; on the other hand, body waves take place at all depths. Surface waves cause serious damage to structures due to their long duration [6]. **Figure 9a** and **b** shows these types of waves.

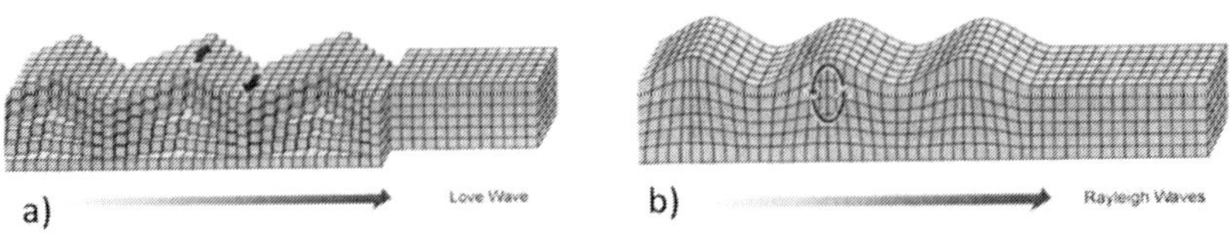

FIGURE 9. (a) Love (L) wave and (b) Rayleigh (R) wave [8].

2.5. Effects Of Earthquakes

Earthquakes can cause devastating effects in terms of life and property. The destructive potential of earthquakes depends on many factors such as focal depth, epicenter distance, and local site conditions. But the causes of fatalities and the extent of damages depend on lack of engineering service, design faults, material quality, and workmanship.

Many researchers studied and evaluated structural damages of reinforced concrete (R/C) and masonry buildings after the past earthquakes in different regions. Watanabe studied the behavior of R/C buildings during the Hyougoken-Nanbu earthquake and evaluated the damage reasons of reinforced concrete buildings [9]. Adalıer and Aydıngün evaluated the structural damages and the geological conditions for 1998 Adana–Ceyhan earthquake in Turkey [10]. Humar et al. evaluated the performance of masonry and reinforced concrete buildings during the Bhuj earthquake in India on January 26, 2001 [11]. Bruneau investigated the seismic behavior of structures after August 1999 Kocaeli earthquake and listed some general lessons obtained from this earthquake [12]. The structural damages of buildings were investigated during the December 26, 2003, Bam earthquake in Iran by Ahmadizadeh and Shakib [13]. Doğangün carried out a study about the reasons of damages of reinforced concrete structures during the May 1, 2003, Bingöl, Turkey, earthquake [14]. Ghobarah et al. conducted a field investigation in Thailand after December 26, 2004, Southeast Asia earthquake and tsunami. For this purpose, they evaluated the effects of the event on buildings, bridges, and infrastructure [15]. Bayraktar et al. evaluated the performance of stone masonry buildings during the Aşkale andErzurum earthquakes in Turkey on March 25 and 28, 2004 [16]. Arslan and Korkmaz carried out a study about R/C buildings during recent earthquakes in Turkey [17]. Zhao et al. focused on description of building performance after the May 12, 2008, Wenchuan earthquake in China [18]. Rossetto and Peiris evaluated the performance of government, commercial, and residential buildings after the October 8, 2005 Kashmir earthquake in Pakistan. Also, seismic design provisions of Pakistan were compared with seismic requirements of UBC (1997) and EC8 (1998) [19]. Maqsood and Schwarz (2010) carried out a study about the damages to rural buildings after Baluchistan earthquake that occurred in Pakistan in 2008 [20]. Augenti and Parisi evaluated the seismic performance of reinforced concrete and unreinforced masonry (URM) buildings during the L'Aquila earthquake that struck Italy in 2009 [21]. Chiou and Wang investigated the traditional Chinese residences after Chi-Chi earthquake that occurred in Taiwan on September 21, 1999 [22]. Celep et al. assessed the damages of concrete and masonry buildings after the March 8, 2010, Elazığ earthquakes in Turkey [23]. Ricci et al. conducted a study about the behavior of R/C buildings after L'Aquila earthquake which occurred on April 6, 2009, in Italy [24]. Braga et al. investigated the behavior of non-structural elements, particularly masonry infills in reinforced concrete frames at the 2009 L'Aquila earthquake [25]. Kam and Pampanin assessed the performance of R/C buildings during February 22, 2011, Christchurch earthquake in New Zealand [26]. Mahmood and Ingham assessed the seismic vulnerability of unreinforced masonry buildings in Pakistan using three empirical (New Zealand, United States, and Indian) methods [27]. Calayır et al. studied the damages of various structures during the March 8, 2010, Elazığ-Kovancılar earthquake in Turkey [28]. Yön et al. investigated the seismic performance of reinforced concrete and masonry buildings after 2011 Simav (Turkey) earthquake [29]. Taskin et al. evaluated strong ground motion and geotechnical and structural damages during the 2011 Van Turkey earthquakes [30]. Sayın et al. studied the failures of masonry and adobe buildings during the June 23, 2011, Maden, Elazığ earthquake in Turkey [31]. Bayraktar et al. and Ates et al. evaluated the reinforced concrete building damages during the 2011 Van earthquakes [32, 33]. Hermanns et al. investigated the performance of buildings with masonry infill walls after the 2011 Lorca, Spain, earthquakes. They described and analyzed the failure patterns in reinforced concrete frame buildings with masonry infill walls [34]. Manfredi et al. focused on damage observations and analyzed the emphasizing typical weaknesses of R/C buildings after the 2012 Emilia, Italy, earthquake [35]. Lemnitzer et al. assessed the behavior of reinforced concrete buildings after 2010 Chile earthquake [36]. Yön et al. conducted a field observation after the 2011 Van, Turkey, earthquakes. They investigated failure mechanisms and damage patterns of reinforced concrete buildings and presented their observations in the study [37]. Pinilla et al. carried out a study after the 2011 Lorca, Spain, earthquake. They determined the earthquake damages in reinforced concrete buildings and presented in their study [38]. Sharma et al. carried out a field investigation after the April 25, 2015, Gorkha earthquake in Nepal. They investigated the damaged buildings and the failure mechanisms in their study [39].

3. RESPONSE OF REINFORCED CONCRETE (R/C) STRUCTURES

3.1. Soft And Weak Storey Mechanism

In some R/C buildings, especially at the ground floor, walls may not be continuous along to height of building for architectural, functional, and commercial reasons. While ground floor generally encloses with glass window instead of brick infill walls, partition walls are constructed above from this storey for separating rooms for the residential usage. This situation causes brittle failures at the end of the columns. In mid-rise reinforced concrete buildings, the most common failure mode is soft-storey mechanism, particularly at the first storey. Failures can be concentrated at any story called as weak storey in which the lateral strength changes suddenly between adjacent stories due to lack of or removing of partition walls or decreasing of cross section of columns. Thus, during an earthquake, partial and total collapses occur in these storeys. These types of damages can be seen in **Figures 10** and **11**.

FIGURE 10. Unexpected inter-storey drift due to soft storey during the Van earthquake.

3.2. Inadequate Transverse Reinforcement In Columns And Beams

Shear forces increase during an earthquake especially at columns and beam–column joints. Consequently, special attention should be paid to construction and design of beam–column joints and columns. Seismic design requires increasing of ductility of structures for performance-based design approach. In particular, columns of buildings can be having insufficient transverse reinforcement in the plastic hinge region. Therefore, structural elements which have such details show low performance against to dynamic loads and lost their shear and axial load carrying capacity. **Figure 12** shows this failure below.

FIGURE 11. Weak storey mechanism during the Bingöl earthquake.

FIGURE 12. Damaged structure due to inadequate spacing between shear reinforcements during (a) Van earthquake and (b) Bingöl earthquake.

3.3. Short Column

This type of mechanism can be developed due to structural adjustments and/or to continuous openings at the top of infill walls between columns. Lateral forces that occurred by an earthquake are carried by columns and shear walls. Length of column is an important factor for dissipation of these loads. When the length of column decreases, the column becomes stiffer and brittle than the other columns and this column attracts more shear forces. Thus, shear failure which is a critical type of concrete column damage occurs at these columns. Short column failure is given in **Figure 13**.

FIGURE 13. Short column damages during the 2003 Bingöl earthquake.

3.4. Inadequate Gaps Between Adjacent Buildings

Buildings are sometimes constructed adjacent because of the lack of building lots. In this layout plan, one or two faces of two buildings are in contact to each other. Consequently, the buildings that have not adequate gaps pound to each other during the earthquake. If the floors of the buildings are not at the same level, pounding effect of the buildings becomes more dangerous. **Figure 14** shows this type of damage during the 2003 Bingöl earthquakes.

FIGURE 14. Collapse of adjacent buildings during the Bingöl earthquake.

3.5. Strong Beam–Weak Column

Deep and rigid beams are used with flexible columns in type of buildings. Therefore, these beams resist more moments, occurred by dynamic loads, than weak columns. In such a design during an earthquake while deep and rigid beams show elastic behavior, shear failure or compression crushing causes plastic hinges at flexible columns. Failure mechanism of strong beam–weak column can be seen in **Figure 15**.

FIGURE 15. Failure of a building due to strong beam–weak column effect during the Van earthquake.

3.6. Failures Of Gable Walls

The most common failure mode at gable walls is out-of-plane collapse in the earthquakes. Although failures of gable walls are not structural damages, these damages may be cause loss of lives and properties. Stability problems and large unsupported wall lengths cause damages at these walls. Failure of gabble wall is presented in **Figure 16**.

3.7. Poor Concrete Quality And Corrosion

The other main reasons of damages are low concrete strength and workmanship. Concrete quality is an important factor for building performance against to earthquakes. Handmade concrete is used to without using vibrator in construction of old buildings. Thus, homogeny mixing was not obtained and expected compressive strength was not provided in these buildings. In addition to this, using of aggregates which have improper granulometry, corrosion which decreases reinforcement bar area, and using of smooth steel reinforcement effected strength of concrete. This type of damages is given in **Figure 17**.

FIGURE 16. Failure of gabble walls on top of the building during the Van earthquake.

FIGURE 17. Failure of column due to poor concrete quality during; (a) Van earthquake and (b) Bingöl earthquake.

3.8. In-Plane/Out-Of-Plane Effect

One of the most important reasons of life and economic loss during the earthquake is combined effect of in-plane and out-of-plane movement of the wall. In-plane and out-of-plane interaction is very complicated and should be analyzed well for this phenomena. For low-rise and mid-rise unreinforced masonry (URM) infilled R/C frames, ground story infill walls are expected to be damaged firstly, because they are subjected to the highest in-plane demands. However, under the effect of bidirectional loading, where the two components of a ground motion are equally significant, infill walls of the upper stories may fail under the combination of in-plane and out-of-plane effects. The in-plane demand reduces at the upper stories, while that of out-of-plane forces increases due to the increase of accelerations [40]. To prevent this problem, in-plane carrying capacity of the wall should increase and out-of-plane ductility should increase with possible and applicable developments like bed-joint reinforcements and wire mesh. These listed applications will prevent detachment of infill wall from reinforced concrete elements and will increase the stiffness of the total structural system. **Figure 18** shows out-of-plane and in-plane damages.

FIGURE 18. (a) Detachment of infill wall during the Bingöl earthquake and (b) In-plane damage of during the Van earthquake.

4. RESPONSE OF MASONRY STRUCTURES

4.1. Earthen Roof Damages

One- and two-story masonry buildings are common in the rural areas because they require easy workmanship. These buildings are constructed with thick adobe and stone walls and mostly vulnerable to strong ground motions. But masonry buildings are preferred because of the advantages such as thermal properties and simple construction by using local materials [31]. These structures are constructed with traditional techniques using locally available materials. Nearly no engineering services are used in these buildings [20]. Adobe blocks are produced from local material contains mixed soil with straw and leave dried under the sun. These units are easily broken into small pieces as they have very low strength. As adobe blocks have low strength, walls of masonry buildings are thick and massive.

Generally, earthen roofs are constructed over wooden logs supported by two main walls of the buildings to provide thermal and water insulation. Also, heavy earthen roofs are constructed by the residents because a thick roof keeps the house cool or warm according to the seasons. However, these roofs lose their effectiveness because of weather conditions, such as rain and snow. Therefore, the residents place a new earthen cover on top of the existing roof to repair these roofs [16]. Thus, the weight and thickness of the roof increase over time. As a result of these heavy roofs, the structures are subjected to larger inertia forces during the earthquakes [41]. During the horizontal displacements, these heavy roofs are very vulnerable since they can easily slide over the walls [30]. **Figure 19** shows the damaged masonry buildings arising from heavy earthen roof at various earthquakes.

FIGURE 19. Damage to adobe structures due to heavy earthen roof during the Van earthquake [42].

4.2. Corner Damages

Corner damages are common in the adobe and masonry buildings. This type of mechanism generally occurs at wall-to-wall and wall-to-roof connections when subjected to out-of-plane displacements. During an earthquake, the stress concentrations increase at intersection of the walls. In this way, vertical or inclined cracks appear in the corners of masonry buildings. If bond beams are not used in the corners or connection, two walls are not properly anchored each other, intensity of the cracks increases and these cracks spread along the height of the wall. Similar cracking may have been observed at adobe buildings [41]. Poor connections between adjacent walls and the absence of bond beams cause serious damages. In addition, there are no appropriate connections at the corner of the walls in damaged buildings. This type of damage for adobe and masonry buildings is shown in **Figure 20** [42]. Due to reduced compression stress and increase seismic acceleration at upper stories, the common failures are seen at the corners of the roof level. When there is no slab with some in-plane rigidity at roof level, top corners are more sensitive to fail because of cantilever-like behavior [43].

FIGURE 20. Corner damage of briquette masonry buildings [42].

4.3. Out-Of-Plane Mechanism

Out-of-plane mechanism may appear from the combination of several deficiencies. A lack of bond beams, poor connections among the walls and the roofs, and large unsupported wall lengths cause the separation of walls and cause damage to occur via the of out-of-plane mechanism. Thus, the whole or the significant parts of the wall fall down during the earthquake. Wooden logs that bear the weight of the floor of the building are generally placed on load-bearing walls in only one direction. Thus, earthquake loads are transferred to perpendicular walls by wooden logs. Therefore, the walls that are not supported by the wooden logs may easily overturn to out-of-plane direction during the earthquake. This failure mechanism can be commonly observed in the earthquake region. **Figure 21** shows the out-of-plane mechanism of the adobe and briquette masonry buildings, respectively.

4.4. In-Plane Mechanism

The seismic performance of the masonry buildings relates to the in-plane stiffness of the walls. In-plane mechanism is generally observed in most of the masonry buildings that are affected by shear cracking. Earthquake loads increase the shear force. It can damage walls and their connections. These damages generally occur near openings, because most of the masonry buildings don't have sufficient and proper bond beams that distribute the lateral forces uniformly and enhance the lateral strength of the walls. During the earthquakes, excessive bending and shear can produce in-plane failures depending on the aspect ratio of the unreinforced masonry elements [41]. In the areas struck by earthquake, three failure modes of the shear damages in masonry buildings are generally observed, namely diagonal shear failures that proceed through masonry units and mortar (**Figure 22a–b**), sliding consisting of straight failure at the horizontal bed joints (**Figure 22c**), and stepped failures from the head to bed or bed to head joints (**Figure 22d**).

FIGURE 21. Out-of-plane mechanism of the briquette buildings [42].

FIGURE 22. Diagonal shear (a-b), horizontal (c) and stepped (d) failures wall [42].

4.5. Disintegration Of Stone Masonry Walls

Most of the masonry buildings are constructed with thick stone walls. As the thickness of the walls is relatively large, these stone walls are constructed by using more than one stone along the thickness direction [28]. In these walls, stones are placed in a random order. These walls have two exterior vertical wythes of large coarse stones. However, smaller rubble stones with mud mortar are used between two exterior layers. Using mud mortar instead of cement mortar causes insufficient adherence between the layers. Thus, the interior and exterior layers of the wall behaved independently and separated each other during the earthquake. Some reasons such as the quality of construction, poor workmanship, and the use of improper materials increase the intensity of the disintegration. **Figure 23** presents this failure in stone masonry buildings.

FIGURE 23. Disintegration in stone masonry buildings [28].

5. CONCLUSIONS AND SUGGESTIONS

In this book chapter, reasons of damages for reinforced concrete and masonry structures arose from earthquakes are presented. According to information that obtained from investigated buildings, the main reasons of failures of are presented below.

For reinforced concrete structures,

- The reason of soft storey collapse is occurred due to low rigidity of reinforced concrete structural members at ground floor. In case of the absence of the infill wall, the rigidity of the ground floor is lower than the upper storeys. Thus, this failure mechanism is triggered by earthquake. This type of failure is prevented during design phase by designing with more detail.
- Inadequate transverse reinforcement and no bending of hooks of ties in structural elements cause damages. This problem can be solved by using close-spaced stirrups and 1350 bended hooks to increase shear resistance of structural elements.
- Failures reason of short column is especially occurred due to partially filled infill walls in R/C frame system. These failures can be prevented by increasing of shear strength of this part of the column.
- During the earthquake, after first shake, different natural vibration periods cause hammer effect and then result in total collapse. To prevent this problem, adequate gaps according to current codes should be left between the adjacent buildings.
- When deep and rigid beams are used with flexible columns, weak column–strong beam failure mechanism is developed. To refrain this type of problem, sum of moments at column connected to any of joint should be bigger than sum of moments at beam connected to the same joint.
- Low concrete strength, workmanship, and corrosion of steel bars decrease the lateral stiffness of the structural system. This important reason can be eliminated by inspecting concrete and workmanship.
- The reason of in-plane, out-of-plane, and gabble wall failures is lower strength of infill materials than reinforced concrete frame. To prevent this type of failure, adequate connection and high-strength mortar between wall and reinforced concrete frame should be used.

For masonry structures,

- Thick and heavy earthen roofs are one of the reasons for the damages. The walls of the buildings could not support heavy mass during an earthquake, and the heavy roof partially or completely collapsed. This type of application should be refrained.
- Corner damages are developed due to insufficient wall-to-wall connections and lack of horizontal and vertical bond beams. This problem can be eliminated by using proper connection defined in current codes. Moreover, bond beams should be used.
- Another reason for the damages is the out-of-plane mechanism. The main reasons for this mechanism are the lack of bond beams, poor connections among the walls and roofs, and large unsupported wall lengths. Also, the gable walls of some masonry buildings are affected negatively by the out-of-plane mechanism. This problem can be eliminated by using proper connection defined in current codes. Constructing long and unsupported walls should be refrained. In addition to this, vertical and horizontal bond beams should be used.
- However, a lack of bond beams and large openings that decrease the stiffness of the walls increase the shear effects and cause in-plane failures, such as diagonal shear failures, sliding consisting of straight cracks, and stepped failures. It should be refrained to construct large window and door openings. In addition to this measure, vertical bond beams should be constructed near the openings.
- Construction of multilayer walls with inadequate connections causes the disintegration at stone masonry buildings along the wall thickness. This application should be prevented while constructing stone masonry wall.

According to listed damages and possible solutions above, it is strongly advised to obey current seismic codes. Furthermore, construction workers should be trained about earthquakes and construction of earthquake-resistant buildings. Process of construction should be controlled by the local government and professional civil engineers.

REFERENCES

1. Canbay E., Ersoy U., Özcebe G., Sucuoğlu H., Wasti S. T., editors. Basic Principles of Earthquake Engineering for Buildigs (In Turkish). 1st ed. Ankara, Turkey: Middle East Technical University Press; 2008. 434 p.
2. Beatty J. K., Chaikin A., editors. The New Solar System. 3rd ed. Massachusetts: Sky Publishing; 1990.

3. Celep Z., Kumbasar N. Introduction to Earthquake Engineering and Earthquake Resistivity Design of Buildings (In Turkish). 3rd ed. Istanbul, Turkey: Beta Publishing; 2004. 700 p.
4. www.johomaps.com [Internet]. Available from: http://www.johomaps.com/world/worldtecton.html [Accessed: August 10, 2016]
5. Vallee M. Source time function properties indicate a strain drop independent of 18 earthquake depth and magnitude. Nature Communications. 2013;4:1–6
6. Elnashai A. S., Sarno L. D. Fundamentals of Earthquake Engineering. 1st ed. England: Wiley; 2008. 347 p.
7. www.groovylabinabox.com [Internet]. [Updated: July 25, 2015]. Available from: http://www.groovylabinabox.com/its-not-my-fault [Accessed: August 10, 2016]
8. www.domus.urbanaction.org [Internet]. Available from: http://allshookup.org/quakes/wavetype.htm [Accessed: August 10, 2016]
9. Watanabe F. Behavior of reinforced concrete buildings during the Hyougoken-Nanbu earthquake. Cement and Concrete Composites. 1997;19(3):203–211.
10. Adalıer K., Aydıngün O. Structural engineering aspects of the June 27, 1998 Adana–Ceyhan (Turkey) earthquake. Engineering Structures. 2001;23:343–355.
11. Humar J. M., Lau D., Pierre J. R. Performance of buildings during the 2001 Bhuj earthquake. Canadian Journal of Civil Engineering. 2001;28:979–991.
12. Bruneau M. Building damage from the Marmara, Turkey earthquake of August 17, 1999. Journal of Seismology. 2002;6(3):357–377.
13. Ahmadizadeh M., Shakib H. On the December 26, 2003, Southeastern Iran earthquake in Bam region. Engineering Structures. 2004;26:1055–1070.
14. Doğangün A. Performance of reinforced concrete buildings during the May 1, 2003 Bingöl earthquake in Turkey. Engineering Structures. 2004;26(6):841–856.
15. Ghobarah A., Saatcioglu M., Nistor I. The impact of the 26 December 2004 earthquake and tsunami on structures and infrastructure. Engineering Structures. 2006;28:312–326.
16. Bayraktar A., Coşkun N., Yalçin A. Performance of masonry stone buildings during the March 25 and 28 2004 Aşkale (Erzurum) earthquakes in Turkey. Journal of Performance of Constructed Facilities. 2007;21(6):432–440.
17. Arslan M. H., Korkmaz H. H. What is to be learned from damage and failure of reinforced concrete structures during recent earthquakes in Turkey?. Engineering Failure Analysis. 2007;14(1):1–22.
18. Zhao B., Taucer F., Rossetto T. Field investigation on the performance of building structures during the 12 May 2008 Wenchuan earthquake in China. Engineering Structures. 2009;31(1):1707–1723.
19. Rossetto T., Peiris N. Observations of damage due to the Kashmir earthquake of October 8, 2005 and study of current seismic provisions for buildings in Pakistan. Bulletin of Earthquake Engineering. 2009;7:681–699.
20. Maqsood S. T., Schwarz J. Building vulnerability and damage during the 2008 Baluchistan earthquake in Pakistan and past experiences. Seismological Research Letters. 2009;81(3):514–525.
21. Augenti N., Parisi F. Learning from construction failures due to the 2009 L'Aquila Italy earthquake. Journal of Performance of Constructed Facilities. 2010;24(6):536–555.
22. Chiou C. W., Wang H. J. Damage to Chinese Courtyard Houses during the September 21 1999 Chi-Chi Earthquake in Taiwan. Journal of Earthquake Engineering. 2011;15(5):711–723.
23. Celep Z., Erken A., Taskin B., Ilki A. Failures of masonry and concrete buildings during the March 8 2010 Kovancılar and Palu (Elazığ) earthquakes in Turkey. Engineering Failure Analysis. 2011;18(1):868–889.
24. Ricci P., Luca F D., Verderame G M. 6th April 2009 L'Aquila earthquake, Italy: reinforced concrete building performance. Bulletin of Earthquake Engineering. 2011;9(1):285–305.

25. Braga F., Manfredi V., Masi A., Salvatori A., Vona M. Performance of non-structural elements in RC buildings during the L'Aquila, 2009 earthquake. Bulletin of Earthquake Engineering. 2011;9(307):307–324.
26. Kam W. Y., Pampanin S. The seismic performance of RC buildings in the 22 February 2011 Christchurch earthquake. Structural Concrete. 2011;12(4):223–233.
27. Mahmood H., Ingham J. M. Seismic vulnerability assessment of Pakistan unreinforced masonry buildings at a national scale. Seismological Research Letters. 2011;82(5):676–685.
28. Calayır Y., Sayın E., Yön B. Performance of structures in the rural area during the March 8 2010 Elazığ-Kovancılar earthquake. Natural Hazards. 2012;61(2):703–717.
29. Yön B., Sayın E., Köksal T. S. Seismic response of buildings during the May 19, 2011 Simav, Turkey earthquake. Earthquakes and Structures. 2013;5(3):343–357.
30. Taskin B., Sezen A., Tugsal U. M., Erken A. The aftermath of 2011 Van earthquakes: evaluation of strong motion, geotechnical and structural issues. Bulletin of Earthquake Engineering. 2013;11:285–312.
31. Sayın E., Yön B., Calayır Y., Karaton M. Failures of masonry and adobe buildings during the June 23 2011 Maden-(Elazığ) earthquake in Turkey. Engineering Failure Analysis. 2013;34:779–791.
32. Bayraktar A., Altunişik A. C., Pehlivan M. Performance and damages of reinforced concrete buildings during the October 23 and November 9, 2011 Van, Turkey, earthquakes. Soil Dynamics and Earthquake Engineering. 2013;53:49–72.
33. Ates S., Kahya V., Yurdakul M., Adanur S. Damages on reinforced concrete buildings due to consecutive earthquakes in Van. Soil Dynamics and Earthquake Engineering. 2013;53(1):109–118.
34. Hermanns L., Fraile A., Alarcón E., Álvarez R. Performance of buildings with masonry infill walls during the 2011 Lorca earthquake. Bulletin of Earthquake Engineering. 2014;12(5):1977–1997.
35. Manfredi G., Prota A., Verderame G. M., De Luca F., Ricci P. 2012 Emilia earthquake, Italy: reinforced concrete buildings response. Bulletin of Earthquake Engineering. 2014;12(5):2275–2298.
36. Lemnitzer A., Massone L. M., Skolnik D. A., Martin J. C. L., Wallace J. W. Aftershock response of RC buildings in Santiago, Chile, succeeding the magnitude 8.8 Maule earthquake. Engineering Structures. 2014;76:324–338.
37. Yön B., Sayin E., Calayir Y., Ulucan Z. C., Karatas M., Sahin H., Alyamaç K. E., Bildik A. T. Lessons learned from recent destructive Van, Turkey earthquakes. Earthquakes and Structures. 2015;9(2):431–453.
38. Pinilla J. G. R., Adam J. M., Cárcel R. P., Yuste J., Moragues J. J. Learning from RC building structures damaged by the earthquake in Lorca, Spain, in 2011. Engineering Failure Analysis. 2016;68:76–86.
39. Sharma K., Deng L., Noguez C. C. Field investigation on the performance of building structures during the April 25, 2015, Gorkha earthquake in Nepal. Engineering Structures. 2016;121:61–74.
40. Mosalam K., Günay M. S. Progressive collapse analysis of RC frames with URM infill walls considering inplane/out-of-plane interaction. Earthquake Spectra. 2015;31(2):921–943.
41. Ural A., Doğangün A., Sezen H., Angın Z. Seismic performance of masonry buildings during the 2007 Bala, Turkey earthquakes. Natural Hazards. 2012;60:1013–1026.
42. Sayın E., Yön B., Calayır Y., Gör M. Construction failures of masonry and adobe buildings during the 2011 Van earthquakes in Turkey. Structural Engineering and Mechanics. 2014;51(3):503–518.
43. Inel M., Ozmen H. B., Akyol E. Observations on the building damages after 19 May 2011 Simav (Turkey) earthquake. Bulletin of Earthquake Engineering. 2013;11(255):255–283.

CHAPTER 6

Earthquake Instrumentation

Jae Cheon Jung*

[1] *KEPCO International Nuclear Graduate School, Seosang-myeon, Ulju-gun, Ulsan, Republic of Korea*

ABSTRACT

Earthquake detectors (seismic instruments) are used to measure low-frequency ground motion caused by earthquakes. They detect the seismic waves created by ground motions and convert the wave motions into electronic signals, which are measurable. Two measurement principles are widely used in the industry sector to detect the strong motion of earthquake that require high sensitivity: force-balanced acceleration and servo acceleration with a feedback loop. In this chapter, the motion of the mass as a function of the ground displacement is discussed with a differential equation resulting from the equilibrium of forces. In addition, the transfer functions of both instruments are investigated by using Matlab® Simulink. This technology is applied in NPP (nuclear power plant) to ensure the safety of the plant in systems, such as the SMS (seismic monitoring system) and the ASTS (automatic seismic trip system). SMS provides monitoring and recording capability, whereas ASTS makes a decision to trip the NPP when the PGA (peak ground acceleration) exceeds the pre-defined value, which is determined based on the ground conditions.

Keywords: earthquake detector, seismic instruments, accelerometer, force balanced acceleration, servo acceleration, transfer function

1. INTRODUCTION

Earthquake detection instrumentation is widely used in many industries to secure the system, structure and components. In addition, early warning of incident seismic wave is critical to safety of human lives. On 12 September 2016, South Korea experienced the most powerful earthquake ever recorded in the country since measurements began in 1978. A 5.8-magnitude earthquake struck the historic city of Gyeongju and the people were subjected to a series of aftershocks affecting their daily lives [1]. Buildings built inside the historic district were not strong enough to sustain seismic incident despite its low destructive power. Luckily, since the frequency of the wave was higher than 16 Hz, a catastrophic falling down of buildings was not experienced. This tells the relationship among magnitude, wave frequency and the destructive power. Earthquake detection instrumentation has been improved to detect the magnitude and frequency effectively. Historically, the first seismic instrument was invented in Han dynasty. Modern instruments are based on electronic sensors which are able to detect tri-axial acceleration.

In order to investigate both theory and application of earthquake instruments, the mathematical expression of the instrument is defined by the motion of the mass-spring-damper system. Then the differential equation resulting from the equilibrium of forces is discussed. In addition, the transfer functions of both force balanced and servo balanced and feedback circuit are analysed. Finally, the application of earthquake instrument in the nuclear power plant is introduced.

2. MATHEMATICAL EXPRESSION

Let us define a single seismic instrument having mass, spring and dashpot as shown in **Figure 1**. A dashpot is a device that provides viscous friction or damping [2]. The dashpot absorbs energy. This spring-mass-dashpot system can be expressed using Newton's second law as represented in Eq. (1):

$$\text{(D.E.)}\ m\ddot{y} = -ky - b\dot{y} + u \tag{1}$$

$$\text{(T.F.)}\ ms^2Y(s) = -kY(s) - bsY(s) + U(s) \tag{2}$$

$$G(s)\frac{Y(s)}{U(s)} = \frac{1}{ms^2 + bs + k} \tag{3}$$

The mathematical expression of the system above can be expressed as Eq. (1). The transfer function of Eq. (1) can be obtained as Eqs. (2) and (3) by taking Laplace transformation.

The motion of the mass as a function of the ground displacement is given by a differential equation resulting from the equilibrium of forces as shown in Eq. (4) [3].

$$Fs + Fr + Fg = 0 \tag{4}$$

where $Fs = -kx$ (k is the spring constant);

$Fr = -bx$ (b is the friction coefficient);

$Fg = -mu$ (m is the spring mass).

When the seismic instrument moves slowly, the acceleration and velocity become negligible, it records ground acceleration. If however, movement is fast enough, the acceleration of the mass dominates and the ground displacement is recorded.

Figure 2 shows the seismic instrument under ground motion. Where the u_g is the ground displacement, x_r is displacement of the seismometer mass and x_0 is the mass equilibrium position.

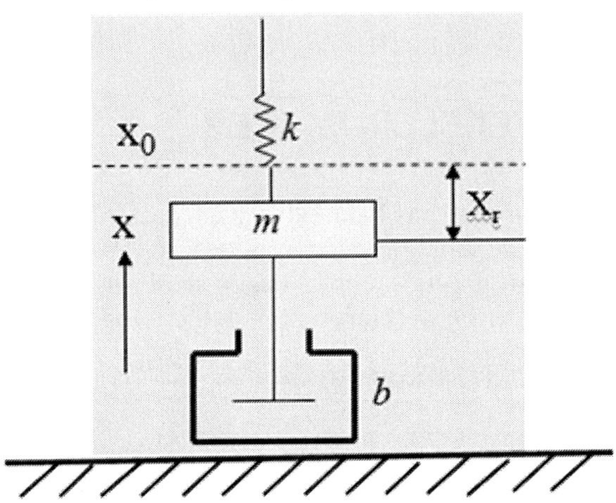

FIGURE 1. Spring-mass-dashpot system.

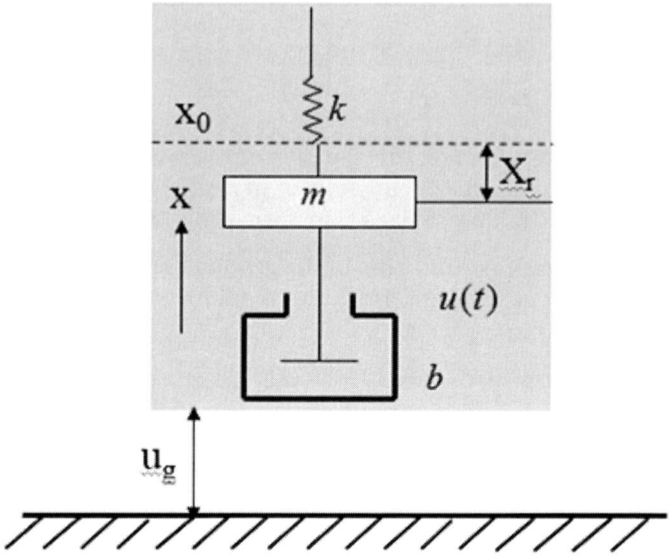

FIGURE 2. Concept of seismic instrument under ground motion.

In general, the frequency of the seismic wave tends to be increased if the magnitude of the wave is getting smaller. The ground motion has very low-frequency profile due to the movement of the earth.

This study is focused on the plant industry application of the earthquake instrument. In this case, the detector covers DC to 50 Hz frequency components.

3. TRI-AXIAL ACCELEROMETER

The tri-axial accelerometer has been widely used in industrial applications. An accelerometer is a sensor which can detect the earthquake-induced motion. The tri-axial accelerometer has a coil in the sensor that is suspended to it. When the accelerometer detects the earthquake-induced motion, the coil is moved and the voltage is induced proportionally to the variation of magnetic force. This voltage is converted to acceleration.

The tri-axial accelerometer detects incident seismic waves from longitudinal (L), transverse (T) and vertical (V) direction using three identical sensors as illustrated in **Figure 3**. In general, each accelerometer has a feedback amplifier in consideration of the stability of the sensor output. Moreover, an added RC filter eliminates the high-frequency component from the output. The accelerometers are installed at free-field where the earthquake ground motion is detected by the surface motion.

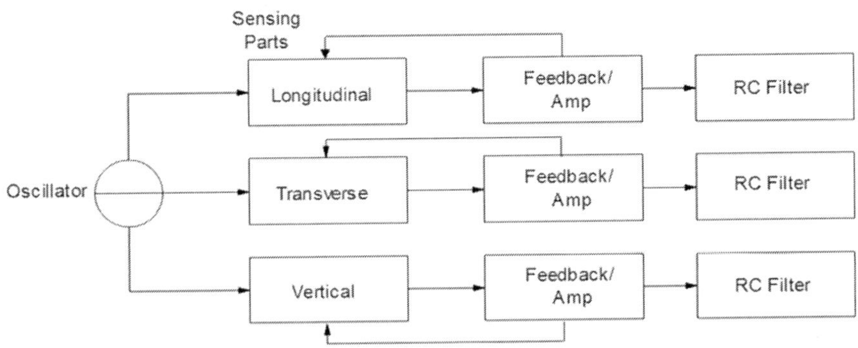

FIGURE 3. Concept of tri-axial accelerometer.

3.1. Types Of Accelerometers

An accelerometer detects low-frequency component from the earthquake ground motion. Two measurement principles are used to detect a strong motion earthquake requiring high sensitivity: open-loop type or closed-loop type with a feedback loop. In the plant industry, two sensor types have been used; force balanced acceleration using a pendulum, and servo acceleration with a feedback loop.

3.1.1. Force Balanced Acceleration Using Pendulum

In this section, the FBA-23 (by Kinemetrics Inc.) accelerometer is analysed to provide a more detailed operational theory of pendulous type servo sensors. As seen in **Figure 4**, FBA-23 consists of three-axis FBAs (force balanced accelerators) for detecting incident seismic waves in (L), (T) and (V) directions. Each accelerometer is identical, however, the vertical one has an active oscillator circuit as an additional function.

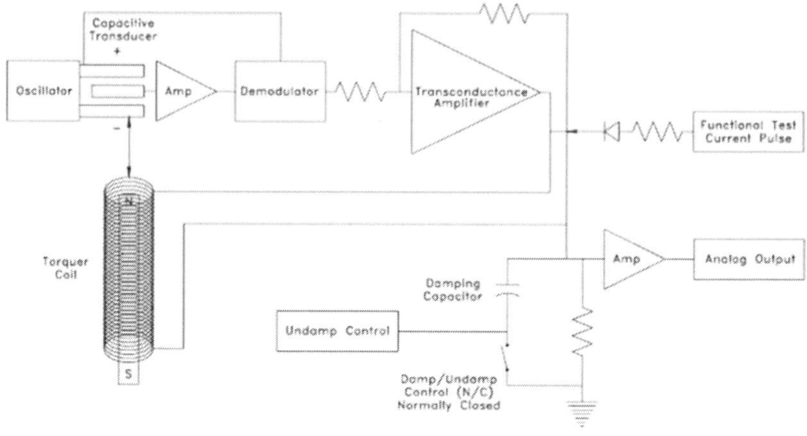

FIGURE 4. Simplified circuit of FBA-23 [4].

The electrical circuit of FBA-23 has a non-inverting amplifying circuit for adjusting the polarity of the output signal. The transfer function of the accelerometer can be expressed by Eq. (5).

$$\frac{E_{\mathrm{OUT}}}{E_{\mathrm{DC}}} = \frac{\omega_n^2}{s^2 + 2\zeta\omega_n s + \omega_n^2} \tag{5}$$

where ωnωn = natural frequency; *s* = Laplacian; ζ ζ = damping factor.

The solution of Eq. (5) can be denoted by the second order equation having two poles (p1, p2p1, p2) as shown in Eqs. (6) and (7).

$$p_{1=} - \zeta\omega_n + \mathrm{j}\omega_n\sqrt{1-\zeta} \tag{6}$$

$$p_{2=} - \zeta\omega_n - \mathrm{j}\omega_n\sqrt{1-\zeta} \tag{7}$$

The circuit has sensitivities of 0.25, 0.5, 1, 2 and 4 g full scale. This corresponds to normal absolute sensitivity at DC.

The second order transfer function in Eq. (5) has natural frequencies of 50, 90 or 100 Hz. In addition, if we consider the damping factor of 0.707, then the two poles (p1, p2)p1, p2) are located in the left half of the *s*-plane. The transfer function of this accelerometer can be rewritten as Eq. (8) having two poles.

$$\frac{E_{OUT}}{E_{DC}} = \frac{\omega_n^2}{s^2 + 2\zeta\omega_n s + \omega_n^2} = \frac{\omega_n^2}{(s - P1)(s - P2)} \tag{8}$$

From Eq. (8), the natural frequency and damping factor of this instrument can be calculated as Eqs. (9) and (10) respectively.

$$\omega_n = 2 \cdot \pi \cdot 50 = 314.159 \tag{9}$$

$$\zeta = 0.2550(\text{for } 1.0 \text{ g}) \tag{10}$$

The poles are then calculated as, *P*1 = -222.1 + j222.1 rad/s and *P*2 = -222.1 - j222.1 rad/s. Therefore, the resultant transfer function of this system becomes Eq. (11)

$$TF = \frac{98696}{s^2 + 160.22s + 98696} \tag{11}$$

Figure 5 shows the Matlab Simulink results when the step input is applied. The corresponding plot described in **Figure 6** shows overshoots and it needs relatively reduced settling time. In this simulation, the poles have 50 Hz of natural frequency and 1 g of absolute gain.

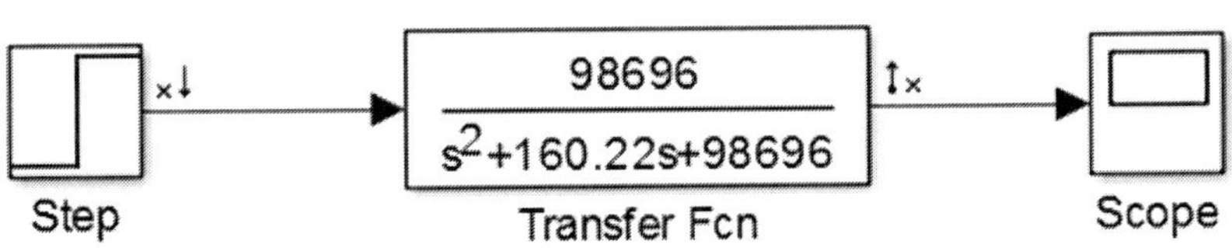

FIGURE 5. Matlab Simulink diagram (without RC filter).

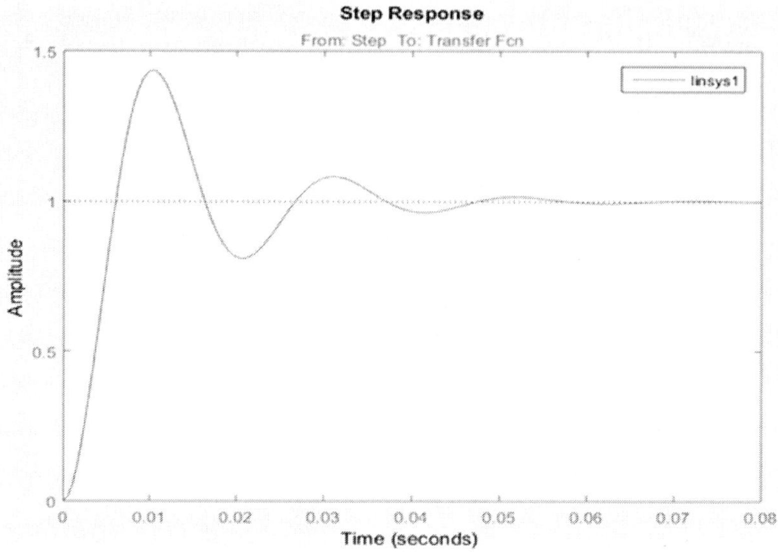

FIGURE 6. Step response (without RC filter).

Since this transfer function may be affected by the zeros, therefore by adding low pass RC filter to the post amplifier, the stability and accuracy of this circuit would be improved. In this case, the RC filter adds one more pole as summarized in **Table 1**. This proves that the calculated poles are placed in the left half of the *s*-plane as well.

TABLE 1. Pole placement of FBA-23 in 50 Hz and 1 g

	P1	*P2*	*P3*
Real (rad/s)	−222.1	−222.1	−1000
Imaginary (rad/s)	222.1	−222.1	0

The full transfer function of the electrical circuit in 50 Hz has three poles as shown in Eq. (12).

$$TF_{\mathrm{RC}} = \frac{98696}{s^3 + 1160.225s^2 + 258921s + 98696000} \tag{12}$$

Figure 7 shows the Matlab Simulink results when a step input is applied. **Figure 8** also shows almost the same plot disregarding the existence of the RC filter.

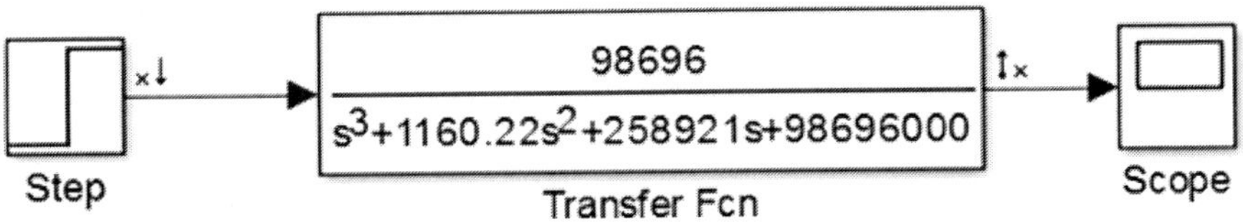

FIGURE 7. Matlab Simulink diagram (with RC filter).

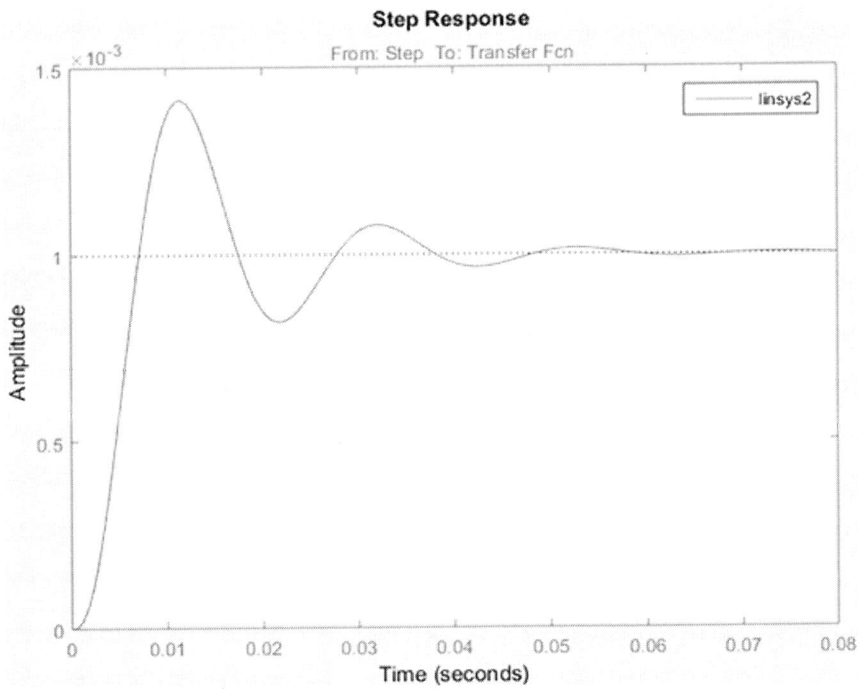

FIGURE 8. Step response (with RC filter).

3.1.2. Servo acceleration with feedback loop

In this section, the servo acceleration with a feedback loop type accelerometer (AC-23 by Geosig), which consists of geophone and electrical circuits, is discussed. The geophone response can be seen from two viewpoints:

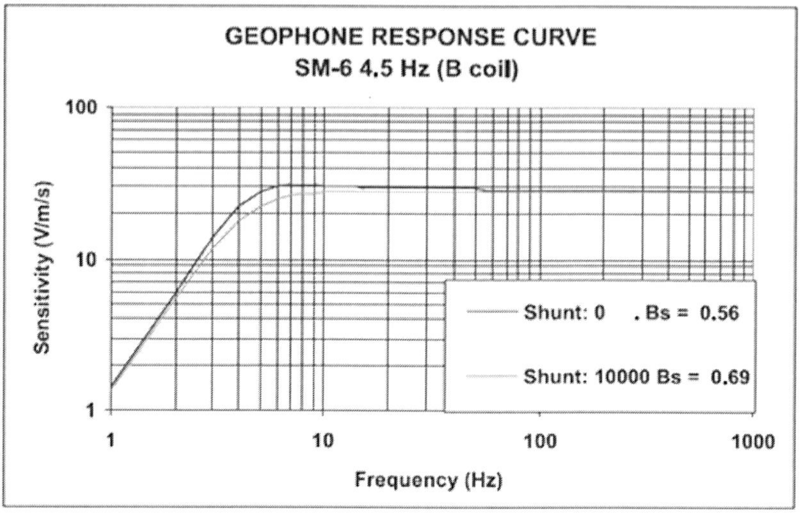

FIGURE 9. Geophone response curve (4.5Hz) [5].

- With a constant velocity, amplitude applied to the geophone.
- With a constant acceleration, amplitude applied to the geophone.

Figure 9 represents the response curve of Geophone at 4.5 Hz.

The relationship between these two points of view is a rotation when plotted with log-log axis scales. In **Figure 10**, the principle of Geophone is drawn. When the motion on the *x*-axis is incident then Geophone generates the open-loop signal which has peak response of 4.5 Hz.

The geophone is a long coil travel version and the extra coil travel offers an advantage for higher tilt requirements where larger amplitude signals may be encountered. A range of natural frequencies is available from 4.5 Hz to 14 Hz, providing a choice of the correct geophone for a wide variety of applications. The geophone is suitable for detecting an extensive range of ground motion.

The AC-23 provides over-damping the geophones with a feedback amplifier in a bridge circuit. The principle of over-damping geophones is done by applying a voltage on the geophone, which has opposite polarity from the voltage, which is induced by the moving geophone coil. Since the voltage induced by the geophone coil is proportional to the velocity, the externally

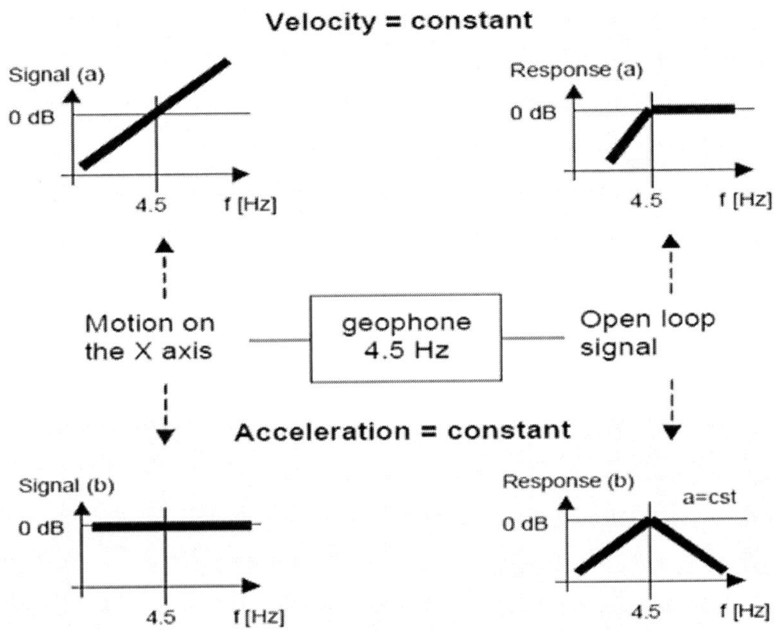

FIGURE 10. Principle of Geophone [6].

applied voltage is also proportional to it. This results in a current in the coil (~force), which is also proportional to the velocity and therefore is a "damping" current, or additional damping. Increasing this damping further will lead to a resulting output that is proportional to acceleration.

The function of the sensor, seen from a constant acceleration point of view is illustrated in **Figure 11**.

The geophone is connected in a resistor bridge, driven by a feedback amplifier and an inverter, which apply the amplified bridge differential signal in opposite polarity [6]. The amplifier has a fixed gain that will define the bandwidth of the accelerometer. The gain *G* and output *Vo* are represented by Eqs. (13) and (14). **Figure 12** shows the equivalent circuit for the linear servo-balanced accelerometer.

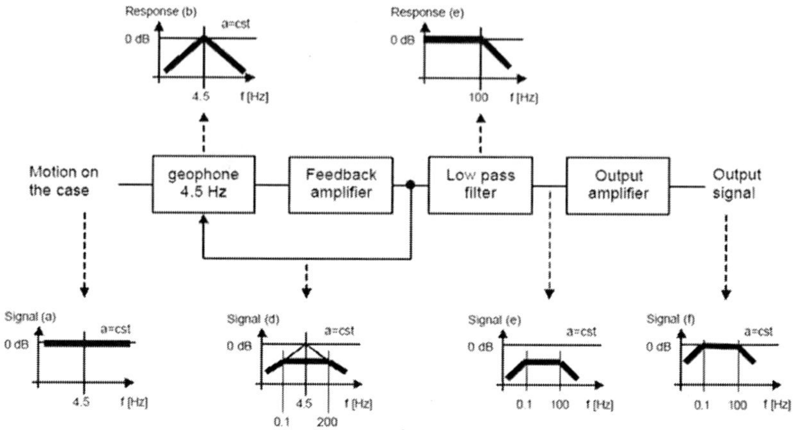

FIGURE 11. Operational principle of AC-23 [6].

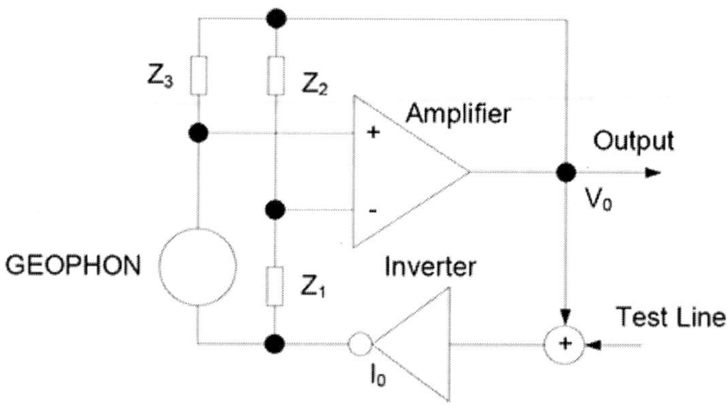

FIGURE 12. Equivalent circuit for AC-23 (linear servo balanced accelerometer) [6].

$$G = 1 + \left(\frac{Z_2}{Z_1} \right) \tag{13}$$

$$V_0 = G \cdot V_i = \left(1 + \frac{Z_2}{Z_1} \right) \cdot V_i \tag{14}$$

In **Figure 12**, the test-line shifts the voltage to one side of the bridge, which produces a current flow in the geophone, resulting in a displacement step of the seismic mass. The movement of the mass generates a voltage across the Geophone, which is detected by the differential amplifier and induces an output signal. The effect of the test signal on the bridge is cancelled by the differential input of the amplifier.

In this part, the Geophone transfer function transfer is analysed in order to define how the sensor generates open-loop transfer function as depicted in Eq. (13).

$$TF = \frac{1.564s}{0.1251s^2 + 1.514s + 1} \tag{15}$$

The transfer function in Eq. (15) has two poles and one zero. In this work, the MATLAB script of Geophone is used to illustrate the response curve under step input condition as shown in **Figure 13**.

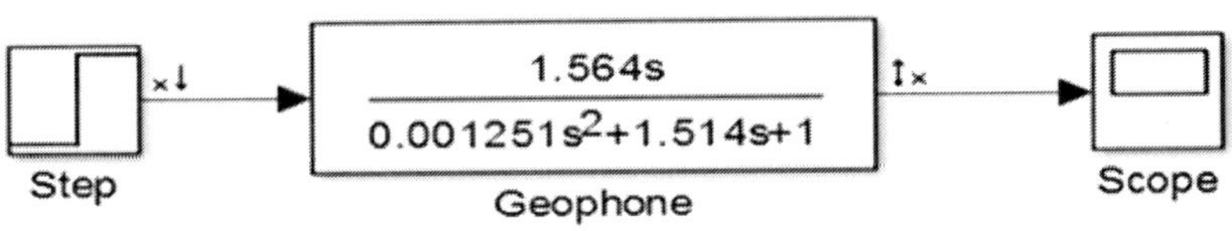

FIGURE 13. Open-loop transfer function of Geophone.

The transfer function plotted as the frequency range vs. relative response to the acceleration is shown in **Figure 14**.

If a unit step function is applied to the transfer function, the response as in **Figure 15** is obtained.

In this case, the magnitude and the phase can be drawn by the Bode plot as seen in **Figure 16**.

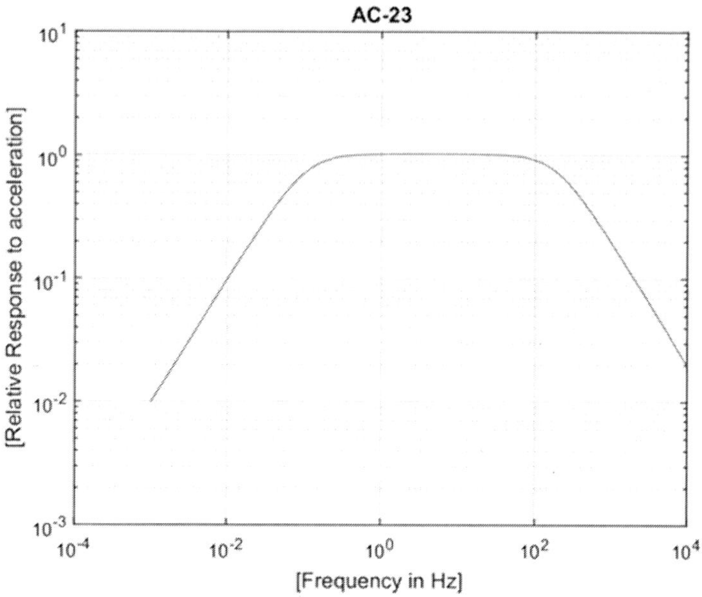

FIGURE 14. Transfer function plotted by the frequency vs. relative response to the acceleration.

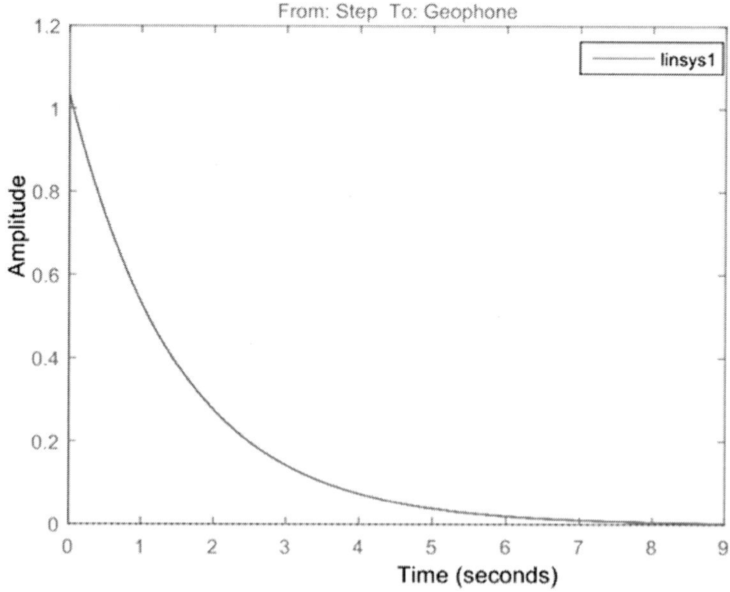

FIGURE 15. Step response.

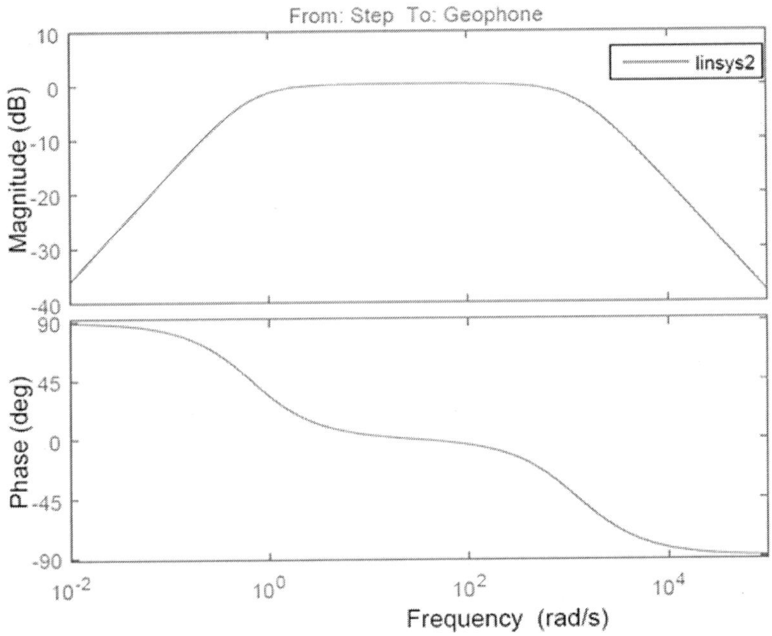

FIGURE 16. Bode plot.

4. APPLICATION OF SEISMIC INSTRUMENT

In this chapter, two application examples for the NPP (nuclear power plant) are discussed with their designs. They are SMS (seismic monitoring system) and ASTS (automatic seismic trip system). SMS detects the seismic wave and provides alarms to the NPP operators. The operator then makes a decision whether or not to trip the nuclear reactor. ASTS has an automatic trip function, if the peak acceleration of the incident wave exceeds, the pre-determined level trigger the reactor trip automatically. In the protection view-point, SMS provides alarms for the manual trip while ASTS provides automatic protection.

4.1. Seismic Monitoring System

The SMS consists of tri-axial accelerometers and a seismic monitoring cabinet which is located in an electrical equipment room in unit one (only one unit). The cabinet contains recorders, a playback and analysis unit, an annunciator, a seismic switch, test units and an uninterruptible power supply.

The signal flow from the accelerometer to the seismic monitoring system is shown in **Figure 17**. From the accelerometer, the voltage signal is transmitted to the seismic monitoring cabinet and is converted by electronics into a signal proportional to the acceleration. The acceleration signal can activate recorders and/or local and MCR annunciators/alarms according to its magnitude.

When the acceleration of earthquake exceeds the seismic trigger or set-points for OBE (Operating Basis Earthquake or Event) or SSE (Safe Shutdown Earthquake), the alarm is provided on the local annunciator panel of the seismic monitoring cabinet and that of large panel display system (LDPS) in main control room or alternatively in the remote shutdown room. That event also activates the time-history recorder(s) in the seismic monitoring cabinet.

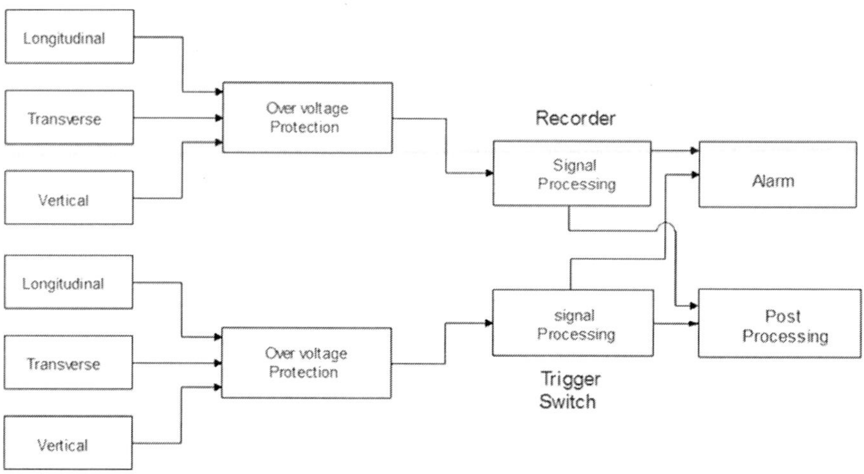

FIGURE 17. SMS block diagram.

4.2. Automatic Seismic Trip System

ASTS continuously monitor PGA (peak ground acceleration) from the seismic wave and automatically generates the trip signal. As shown in **Figure 18**, the seismic signal detected by the accelerometer is filtered, rectified and converted to the current. The sensor module is to cut off the frequency ranges over 10 Hz in order to pick out the strong motion of the earthquake. Since the seismic signal has a positive and negative signal, a rectifier circuit is added. The rectifier converts the bipolar signal to the unipolar.

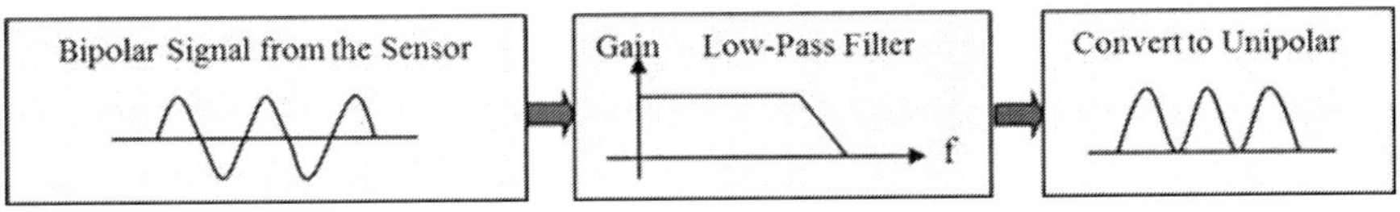

FIGURE 18. Conceptual diagram of sensor module for signal processing [7].

The overall scheme of ASTS is shown in **Figure 19**. The sensor output is interfaced with digital input card at the ASTS cabinet. Even the decision logic is set as two out of four but the system is composed of two independent channels. They are N1 and N2. For isolation between the channels, the digital input/output card is applied. The dotted line on the ASTS cabinet is the scope of trip logic channel N1 and N2. The ASTS cabinet is implemented by digitalized system such as PLC (programmable logic controller), FPGA (field programmable gate array), or DCS (distribute control system). The bi-stable and decision logic are configured with software inside each channel.

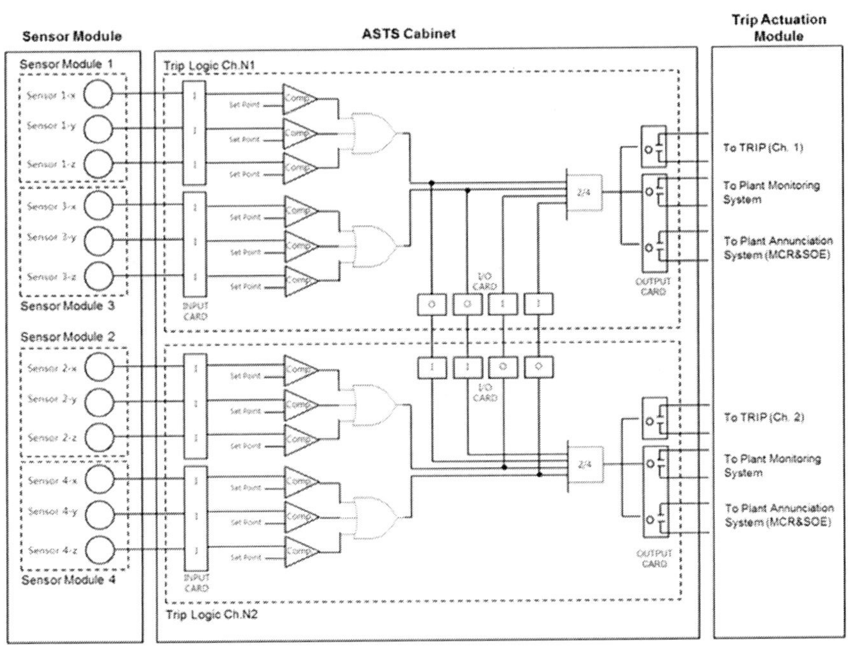

FIGURE 19. Overall scheme of digitalized ASTS [7].

The operational principles are as follows. When the measured signal exceeds the set-point, the edge triggering happens and the status is set to '0' by comparator actuation. In this case, the latch is engaged due to the edge triggering, and keep the status for 10 s. The purpose of 10-s latch is to synchronize the individual channel for the coincidence logic. If the signals are not properly synchronized, the trip initiation signal may not be actuated during a strong earthquake. ASTS consists of four diverse channel applications. The bi-stable signal from each channel feeds into two-out of-four coincidence logic for generating trip initiation signal.

5. CONCLUSIONS

In this work, two measurement principles are analysed to obtain the transfer function plotted by the frequency vs. relative response to the acceleration as well as the step responses. In order to define the sensitivity of open-loop type or closed-loop type with a feedback loop, two sensor types have been used; the force-balanced acceleration and the servo acceleration. The open-loop and closed-loop response when the step input describes the transfer function by plotting it with amplitude and time. The open-loop geophone shows exponentially decreasing function while the force balanced type illustrates overshoot before it is stabilized.

Two application examples for the NPP are discussed with their designs in this work. They are SMS and ASTS. When the Gyeongju earthquake occurred, the main concern in the NPP, whether the peak ground acceleration exceeded in OBE or SSE, was highlighted. The second issue was the frequency range determining the manual or automatic reactor trip. ASTS eliminates frequency over 10 Hz for generating a trip signal. Through Gyeongju earthquake the adequacy of this theory has been proved. The third issue that arose in ASTS design was a latching time of 10 s to synchronize the individual channel for the coincidence logic. This issue needs to be resolved by analysing the seismic waves in Gyeongju.

REFERENCES

1. "Strongest-ever earthquake hits Korea, tremors felt nationwide". The Korea Times. September 12, 2016. [Accessed 2016. 10].
2. Katsuhiko Ogata, "Modern Control Engineering", 3rd ed. Chapter 3, New Jersey: Prentice-Hall; 1997. pp. 83–84
3. Dept. of earth and environmental, Ludwig Maximilians Universiat https://www.geophysik.uni-muenchen.de/.../2_ms_seismic_instruments.pdf [Accessed 2016. 08]
4. Kinemetrics FBA-23 User Manual, https://nees.org/data/get/facility/sensor Models/180/Documentation/FBA23.pdf [Accessed 2016. 08]

5. Input/Output, SM-6 Geophone, http://www.iongeo.com/content/ documents /.../DS_SEN_121026SM6.pdf [Accessed 2016. 08]
6. GeoSig, AC-23 User Manual, http://www.geosig.com/files /GS_AC-23 _UserManual_V12.pdf [Accessed 2016. 08]
7. JC. Jung, Design of the Digitalized Automatic Seismic Trip System for Nuclear Power Plants, Nuclear Engineering and Technology, 46(2), 2014, pp. 235–246,

CHAPTER 7

The South-Western Alboran Earthquake Sequence of January-March 2016 and Its Associated Coulomb Stress Changes

Fida Medina[1], Taj-Eddine Cherkaoui[2]

[1]*Moroccan Association of Geosciences, Rabat, Morocco*

[2]*Mohammed V University in Rabat, Rabat, Morocco*

ABSTRACT

We expose the results of the study of the south-western Alboran seismic sequence of January-March 2016 and the stress perturbations it caused by means of Coulomb modeling. The use of data from numerous stations allowed us to relocate the largest events ($M_w \geq 3.8$) south of those determined by IGN (Spain). The main shock of 25 January 2016 ($M_w = 6.3$) was relocated at 35.6133°N × 3.6888°W, at a hypocentral depth of 15.5 ± 6.0 km. The epicenters are aligned along two distinct clusters: The first runs N-S towards the Moroccan coast along Al-Idrissi fracture zone, while the second is centered on the Alboran ridge northern thrust fault. The focal mechanisms determined by different agencies correspond to a strike-slip/normal/reverse motion in the western cluster and thrusting in the eastern one. Coulomb stress change modeling shows that the main earthquake: (1) may have been triggered by stress accumulation by the

1994 and 2004 earthquakes; (2) has induced stress perturbations that provide a suitable explanation for the distribution of the aftershock clusters, including the eastern cluster. The prediction of the next event shows that it is likely to occur near the Moroccan coast.

Keywords: Morocco, Seismicity, Alboran, Al Hoceima, Coulomb Stress Change, Seismotectonics, Seismic Hazard

1. INTRODUCTION

From January to March 2016, most of the coastal cities around the Alboran Sea were shaken by a series of earthquakes which started on 21 January at 13 h 47 min by a Mw = 5.1 event, followed on 25 January at 4 h 22 min by a major earthquake of Mw = 6.3. This major event was felt all around the Alboran region with a maximum intensity of VI-V especially at Melilla, where there were several damages and, as in Al Hoceima, the inhabitants remained outside their homes for several days. This event was followed by more than 2000 aftershocks recorded in the area 35.0°N - 36.0°N; 3.5°W - 4.5°W, 25 of which reached magnitudes Mw ≥ 4.0 until 16 March 2016 according to Instituto Geográfico Nacional, Madrid (IGN) (32 according to European-Mediterranean Seismological Centre, EMSC). The seismic activity was still being recorded during the preparation of the present manuscript (July 2016), but with much weaker and less frequent events.

Seismicity in the Alboran area is known to occur at shallow depths (5 - 20 km), mainly along a NNE-SSW trending lineament extending from Almeria to Al Hoceima [1] . This lineament was later named "Trans-Alboran Shear Zone" (TASZ) [2] , a term that was adopted by all following authors. Along the TASZ, the largest earthquakes of the instrumental period were the Al Hoceima earthquakes of 26 May 1994 and 24 February 2004, the magnitudes of which reached Mw = 6.0 and Mw = 6.3 respectively. However, the unusual seismic activity of January–March 2016 arises two main questions: (1) Since the Al Hoceima area was shaken by a similar event (Mw = 6.3) only 12 years ago, did the 2004 event, or even the previous 1994 one, trigger this activity given the distance between the epicentres? (2) why did 25 events considered as aftershocks attain magnitudes more than 4, other 5 had magnitudes Mw ≥ 5.0 and one Mw > 6.0 in such a short interval, a fact that is almost rare in the Alboran area?

In order to understand the causes of the January-March 2016 seismic crisis in the Alboran area north of Al Hoceima, the seismic sequence and the Coulomb stress changes (CSC) were studied with the objective to determine how the major events did jump in space (and when possible in time), taking into account the results of a previous study conducted for the large events of 1994 and 2004 [3] and the numerous recent geological and geophysical studies of the Alboran basin, which provide a new and more precise knowledge on the relations between seismicity and tectonics in the area. A similar study on the subject has been presented recently [4] based on data by Instituto Andaluz de Geofísica (IAG, Granada, Spain); however, our study, based on a long field experience since 1988, includes new relocations, new determinations of the stress tensor, the inclusion of the effects of previous events and more detailed geological and CSC analyses.

2. GEOLOGICAL AND SEISMOLOGICAL SETTING

2.1. Tectonic Context

The Alboran basin (Figure 1), which is the westernmost basin of the Mediterranean Sea, was created during the Early Miocene when the Alboran block or micro-plate collapsed radially and thinned to 15 km after an orogenic uplift phase related to the collision of Africa and Iberia, e.g. [5] . The causes and pre-Pliocene tectonic evolution of the Alboran basin are beyond the scope of this paper and can be found elsewhere in the very numerous papers published on the subject, e.g. [6] [7] [8] . The present-day (or neotectonic) structure, which is the focus of our interest, was inherited from Pliocene times when the Atlantic waters re-flooded the basin which was previously affected by the Messinian Salinity Crisis and associated sea-level fall [9] .

Present-day kinematics determined by GPS studies of the area is characterized by a WNW-ESE convergence of Nubia and Iberia at a rate of 5 $mm \cdot yr^{-1}$, e.g. [10] . However, escape of the Rif to the SW has also been observed [11] .

2.2. Main Structural Units

The Alboran area affected by the January–March 2016 earthquakes consists of a complex mosaic of small structures (Figure 1) which have been studied in detail during the last decade, e.g. [12] [13] [14] . The main physiographic lows are: the almost-radial West Alboran basin, the narrow Alboran channel (ENE-WSW), the South Alboran basin (ENE-WSW) and the eastward-open East Alboran basin. The most conspicuous highs are the volcanic Alboran Ridge (ENE-WSW) prolonged by the also volcanic Tofiño bank and the sedimentary Xauen bank [6] [12] [15] [16] , the Algarrobo-Herradura-Djibouti banks off the Spanish margin [17] , the Pytheas-Cabliers-Provençaux banks [12] , and the WNW-ESE Yusuf-Habibas lineament consisting of a ridge and a scarp [18] .

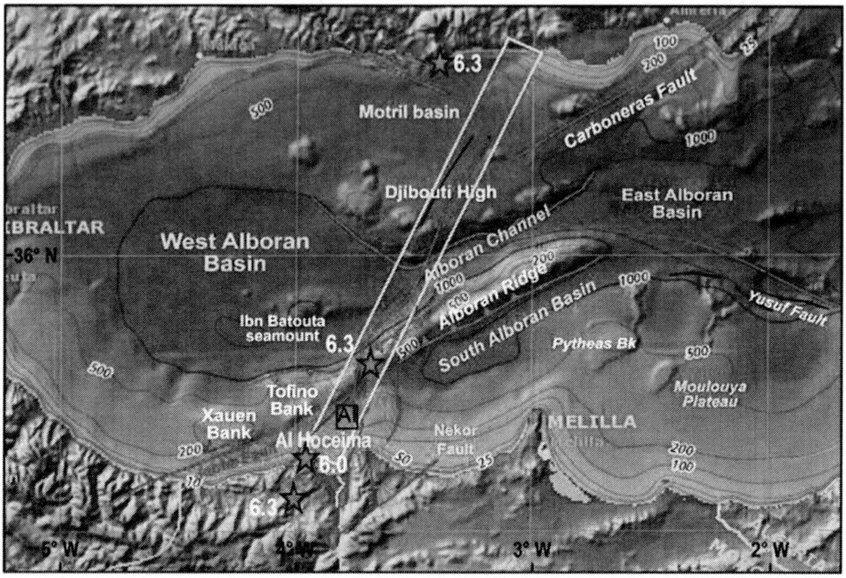

Figure 1. Main structures of the Alboran area superimposed on NOAA bathymetry and physiography (http://maps.ngdc.noaa.gov/ viewers/ bathymetry/). AI: Al-Idrissi fault zone; stars : main earthquakes of the area. Half-arrows indicate the motion of the fault block on which they are drawn. Triangles correspond to the hanging walls of reverse faults. The rectangle extending across the Alboran sea form the Spanish coast to Al Hoceima is the shear zone suggested by [1 .

2.3. Main Faults

In addition to previous commercial and academic seismic-reflection surveys, e.g. [12] [15] [16] , the recent use of swath bathymetry and high-resolution seismic profiling performed during several oceanographic surveys conducted in the last decade [8] have led to a highly improved knowledge on the geometry, trend and surface expression of the faults affecting the Plio-Quaternary formations (Figure 1).

Thus, the most important recent (Plio-Quaternary) faults are those delimiting in part the Alboran Ridge at both flanks, those oriented ENE-WSW bounding the Xauen and Tofiño banks [15] [16] , and the WNW-ESE Yusuf ridge and scarp [8] [18] . Along the Spanish margin, there are conspicuous structures such as the NE-SW Adra and Serrata-Carboneras faults [17] and smaller structures described in detail [19] , while along the Moroccan margin, some surface expressions of the Bousekkour-Aghbal and Trougout faults were described by [13] [14] [20] .

The NNE-SSW fracture zone south of the central part of the Alboran Basin was mapped more precisely during the most recent surveys, and was named Al-Idrissi Fault Zone in the Moroccan margin [12] . This fault zone consists of numerous almost-vertical normal faults which are relayed to the north by a major sinistral strike-slip fault offsetting the western termination of the Alboran Ridge and the Tofiño bank for about 28 km [12] . The northern continuation of the TASZ is not as simple as suggested by [1] ; for instance, other authors [8] [12] extend it northwards to the Djibouti bank area, but detailed analysis by [17] show that it continues in a diffuse manner as linear depressions, and more northwards by small segments until it reaches the land east of Motril. The relationship with the NE-SW trending Serrata-Carboneras segment is not obvious as they are separated from each other by the Adra anticline [17] .

Some deep crustal faults have been visualized by means of seismic reflection [6] [19] . One of these is the Intra Crustal Reflector (ICR) but these planes belong to the Alboran basement and seem to be inactive for the moment.

2.4. Seismological end Seismotectonic Setting

Seismicity of the Alboran basin was thought to be distributed because of the continental composition of the basement. However, increase in the earthquake dataset since the installation and development of the seismological networks of Spain and Morocco showed that several epicenters are clearly aligned. Two strips appear to be the most conspicuous: the eastern NNE-SSW alignment, with shallow hypocenters (<20 km) [21] , belongs to the TASZ and runs from Almeria to Al Hoceima entering the Rif, whereas the western alignment (N-S), comprising intermediate-depth hypocenters, runs from Malaga to Jebha [21] [22] (Figure 2). Fault-plane solutions of the earthquakes of the Alboran basin were determined and/or compiled from international centres (IGN, INGV, EMSC, Harvard, NEIC) by numerous authors [21] [22] [23] [24] [25] and show all types of regimes (strike-slip, normal and reverse), with horizontal NW-SE trending to vertical pressure axes and horizontal ENE-WSW to vertical tension axes. However, the largest earthquakes (26 May 1994, 24 February 2004 and 25 January 2016) are strike-slip.

The regional state of stress determined from available fault-plane solutions is a N140-N160 horizontal maximum compressional stress and an ENE-WSW to vertical minimum compressional stress [26] . Local states of stress may differ from the regional one because of local effects reflected by Coulomb stress changes at fault tips [3] .

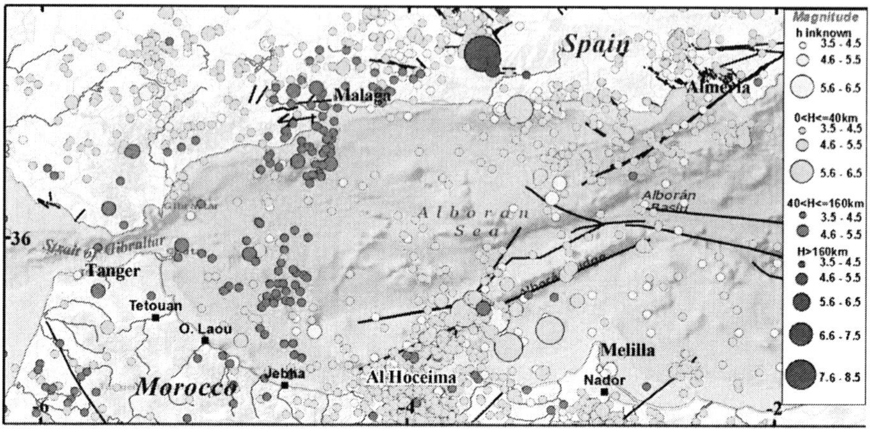

Figure 2. Seismicity of the Alboran Sea from 1901 until March 31, 2016 (Mw ≥ 3.5). Original file from T.E. Cherkaoui.

3. THE EARTHQUAKE SEQUENCE OF 2016

3.1. Distribution

Because of the absence of available public seismological data in Morocco, parameters of the earthquakes were initially taken from the open-access site of IGN (http://www.ign.es/ign/layoutIn/sismoFormularioCatalogo.do), which published the coordinates and the date/time of the events almost one to two hours after their occurrence (origin time). Due to the large number (2200) of small events (1.9 < M < 3.5), only those with magnitude M ≥ 3.8 IGN were initially selected. Thirty-two events were relocated using the arrival times to the seismological stations of Spain, Portugal and Moroccan stations AVE, IFR and MDT. The most important event (Mw = 6.3; on 25 January 2016 at 04:22:02.48 UTC), which triggered most subsequent events, was relocated at 35.6133°N × 3.6888°W, at a hypocentral depth of 15.5 ± 6.0 km.

Figure 3 and Table 1 show that temporally, and in conformity with the observations of [4] , the seismic activity started on 21 January in the western termination of the Alboran ridge with the two main shocks and continued along a rough N-S trend between latitudes 35.45°N and 35.65°N and longitudes 3.75°W and 3.85°W. The events of the initial cluster (4.0 ≤ Mw≤ 6.3) were quite close to each other, with small map separations; then a jump to the south occurred between epicenters 5 and 6, where the next events were recorded. On 25 January at 14 h 52 min, the activity jumped again to an area located 0.1° - 0.2° north-eastwards. The observed trend of the new cluster was clearly NNW-SSE.

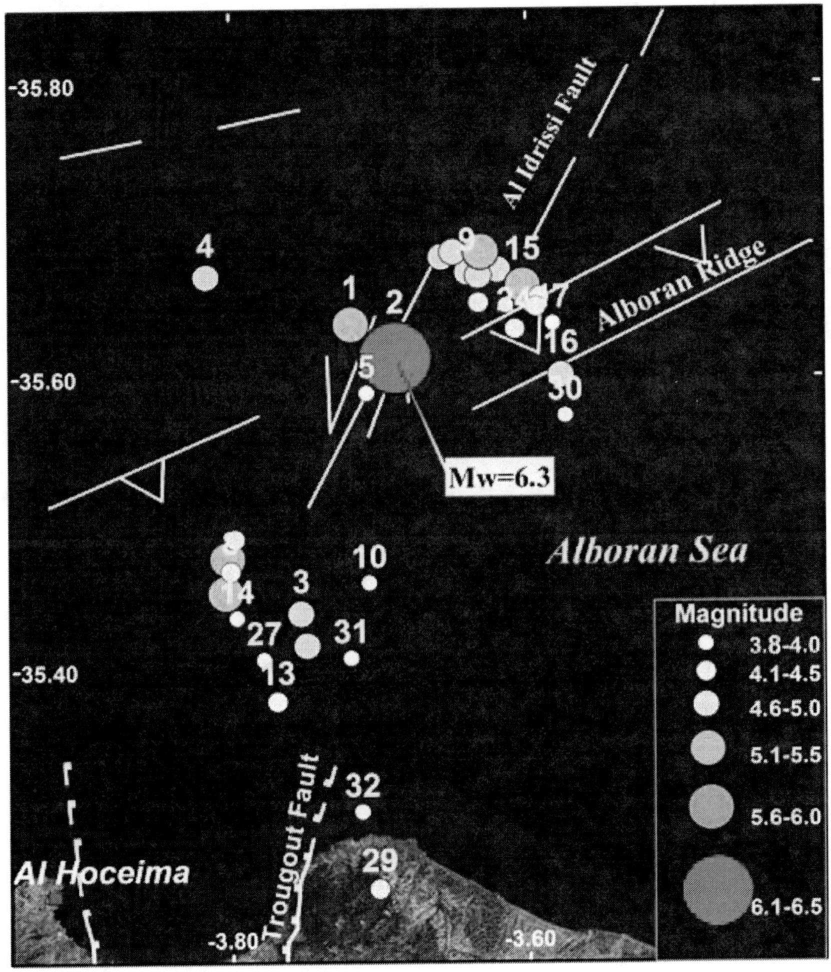

Figure 3. Seismic sequence of the 2016 Alboran events (M ≥ 4) from 21 January to 16 March.

The hypocentral depths of the earthquakes given by IGN and EMSC were probably fixed for calculating the locations, so that most of them were fixed at 10 km. Only moment tensor solutions provided different values ranging from 3 km to 24 km. Relocation of the events shows depths ranging from 0.5 km to 26.2 km, with 9 hypocentres between 5 km and 10 km. Anyway, it clearly appears that the earthquakes are located within the thin crust, in contrast to the deeper ones which are regularly recorded at ~100 km depth between Malaga and Jebha [22] .

3.2. Total Seismic Moment and Seismic Flux

Given the relatively large number of events with magnitudes M ≥ 4, it appeared of interest to calculate the released energy and the seismic flux, as previously done by [10] for the Western Africa-Eurasia plate boundary. The sum of the seismic moments of the largest events leads to a value of ~37 × 10^{24} dyn·cm. As the surveyed area is about 1000 km^2, the seismic flux is about 37 × 10^{21} dyn·cm·km^{-2} for just 3 months. This value is much greater than that calculated by [10] from the database taken from several centers. Authors [22] suggested a value of 1.4 × 10^{16}N·m·yr^{-1} (14×10^{22} dyn·cm·yr^{-1}) for the Alboran-northernmost Rif and South Iberia area, which becomes low when divided by the chosen area.

Table 1. Parameters of major earthquakes recorded by IGN and IMP (Portugal) from 25 January to 31 March 2016 and relocated for this study (map location see Figure 3).

N°	*Date*	*Time*	*Lat.*	*Long.*	*Depth*	*mbIg*	*mb*	*Mw*	*Io*	*NS*
1	21/01/2016	13:47:20.79	35.6360	−3.7185	9.8	4.9		5,1	III	53
2	25/01/2016	04:22:02.48	35.6133	−3.6888	15.5			6.3	V	55
3	25/01/2016	04:30:55.46	35.4387	−3.7525	16.7	4.0	5.0		IV	52
4	25/01/2016	04:34:19.76	35.6690	−3.8163	1.7	4.5	5.3	4.9	IV	54
5	25/01/2016	05:19:35.13	35.5892	−3.7082	18.8	3.8				39
6	25/01/2016	05:54:05.41	35.4760	−3.8023	8.5	4.6	5.2	5.2	IV	53
7	25/01/2016	06:10:41.85	35.4527	−3.8037	8.2	4.7		5.1	V	53
8	25/01/2016	08:25:04.24	35.4675	−3.8002	0.5	4.1	4.5	4.4	II	52
9	25/01/2016	14:52:42.56	35.6712	−3.6400	8.2	4.4	4.5	4.8	IV	55
10	25/01/2016	16:02:44.16	35.4593	−3.7067	18.5	3.5	4.3	3.9		47
11	25/01/2016	18:17:33.94	35.4895	−3.7972	15.1	4.0	4.2	4.1	III	49
12	27/01/2016	06:32:09.50	35.6730	−3.6200	21.8	4.7	4.9		IV	56
13	27/01/2016	21:57:48.21	35.3788	−3.7688	11.5	3.8		4.2	III	41
14	28/01/2016	19:48:52.14	35.4358	−3.7960	9.4	3.7	4.4	4.0	IV	54
15	22/02/2016	03:46:02.57	35.6622	−3.6030	8.8	4.5	5.3	5.1	IV	55
16	22/02/2016	04:14:30.56	35.6013	−3.5773	12.0	3.9	4.7		III	53

17	22/02/2016	06:43:03.14	35.6362	−3.5827	11.4	3.9			III	50
18	23/02/2016	08:46:01.75	35.6547	−3.5918	15.8	4.2	4.6	4.3	III	55
19	23/02/2016	10:12:33.23	35.6485	−3.6143	9.9	4.1	4.0		III	54
20	03/03/2016	11:36:26.71	35.4170	−3.7485	12.1	4.6	4.9	4.7	IV	57
21	09/03/2016	23:46:08.13	35.6477	−3.5942	12.0	4.5	4.6	4.5	III	51
22	11/03/2016	04:16:48.05	35.6818	−3.6578	1.6	4.8	4.8	5.0	IV	54
23	11/03/2016	04:19:48.58	35.6505	−3.6330	8.0	4.1			III	55
24	11/03/2016	09:40:19.84	35.6327	−3.6083	12.0	4.2			II	54
25	12/03/2016	15:04:06.12	35.6692	−3.6330	3.3	4.9	4.9	4.8	IV	55
26	12/03/2016	15:17:12.20	35.6852	−3.6500	1.9	4.9	4.8	4.8	IV	54
24	13/03/2016	20:54:01.93	35.4072	−3.7775	14.0	4.0			III	55
28	15/03/2016	04:40:39.73	35.6852	−3.6312	8.3	5.2		5.2	IV	54
29	16/03/2016	16:27:33.98	35.2500	−3.7007	3.1	4.2	4.4		IV	52
30	25/03/2016	22:28:47.68	35.5733	−3.5743	21.4	3.8			III	51
31	02/04/2016	14:48:05.01	35.4083	−3.7190	21.0	3.9			III	39
32	08/04/2016	07:10:07.98	35.3032	−3.7120	26.2	3.9				50

3.3. Focal Mechanisms

Twenty focal mechanisms determined by moment tensor inversion (Figure 4 and Figure 5; Table 2) were retrieved from IGN. Mechanisms from EMSC in cluded solutions from IGN, Geo Forschungs Zentrum, Potsdam (GFZ) and Istituto Nazionale di Geofisica e Vulcanologia, Roma (INGV). IGN mechanisms are in fact rapid moment tensors determined automatically from 3 stations, so they could be inaccurate with respect to those determined later with more stations. However, we used them because of their availability and their large number.

As also noted by [4] , who determined mechanisms from IAG unpublished data, the most striking observation is that the solutions obtained for the western and for the eastern clusters have very different patterns. The mechanisms of the western cluster (Figure 4) show various regimes depending on the position of the epicenter within the cluster: the northernmost events (#1; 2) and one in the south (#6) correspond to strike-slip faults with normal or reverse components; two events (#8; 13) show almost-pure reverse faults and other four (#7; 10; 14; 20) are normal faults. However, all the P axes are oriented NW-SE to NNW-SSE whereas the

T axes have an ENE-WSW orientation with variable plunge values. In contrast, mechanisms of the eastern cluster exclusively show thrust faulting across the Alboran Ridge (Figure 5) with NW-SE horizontal pressure axes and almost vertical tension axes.

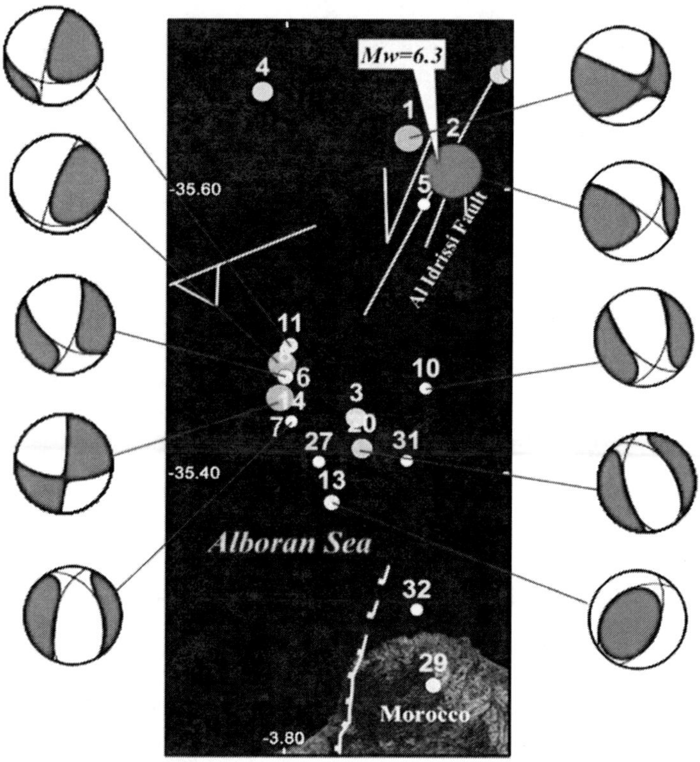

Figure 4. IGN fault-plane solutions for the western cluster for earthquakes of magnitude M ≥ 3.8 during the Alboran seismic crisis from 21 January to 31 March.

4. REGIONAL STRESS ORIENTATION AND MAGNITUDE FROM FOCAL MECHANISMS

Given the NNE-SSW to N-S trend of the first cluster, it is intuitive to consider the NNE-SSW planes as the main fault planes. However, modeling Coulomb stress changes (CSC) requires a more precise knowledge of the main plane in order to trace the faults. To this end, the stress determination software R4DT [27] was used to obtain the regional stress and, as an important output, the angle difference between the observed and the theoretical slip vector on the fault plane. For the western cluster, the obtained state of stress shows that σ_1 is oriented N344°, 20°; σ_2 is oriented N180°; 69° and σ_3 is oriented N76°; 05°. The stress ratio

$$R = \sigma_2 - \sigma_3 / \sigma_1 - \sigma_3 \tag{1}$$

is 0.45 ($\sigma_1 > \sigma_2 > \sigma_3$). However, this particular stress ratio is only obtained when assigning a weight to the fault (nodal) planes depending on the magnitude of the earthquake (weight 9 for M ≥ 6; 8 for M ≥ 5 etc.). Events #6, 7, 10 and 20 show small deviations, while events 8 and 13 show too large deviations with respect to the calculated state of stress. The stress tensor obtained from the eastern cluster shows that, using the same weights because of similar magnitudes, σ_1 is oriented N326°, 17°; σ_2 is oriented N233°; 11° and σ_3 is oriented N113°; 70°. The stress ratio R is 0.375. Only one plane shows a high deviation.

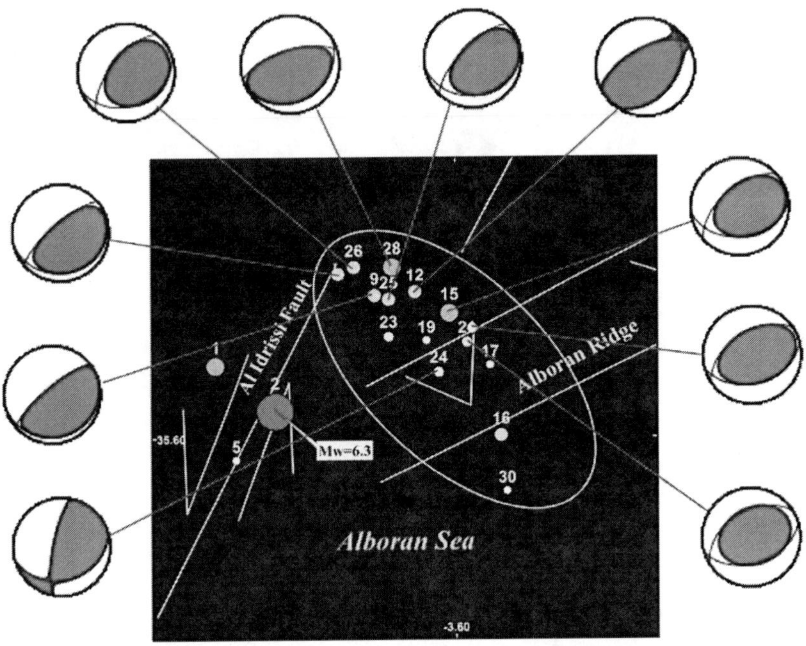

Figure 5. IGN fault-plane solutions for the eastern cluster for earthquakes of magnitude M ≥ 3.8 during the Alboran seismic crisis from 21 January to 31 March.

Table 2. Parameters of moment tensor solutions determined by several seismological agencies from 25 January to 31 March 2016. IGN data were used for the calculation of the state of stress.

Date and time	Lat.	Long.	Z	M	Plane A			Plane B			Agency
					Strike	Dip	Rake	Strike	Dip	Rake	
2016-01-21 13:47:22.0	35.3	−3.7	14	5.2	304	69	−177	213	87	−21	INGV
2016-01-21 13:47:20.0	35.7	−3.7	10	5.0	299	85	153	31	64	6	GFZ
2016-01-21 13:47:00.0	35.6	−3.8	3	5.1	298	84	136	34	46	9	IGN
2016-01-25 04:22:07.0	**35.7**	**−3.6**	**12**	**6.4**	**120**	**72**	**168**	**214**	**78**	**19**	**GCMT**
2016-01-25 04:22:04.0	**35.7**	**−3.5**	**15**	**6.5**	**217**	**81**	**−1**	**307**	**89**	**−171**	**CPPT**
2016-01-25 04:22:02.0	**35.7**	**−3.7**	**4**	**6.3**	**112**	**61**	**175**	**204**	**85**	**28**	**USGS**
2016-01-25 04:22:02.0	**35.7**	**−3.6**	**18**	**6.5**	**34**	**73**	**2**	**303**	**88**	**163**	**IPGP**
2016-01-25 04:22:02.0	**35.6**	**−3.6**	**12**	**6.3**	**211**	**75**	**10**	**119**	**80**	**165**	**INGV**
2016-01-25 04:22:02.0	**35.7**	**−3.7**	**15**	**6.3**	**122**	**88**	**152**	**213**	**62**	**2**	**GFZ**
2016-01-25 04:22:00.0	**35.6**	**−3.9**	**3**	**6.3**	**303**	**80**	**140**	**41**	**51**	**13**	**IGN**
2016-01-25 04:34:20.0	35.7	−3.8	15	4.9	308	87	175	39	85	3	GFZ
2016-01-25 05:54:05.0	35.5	−3.8	15	5.1	92	76	−173	1	84	−13	GFZ
2016-01-25 05:54:00.0	35.5	−3.8	9	5.2	242	61	119	13	40	49	IGN
2016-01-25 06:10:42.0	35.5	−3.9	15	4.9	298	84	164	30	75	6	GFZ
2016-01-25 06:10:00.0	35.3	−3.9	3	5.0	19	89	−2	109	88	−179	IGN
2016-01-25 08:25:04.0	35.5	−3.8	2	4.4	203	84	76	88	15	155	IGN
2016-01-25 14:52:41.0	35.7	−3.7	3	4.8	288	71	114	54	30	40	IGN
2016-01-25 16:02:42.0	35.5	−3.8	6	3.9	134	61	−134	16	51	−39	IGN
2016-01-25 18:17:32.0	35.5	−3.8	4	4.1	195	81	47	96	44	167	IGN
2016-01-27 06:32:07.0	35.7	3.7	5	4.5	241	58	105	35	35	68	IGN
2016-01-27 21:57:00.0	35.5	−3.8	3	4.2	23	54	73	231	39	112	IGN
016-01-28 19:48:00.0	35.5	−3.9	3	4.0	197	60	−62	331	40	−129	IGN
2016-01-31 16:25:00.0	36.6	−3.1	3	4.5	308	81	−128	207	39	−15	IGN
2016-02-22 03:46:02.0	35.7	−3.5	5	5.0	63	29	83	250	60	93	USGS
2016-02-22 03:46:02.0	35.7	−3.6	10	5.1	54	33	69	258	59	103	GFZ
2016-02-22 03:46:01.0	35.6	−3.7	10	5.2	236	32	62	87	62	106	INGV
2016-02-23 08:46:00	35.7	−3.6	4	4.3	236	51	78	075	41	105	IGN
2016-03-03 11:36:27.0	35.4	−3.9	14	4.8	130	31	−104	327	61	−81	GFZ

2016-03-09 23:46:07.0	35.5	−3.6	4	4.5	238	46	079	73	45	101	IGN
2016-03-11 04:16:48.0	35.6	−3.6	11	4.9	59	29	76	256	62	98	GFZ
2016-03-11 04:16:47.0	35.7	−3.6	10	4.9	237	40	69	84	53	107	INGV
2016-03-11 09:40:18.0	35.6	−3.6	04	3.8	191	75	62	75	32	150	IGN
2016-03-12 15:04:06.0	35.6	−3.6	10	4.8	59	31	69	262	61	102	GFZ
2016-03-12 15:04:06.	35.6	−3.8	10	4.9	245	31	70	88	61	101	INGV
2016-03-12 15:04:05	35.5	−3.6	3	4.8	224	70	75	81	25	124	IGN
2016-03-12 15:17:12.0	35.6	−3.7	10	4.9	235	26	63	84	67	103	INGV
2016-03-12 15:17:12.0	35.7	−3.7	10	4.8	48	31	68	253	61	102	GFZ
2016-03-12 15:17:11.0	35.6	−3.6	4	4.8	219	63	74	71	31	118	IGN
2016-03-15 04:40:40.0	35.8	−3.6	4	5.2	78	25	100	246	65	84	USGS
2016-03-15 04:40:39.0	35.6	−3.7	10	5.2	58	42	76	257	50	102	GFZ
2016-03-15 04:40:38.0	35.6	−3.8	10	5.3	236	42	68	84	52	109	INGV
2016-03-15 04:40:38.0	35.6	−3.6	04	5.2	252	66	96	58	25	77	IGN

5. COULOMB STRESS CHANGES

5.1. Methodology

The study of Coulomb stress changes (CSC), which has become widespread since the early 1990s [28] [29] , is a very useful tool to understand the distribution of seismicity around active faults and even in unfaulted areas. CSC is expressed by the equation e.g. [28] :

$$\Delta\sigma_c = \Delta\tau - \mu'\Delta\sigma_N \tag{2}$$

where $\Delta\sigma_c$, $\Delta\tau$ and $\Delta\sigma_N$ are, respectively, the coseismic changes in Coulomb stress, shear stress and normal stress, and μ' is the effective coefficient of friction, e.g. [30] and references therein, which corresponds to $\mu' = \mu(1-B)$ where B is Skempton's coefficient, comprised between 0.5 and 0.9. Slip on a seismic fault produces changes in Coulomb stresses in the surrounding areas. These changes depend on the orientation of the planes and on the direction and magnitude of the regional stresses. Detailed analytical equations of stress transfer are exposed by [28] , but a more simple way of output is the use of Coulomb 3.2 software [31] which allows drawing maps of Coulomb stress changes resolved on various fault orientations and regimes, and also superimposing topographic and tectonic features or splitting faults.

5.2. Input Parameters

Epicenter location and focal depth. As exposed in section 3.1, the epicenter of 25 January 2016 was relocated at 35.6133°N × 3.6888°W. The calculated final depth was 15.5 km.

Fault parameters. A major issue is the determination of the main fault plane. For events 1 and 2, which are respectively the initial (foreshock) and major events, the B plane (N34 and N41 respectively of the solution IGN in Table 2) show the smallest angles to the theoretical fault plane obtained from the R4DT stress determinations, so they can be considered as the main fault-planes as expected from the distribution of epicenters. Therefore, CSC was initially calculated using IGN parameters (Table 2); length 10 km; width 10 km (5 - 15 km depth); mean displacement 1 m left-lateral (when uniform), which lead to a seismic scalar moment of about 3.2×10^{25} dyn.cm (3.2×10^{18} N.m) and a moment magnitude of ~6.3. However, subsequent determinations of focal mechanisms listed in EMSC files led us to rather use the solution of INGV (N211 and N213 respectively; dip 75°) with the same parameters (Figure 6). In order to avoid in excess concentration of stress at the tips of the fault, the plane was tapered into 5 patches as recommended by the Coulomb 3.2 tutorial.

Stress orientation and magnitude. CSC calculations were carried out taking into account the regional state of stress, derived from our previous R4DT determination for the Al Hoceima area [26] : σ_1 and σ_3 horizontal at 150°E and N60°E respectively. The stress magnitudes used by software Coulomb are 100 bar for σ_1; 30 bar for σ_2; and 0 bar for σ_3. No vertical gradient was taken into account. The CSC calculation depth is 15 km. We use this pre-earthquake regional tensor, in which the trend of σ_1 differs by 14° counterclockwise from that determined from the western cluster and 4° clockwise from the eastern cluster, because the latter was determined after the earthquake sequence which lasted many weeks after the main events. However, the new tensor should be the one to be used in the next studies.

Fault strength parameters. We used the default values recommended by the Coulomb 3.2 software. These are 8×10^5 bar for Young modulus; 0.25 for Poisson's ratio, and 0.8 for the coefficient of friction.

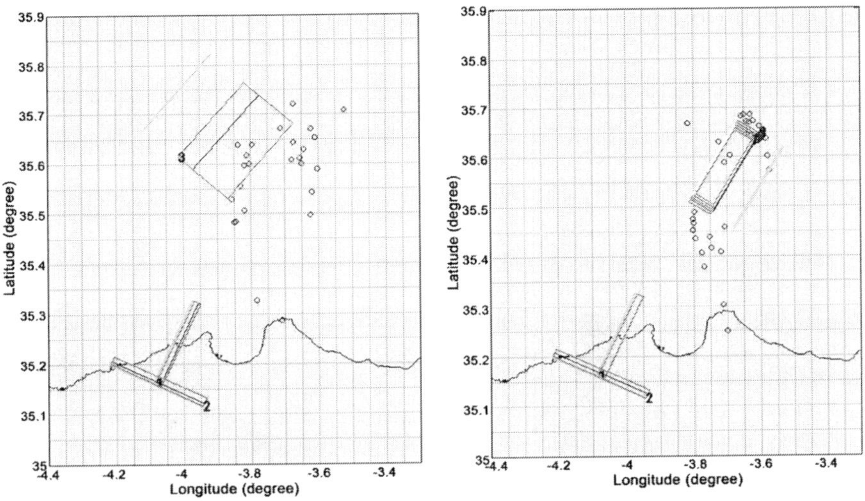

Figure 6. Initial state of the source fault plane model, corresponding to the 25 January 2016 shock, of Mw = 6.3 (tapered fault 3-7). Left panel: initial determinations from IGN, right: relocation of epicentres (this paper) and plane determined by INGV. Red rectangles are the fault planes of the events of 1994 (1) 2004 (2) and 2016 (3-7). Green lines are the extended surface traces of the faults.

5.3. Coulomb Stress Perturbations Caused by the 25 January Quake

From all seismic data, it appears that events 1, 2 and 5 may belong to ruptures on the same plane or on close planes. As mechanism 1 is almost similar to mechanism 2, this one will be considered for CSC modeling because of its larger magnitude (M = 6.3) and effects. In this case, plane B has been considered as the main fault plane because of the smaller angular deviation of the slip vector with respect to the theoretical striation obtained from R4DT software.

Coulomb 3.2 was run for fault of event #2 (Table 1 and Table 2) taken as the source fault using the additional parameters indicated in Table 2 (Figure 6). The effect of the main fault on CSC was calculated on optimum strike-slip, normal and reverse faults (Figures 7(A)-(C) respectively). In the case of optimum normal faults, it clearly appears that the southernmost events of the western cluster, the mechanisms of which are normal, match fairly well the N-S optimum planes of normal faults, on which there is a rise in CSC. This is the case of events #7; 10; 14 and 20. In the case of events with reverse faults of the eastern cluster, the CSC map shows that they fall in the CSC increase sector, exactly at the north-eastern tip of the source fault.

A cross-section was constructed in order to better visualize the zones of CSC rise on reverse faults at 3 bar and 10 bar saturation (Figure 8(A) and Figure 8(B)). This cross section shows that the events of the eastern cluster clearly plot in the CSC rise zone.

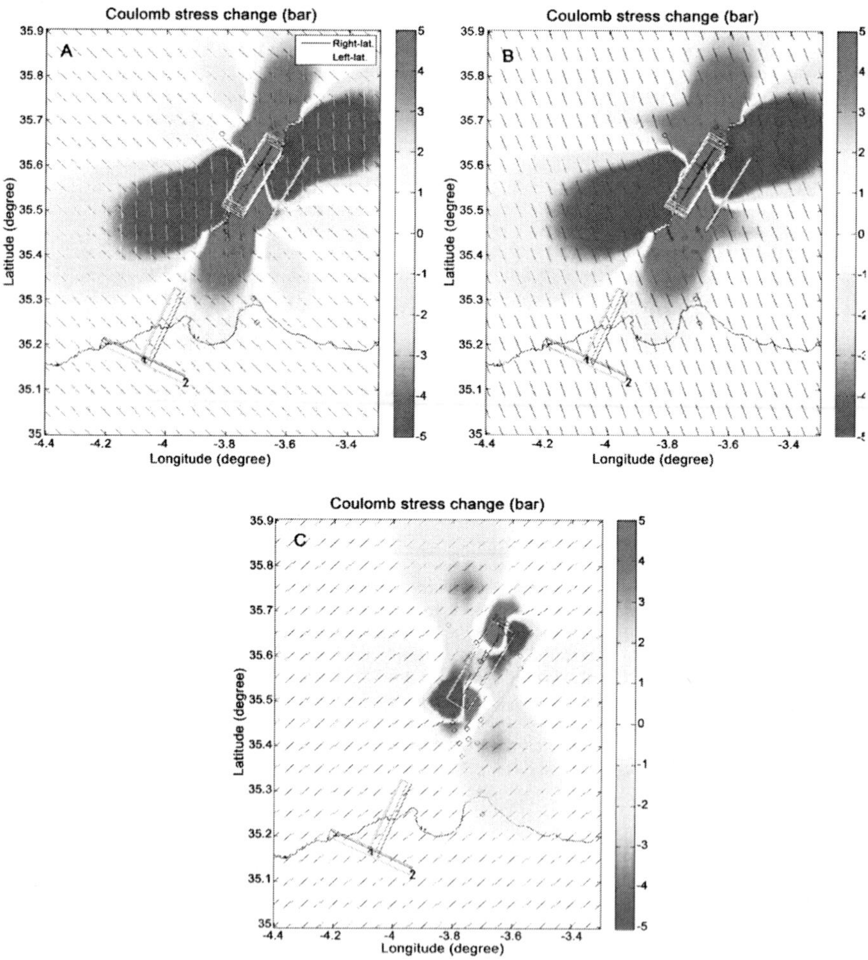

Figure 7. CSC produced by the 25 January 2016 around the epicentral area resolved on optimum strike-slip planes (A), optimum normal planes (B), and optimum thrust planes (C). In red: CS increase; in blue, CS decrease.

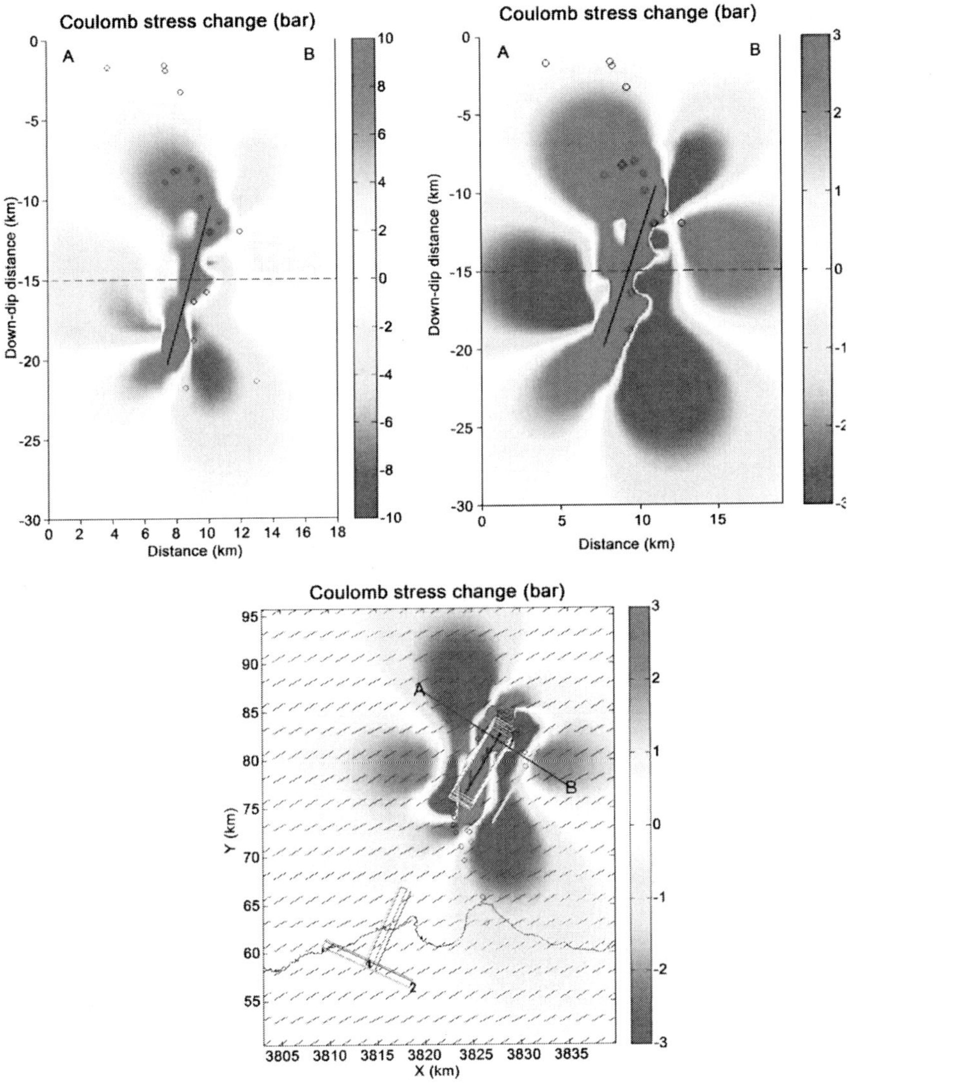

Figure 8. Top: Cross-sections showing CSC induced by slip on the source fault plane of 25 January 2016 (INGV solution) resolved on thrust planes at 10 and 3 bar saturation. In red: CS increase; in blue, CS decrease. Bottom: CSC resolved on thrusts with NE-SW trend at 15 km with 3 bar saturation. (A)-(B), location of the cross section.

6. DISCUSSION

First, it should be stated that the determination of the depth of the hypocenters of small aftershocks is obviously less accurate offshore than onshore where it is generally carried out with a mobile array of stations around the main event, e.g. [32] [33] [34] for the Al Hoceima surveys of 1988, 1994 and 2004. However, the azimutal coverage around the Alboran Sea appears to be quite satisfactory, especially on the Iberian side. This point is discussed in section 6.5.

6.1. Earthquakes and Tectonic Features

From the trend of the two clusters, it clearly appears that the western cluster, trending NNE-SSW, is related to the faults that run from Al Hoceima to the western termination of the Alboran Ridge, i.e. the TASZ and Al-Idrissi FZ. These are either normal or strike-slip faults as interpreted from seismic profiles and high-resolution bathymetry.

Similarly, the eastern cluster, although less linear, is clearly related to the high-angle reverse faults that bound the Alboran Ridge to the NW and to the SE. An important question is whether the motion along the Al-Idrissi FZ or more generally the TASZ can generate large earthquakes given its recent and shallow character. Indeed, seismic profiles published by [1] and following authors, and the maps published by [8] , show that this system is quite recent and shallow and its extension downwards to 10 km depth is not established with other means than seismology. However, it may reflect surface expression of older, deeper faults inherited from Mio-Pliocene times.

6.2. Focal Mechanisms

It is interesting to note that since the 2004 seismic crisis there was only a weak seismic activity in the Alboran area before the 2016 crisis. Authors [25] , who installed an array of ocean bottom seismographs (OBS) from 13 August 2009 to 15 January 2010, did not record shocks of magnitude higher than 3.8. The only focal mechanism they determined in the 2016 crisis area is located at Ras Tarf and consists of a reverse fault similar to those of the eastern cluster of 2016. Authors [24] also conducted a study on moment tensor solutions for the Iberian-Maghreb region during the Iber Array deployment from 2009 to 2013, but no one was determined in the studied area, probably because of small magnitudes.

6.3. Influence of the Previous Al Hoceima Earthquakes of 1994 and 2004

In order to assess the influence of the previous Al Hoceima earthquakes, CSC after the 1994 and 2004 earthquakes were resolved on planes parallel to the NNE-SSW fault plane of 2016 determined by INGV (see also [3] ; his Figure 6(A) and Figure 7(A) respectively). Adding the 2016 earthquake data to these maps clearly shows that the swarms are located in the sector where a rise of Coulomb stress has been obtained after the occurrence of both earthquakes (Figure 9(A) and Figure 9(B)), although the stress magnitude becomes higher when removing the arbitrary values of the regional stresses suggested by the model. This clearly testifies for the stress transfer to the Al-Idrissi Fault Zone and to the Alboran ridge, so it is likely that the faults in this area were reactivated or at least fostered by these events. However, only the southern cluster falls in the 1 - 2 bar rise sector, while the northern events, the major ones, are located in the 0 - 1 bar rise sector. Another important observation is that the 2004 earthquake had weaker CSC rise with respect to the 1994 one for the area where the 2016 earthquakes were recorded. This is due to the southerly location of the 2004 event with respect to the 1994 shock.

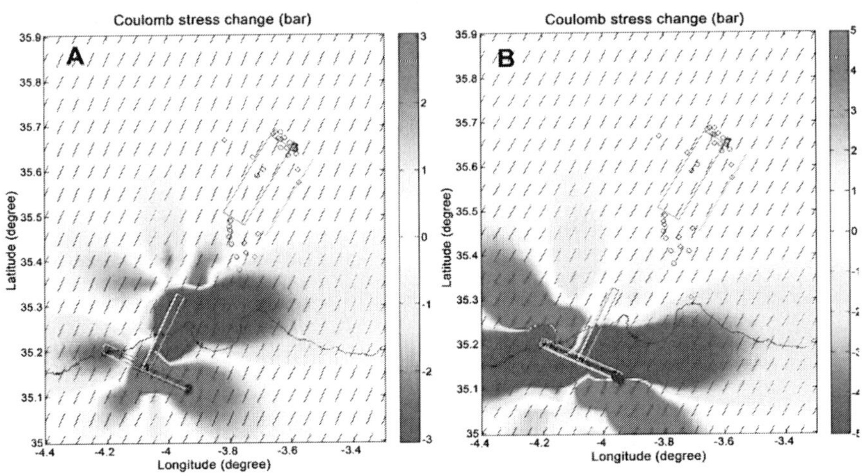

Figure 9. (A) CSC effect of the Al Hoceima earthquakes on 1994 at 10 km and 3 bar saturation and 2004 (B) at 10 bar and 10 km on the location of the 2016 events with M ≥ 4. In red, CS increase; in blue, CS decrease.

6.4. Jumps in Seismic Activity

Another major question is how the jumps in seismicity occurred during the 2016 crisis. The location of the western cluster in the area where an increase in CS was obtained using INGV parameters is almost perfect (Figure 7). The location of the aftershocks of the eastern cluster is more difficult to match because they are slightly offset from the CS rise area when using (unrealistic) uniform slip along the whole fault. However, in the tapered fault plane model, they match the CS rise at the NE fault tip so that there is no need for a new fault to account for their occurrence. It remains that the NNW-SSE orientation of the epicenters along theoretically NE-SW fault planes (Alboran Ridge) may arise some questions such as the existence of a pop-up structure. Although the cross-section of Figure 8 clearly shows that the epicenters appear to be aligned along a plane that could be interpreted as a reverse fault, this assumption needs more accurate determinations with smaller horizontal/vertical errors.

6.5. Influence of Errors on Hypocentral Depths

One of the most unconstrained parameters in this study is the hypocentral depth, because most shocks were in the Alboran Sea and the closest stations are relatively far (80 km for EALB and EMEL). This leads to difficulties in the determination of the depth of hypocenters, as generally observed on IGN earthquake catalogues, where most shallow earthquakes are (fixed) at 10 km, followed by those at 5 km and 2 km, e.g. [21] [22] . Our relocations show that the errors on depth range from 3.1 to 17.3 km with most events with ERZ between 2.5 and 7.5 km. Anyway, as illustrated in Figure 8, the vertical error bars of the eastern cluster of the 2016 crisis remain situated in an area of a wide CSC rise.

6.6. What about the Next Large Earthquake (and Tsunami)?

One of the possibilities provided by the study of the CSC is to assess the seismic hazard and to determine the faults that have undergone CS rise. Previous investigations have been successful in predicting the location of the occurrence of the next events within CS increase areas, although it was not possible to determine when. For instance, the CSC during the seismic crisis of 2016 (Figure 7) shows that there is an increase in CS to the north and to the south, spanning the Al-Idrissi FZ and the Algarrobo–Herradura–Djibouti banks off the Spanish margin (Figure 1). An interesting feature is that the area NE of Al Hoceima, which was in a CS decrease zone [3] , has undergone an increase of CS after 2016, and should be considered in the next years. The same can be said about the Alboran ridge, which has been activated in 2016.

In parallel, recent research on tsunami hazard in Alboran area [21] [35] has simulated fault slip on several potential seismogenic/tsunamigenic faults (Figure 6 in [21]), which can also be determined by Coulomb 3.2, but no CSC was studied. As tsunami excitation and hazard are intimately linked to fault slip, we can also state that the faults of the affected area may accumulate stress, especially those displaying vertical motions, such as the thrust faults bounding the Alboran Ridge, the Algarrobo-Herradura-Djibouti and the Tofiño banks because the strike-slip faults with small thrust or normal components do not generate large tsunamis [36] .

7. CONCLUSIONS

In this paper, we have exposed the results of the study of the south-western Alboran seismic sequence of January-March 2016 and the stress perturbations it caused by means of Coulomb modeling. The main results are:

1) The epicenters of the largest events, which were relocated using data from numerous stations, show two distinct clusters: one running NNW-SSE towards the Moroccan coast along Al-Idrissi fracture zone, the other centered on the Alboran ridge northern thrust fault.
2) The focal mechanisms determined by different agencies correspond to strike-slip/normal/reverse motion in the western cluster and thrusting in the eastern one.
3) The state of stress determined from the published focal mechanisms show a horizontal compressional stress oriented NNW-SSE in conformity with previous determinations.
4) Coulomb stress change modeling shows that the main earthquake of 25 January 2016 (Mw = 6.3) may have been triggered by stress accumulation by the 1994 and 2004 earthquakes.
5) The main earthquake of 25 January 2016 has induced stress perturbations that provide a suitable explanation for the distribution of the aftershock clusters, including the eastern cluster. The prediction of the next event shows that it is likely to occur near the Moroccan coast.

CITE THIS PAPER

Medina, F. and Cherkaoui, T.-E. (2017) The South-Western Alboran Earthquake Sequence of January-March 2016 and Its Associated Coulomb Stress Changes. Open Journal of Earthquake Research, 6, 35-54.

REFERENCES

1. Dillon, W.P., Robb, J.M., Greene, H.G. and Lucena, J.C. (1980) Evolution of the Continental Margin of Southern Spain and the Alboran Sea. Marine Geology, 36, 205-226.
2. De Larouzière, F.D., Bolze, J., Bordet, P., Hernandez, J., Montenat, C. and Ott d'Estevou, P. (1988) The Betic Segment of the Lithospheric Trans-Alboran Shear Zone during the Late Miocene. Tectonophysics, 152, 41-52.
3. Medina, F. (2015) Coulomb Stress Perturbations Related to the Al Hoceima (Morocco) Earthquakes of 1994 and 2004. Open Journal of Earthquake Research, 4, 37-54.
4. Alvarez-Gómez, J.A., Martín, R., Pérez-López, R., Stich, D., Cantavella, J.V., Martínez-Díaz, J.J., Morales, J., Martínez García, P., Soto, J.I. and Carreno, E. (2016) La Série Sísmica de Alhucemas 2016. Partición de la deformación e Interacción de Estructuras en un Límite de Placas Difuso. Geo-Temas, 16, 491-494.
5. Dewey, J.F. (1988) Extensional Collapse of Orogens. Tectonics, 7, 1123-1139.
6. Watts, A.B., Platt, J.P. and Buhl, P. (1993) Tectonic Evolution of the Alboran Sea Basin. Basin Research, 5, 153-177.
7. Gueguen, E., Doglioni, C. and Fernandez, M. (1998) On the Post-25 Ma Geodynamic Evolution of the Western Mediterranean. Tectonophysics, 298, 259-269.
8. Martínez-García, P., Soto, J.I. and Comas, M. (2011) Recent Structures in the Alboran Ridge and Yusuf Fault Zones Based on Swath Bathymetry and Sub-Bottom Profiling: Evidence of Active Tectonics. Geo-Marine Letters, 31, 19-36.
9. Estrada, F., Ercilla, G., Gorini, C., Alonso, B., Vázquez, J.T., García-Castellanos, D., Juan, C., Maldonado, A., Ammar, A. and Elabbassi, M. (2011) Impact of Pulsed Atlantic Water Inflow into the Alboran Basin at the Time of the Zanclean Flooding. Geo-Marine Letters, 31, 361-376.
10. Serpelloni, E., Vannucci, G., Pondrelli, S., Argnani, A., Casula, G., Anzidei, M., Baldi, P. and Gasperini, P. (2007) Kinematics of the Western Africa-Eurasia Plate Boundary from Focal Mechanisms and GPS Data. Geophysical Journal International, 169, 1180-1200.
11. Vernant, P., Fadil, A., Mourabit, T., Ouazar, D., Koulali, A., Davila, J.M., Garate, J., McClusky, S. and Reilinger, R. (2010) Geodetic Constraints on Active Tectonics of the Western Mediterranean: Implications for the Kinematics and Dynamics of the Nubia-Eurasia Plate Boundary Zone. Journal of Geodynamics, 49, 123-129.
12. Ammar, A., Mauffret, A., Gorini, C. and Jabour, H. (2007) The Tectonic Structure of the Alboran Margin of Morocco. Revista Sociedad Geologica de Espana, 20, 247-271.
13. D'Acremont, E., Gutscher, M.A., Rabaute, A., Mercier de Lépinay, B.M., Lafosse, M., Poort, J., Ammar, A., Tahayt, A., Le Roy, P., Smit, J., Do Couto, D., Cancouet, R., Prunier, C., Ercilla, G. and Gorini, C. (2014) High-Resolution Imagery of Active Faulting Offshore Al Hoceima, Northern Morocco. Tectonophysics, 632, 160-166.
14. Lafosse, M., d'Acremont, E., Rabaute, A., Mercier de Lépinay, B., Tahayt, A., Ammar, A. and Gorini, C. (2016) Evidence of Quaternary Transtensional Tectonics in the Nekor Basin (NE Morocco). Basin Research, 1-20.

15. Bourgois, J., Mauffret, A., Ammar, A. and Demnati, A. (1992) Multichannel Seismic Data Imaging of Inversion Tectonics of the Alboran Ridge (Western Mediterranean Sea). Geo-Marine Letters, 12, 117-122.
16. Chalouan, A., Saji, R., Michard, A. and Bally, A.W. (1997) Neogene Tectonic Evolution of the Southwestern Alboran Basin as inferred from Seismic Data off Morocco. AAPG Bulletin, 81, 1161-1184.
17. Ballesteros, M., Rivera, J., Munoz, A., Munoz-Martín, A., Acosta, J., Carbó, A. and Uchupi, E. (2008) Alboran Basin, Southern Spain—Part II: Neogene Tectonic Implications for the Orogenic Float Model. Marine and Petroleum Geology, 25, 75-101.
18. Martínez-García, P., Comas, M., Soto, J.I., Lonergan, L. and Watts, A.B. (2013) Strike-Slip Tectonics and Basin Inversion in the Western Mediterranean: The Post-Messinian Evolution of the Alboran Sea. Basin Research, 25, 361-387.
19. Comas, M.C., Platt, J.P., Soto, J.I. and Watts, A.B. (1999) The Origin and Tectonic History of the Alboran Basin: Insights from Leg 161 Results. Proceedings of the Ocean Drilling Program Scientific Results, 161, 555-580.
20. Calvert, A., Gomez, F., Seber, D., Barazangi, M., Jabour, N., Ibenbrahim, A. and Demnati, A. (1997) An Integrated Geophysical Investigation of Recent Seismicity in the Al-Hoceima Region of North Morocco. Bulletin of the Seismological Society of America, 87, 637-651.
21. álvarez-Gómez, J.A., Aniel-Quiroga, í., González, M., Olabarrieta, M. and Carreno, E. (2011) Scenarios for Earthquake-Generated Tsunamis on a Complex Tectonic Area of Diffuse Deformation and Low Velocity: The Alboran Sea, Western Mediterranean. Marine Geology, 284, 55-73.
22. Buforn, E., Bezzeghoud, M., Udias, A. and Pro, C. (2004) Seismic Sources on the Iberia-African Plate Boundary and Their Tectonic Implications. Pure and Applied Geophysics, 161, 623-646.
23. Stich, D., Serpelloni, E., Mancilla, F. and Morales, J. (2006) Kinematics of the Iberia-Maghreb Plate Contact from Seismic Moment Tensors and GPS Observations. Tectonophysics, 426, 295-317.
24. Martín, R., Stich, D., Morales, J. and Mancilla, F. (2015) Moment Tensor Solutions for the Iberian-Maghreb Region during the Iberarray Deployment (2009-2013). Tectonophysics, 663, 261-274.
25. Grevemeyer, I., Gràcia, E., Villasenor, A., Leuchters, W. and Watts, A.B. (2015) Seismicity and Active Tectonics in the Alboran Sea, Western Mediterranean: Constraints from an Offshore-Onshore Seismological Network and Swath Bathymetry Data. Journal of Geophysical Research: Solid Earth, 120, 8348-8365.
26. Medina, F. and El Alami, S.O. (2006) Focal Mechanisms and State of Stress in the Al Hoceima Area. Bulletin de l'Institut Scientifique, Section Sciences de la Terre, 28, 19-30.
27. Angelier, J. (1990) Inversion of Field Data in Fault Tectonics to Obtain the Regional Stress—III: A New Rapid Direct Inversion Method by Analytical Means. Geophysical Journal International, 103, 363-376.
28. King, G.C., Stein, R.S. and Lin, J. (1994) Static Stress Changes and the Triggering of Earthquakes. Bulletin of the Seismological Society of America, 84, 935-953.
29. Verdecchia, A. and Carena, S. (2015) One Hundred and Fifty Years of Coulomb Stress History along the California-Nevada Border, USA. Tectonics, 34, 213-231.
30. Beeler, N.M., Simpson, R.W., Hickman, S.H. and Lockner, D.A. (2000) Pore Fluid Pressure, Apparent Friction, and Coulomb Failure. Journal of Geophysical Research, 105, 25,533-25,542.
31. Toda, S., Stein, R., Lin, J. and Sevilgen, V. (2008) Coulomb 3.2. User Guide. http://earthquake.usgs.gov/research/software/coulomb/
32. Cherkaoui, T.E., Hatzfeld, D., Jebli, H., Medina, F. and Caillot, V. (1990) Etude microsismique de la région d'A1 Hoceima. Bulletin de l'Institut Scientifique, 14, 25-34.

33. El Alami, S.O., Tadili, B.A., Cherkaoui, T.E., Medina, F., Ramdani, M., Ait Brahim, L. and Harnafi, M. (1998) The Al Hoceima Earthquake of May 26, 1994 and Its Aftershocks: A Seismotectonic Study. Annals of Geophysics, 41, 519-537.
34. Van Der Woerd, J., Dorbath, C., Ousadou, F., Dorbath, L., Delouis, B., Jacques, E., Tapponnier, P., Hahou, Y., Menzhi, M., Frogneux, M. and Haessler, H. (2014) The Al Hoceima Mw 6.4 Earthquake of 24 February 2004 and Its Aftershocks Sequence. Journal of Geodynamics, 77, 89-109.
35. álvarez-Gómez, J.A., González, M. and Otero, L. (2011) Tsunami Hazard at the Western Mediterranean Spanish Coast from Seismic Sources. Natural Hazards and Earth System Sciences, 11, 227-240.
36. Okada, Y. (1985) Surface Deformation due to Shear and Tensile Faults in a Half-Space. Bulletin of the Seismological Society of America, 75, 1135-1154.

CHAPTER 8

Seismic Sequences' Branching Structures: Long-Range Interactions and Hazard Levels

Giulio Riga[1], Paolo Balocchi[2]

[1]*Geologist, Independent Researcher, Lamezia Terme, Italy*

[2]*Geologist, Independent Researcher, Modena, Italy*

ABSTRACT

Branching structures can provide early information on earthquakes' preparation process, trigger stage, different breaking patterns that can occur before strong earthquakes and hazard levels reached in the area to be analyzed. In this study, we aim to understand the earthquakes long-range interactions which constitute the nodes of higher order seismic rods in the upper branching structure, and the hazard level reached in each developmental stage as well as to provide a warning time frame for the most energetic seismic events and a sound method to obtain information on the epicentral area. To this end, we have analyzed several branching structures by using both local and global seismicity. The analysis of different branching structures both on global and local scale highlights long-range interactions between the most energetic earthquakes and their triggering by smaller shocks, thus suggesting that the triggering can occur few minutes to decades before the earthquake, depending on a seismic sequence's development speed and its structure.

Keywords: Branching Structure, Earthquake, Microsequence, Hierarchization, Foreshock, Hazard

1. INTRODUCTION

Several studies show how the aftershocks observed on time scales of minutes to months after a mainshock can be triggered by static and dynamic stresses imparted by the stress change and the mainshock waves' passage [1] .

Branching structures [2] , associated with earthquakes and their development over time, have proven to be a helpful instrument to obtain crucial information about the occurrence of future events, the interaction between the earthquakes they consist of and the local faults' critical and hazard state.

The purpose of this study is to describe the analysis method of seismic sequences in order to locate the nodes of higher order seismic rods in the upper and lower branching structure. Further, by analyzing the seismic sequence and using the nodes in the upper branching structure, we will propose a novel method for identifying the zone within which a future strong earthquake's epicenter may fall.

Through the analysis of the seismic rods in the upper branching structure, we will also introduce a sequence seismic hazard classification in order to define the warning signs during the seismicity's time evolution of the analyzed area.

Finally, the joint analysis of the seismic rods in the upper branching structure allows us to obtain relevant information on possible interactions between different earthquakes on local and global scale, probably due to the effects of the changes caused by static and dynamic stresses.

The analysis of the seismic sequence structure is performed by using: 1) a datasets spatial coverage involving parts of a region or several countries (from a few tens to thousands of km), 2) a range of magnitude values of 2.0 - 10 m, 3), a range of depths of 1 - 50 km.

2. METHOD

2.1. Upper and Lower Branching Structure

The sequences of higher order seismic rods in the upper and lower branching structures are patterns that allow the graphical evaluation of the earthquake preparation process's magnitude and therefore contribute to outline the past and future seismicity evolution of an area. These are identified through the consideration that after an energy shock the magnitude values gradually decrease and increase in the upper and lower structure, respectively, up to the Trigger Point (TP) by following a hierarchization process.

The entire preparation time of the upper branching structure can be divided into three stages: 1) energy accumulation, 2) trigger, 3) energy release [2] .

Figure 1 shows the patterns of the upper and lower branching structures: as it can be seen, the first developmental stage (the largest) in the upper branching structure starts to develop after a strong earthquake (point source), while the subsequent ones are related to the decrease in the magnitude values over time. In the lower branching structure, however, a trend to increase the magnitude values over time can be observed.

Furthermore, we note a decrease in the recurrence rates or the number of events in the final portion of the branching structures and a common trigger point (TP) determining the transition from the energy accumulation to the energy release stage. The energy release stage develops within a shorter time frame compared to the accumulation stage. The distance between the TP and the mainshock is the warning time frame of an energetic seismic event. Usually, a first warning signal results from the convergence of a third order lower branch with a second order higher one (point A). For example, in the case of the Samoan earthquake on 29 September 2009, a second abnormal warning sign resulted from the convergence of a lower fourth order branch with a higher second order one (point B).

2.2. Upper Branching Structure Hazardousness

The classification process provides not only information on the upper branching structure's developmental state, but also a more consistent hazard classification related to the upper branching structure, which can be represented with colors.

The branching structure's hazard level is classified according to the seismic branches magnitude [2] , by establishing seven hazard classes ranging from extremely high to low. The different hazard levels were encoded with colors that let us know, in a simple and quick fashion, the critical level reached by the upper branching structure that is being developed in an area either on local or global scale.

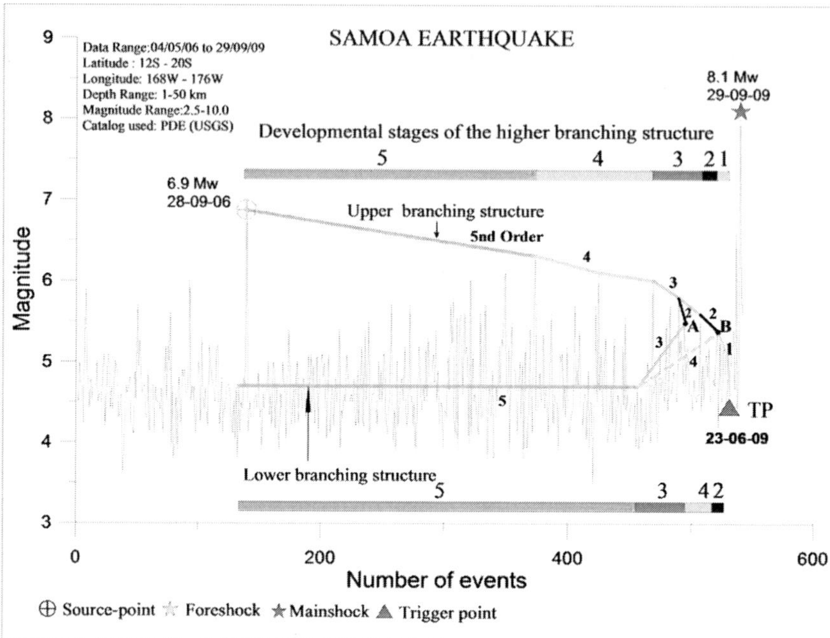

Figure 1. Energy accumulation phase's developmental stages in the upper and lower branching structure. The vertical axis shows the magnitude values, while the horizontal axis displays the number of events recorded in the period analyzed.

We chose the magnitude as a criterion for the division into hazard classes in order to identify the branching structure hazardousness at any point in its evolution.

The branching structures classification process involves several steps, including the following:

- Upper branching structure detection;
- Identification of the magnitude values of the seismic branches nodes it consists of and of the energy release stage;
- Determining the hazard class of each stage of the energy accumulation and release phases;
- Determining the hazard class of the energy release last stage.

As it uses seven hazard classes, this method can be successfully adopted to determine the appropriate warnings during the energy release stage, which can lead to a strong earthquake.

Table 1 lists the different hazard classes of the upper branching structure per magnitude range, while Figure 2 shows the hazard classes, determined according to the magnitude values of the energy accumulation and release stages during the Samoan earthquake on 29 September 2009.

Initially, we note a gradual reduction of the upper branching structure's hazard level up to the TP, thereafter the hazard rises until it reaches an extremely high value coinciding with the mainshock. The midpoint of the fifth order branch, which defines the alleged value of the expected magnitude [2] , enables us to estimate the energy release stage's minimum hazard, which in this case was estimated to be high.

2.3. Epicentral Area Identification-Aimed Procedure

This procedure, which can be of help in identifying the epicentral area of a strong earthquake, consists in plotting on a map the nodes' epicenters in the upper branching structure along with the trigger signals. Figure 3 illustrates the position of the nodes' epicenters in the Japan's upper branching structure (Table 2), whose release stage was characterized by two foreshocks measuring 5.7 Mw and 7.3 Mw magnitudes respectively, recorded on 06/12/2010 and 09/03/2011 and by the earthquake occurred on 11 March 2011, with a magnitude of 9.0 Mw.

If we observe carefully the map, we will see how the epicenters tend to group over time depending on their type, i.e. nodes' epicenters in the energy accumulation and energy release stages and TP-related ones. One of the epicenters of the shocks which the TP or the energy release stage consists of is almost always close to the mainshock.

Therefore, in order to predict the mainshock epicenter, it is considered more appropriate to use the clusters including the shocks epicenters which the TP or the energy release stage consists of.

Table 1. Hazard classes.

Class	Magnitude	Hazard
1	8 or more High	Extremely High
2	7 - 7.9	Very High
3	6 - 6.9	High
4	5 - 5.9	Moderately High
5	4 - 4.9	Moderate
6	3 - 3.9	Moderately Low
7	1 - 3.0	Low

Table 2. Coordinates and magnitude of nodes and warning sign in the upper branching structure of Japan earthquake, 2011.

No Node	Date	Latitude	Longitude	Magnitude	Stage/Node
1	19/07/2008	37.55	142.21	7.0 Mw	Source-point
2	14/03/2010	37.74	141.59	6.5 Mw	Energy accumulation
3	04/07/2010	39.70	142.37	6.3 Mw	Energy accumulation
4	29/09/2010	37.26	139.88	5.5 Mw	Energy accumulation
5	10/10/2010	31.07	141.61	5.1 mb	Energy accumulation
6	17/10/2010	36.39	141.18	4.8 mb	Energy accumulation
7	17/10/2010	30.02	130.92	4.0 mb	Trigger
8	19/10/2010	36.33	142.26	4.7 mb	Trigger
9	22/10/2010	32.39	141.39	4.2 mb	Trigger
10	06/12/2010	40.90	142.97	5.7 Mw	Foreshock
11	09/03/2011	38.44	142.84	7.3 Mw	Source-point-Foreshock
12	09/03/2011	38.30	142.81	6.4 Ms	Energy accumulation
13	10/03/2011	38.60	142.27	5.6 Mw	Energy accumulation
14	10/03/2011	38.05	143.30	5.2 Mw	Energy accumulation
15	10/03/2011	38.53	143.04	5.2 Mw	Energy accumulation
16	10/03/2011	38.49	143.11	4.7 mb	Trigger
17	10/03/2011	38.78	143.05	4.9 mb	Trigger
18	11/03/2011	38.19	143.15	4.2 mb	Trigger
19	11/03/2011	38.30	142.37	9.0 Mw	Mainshock

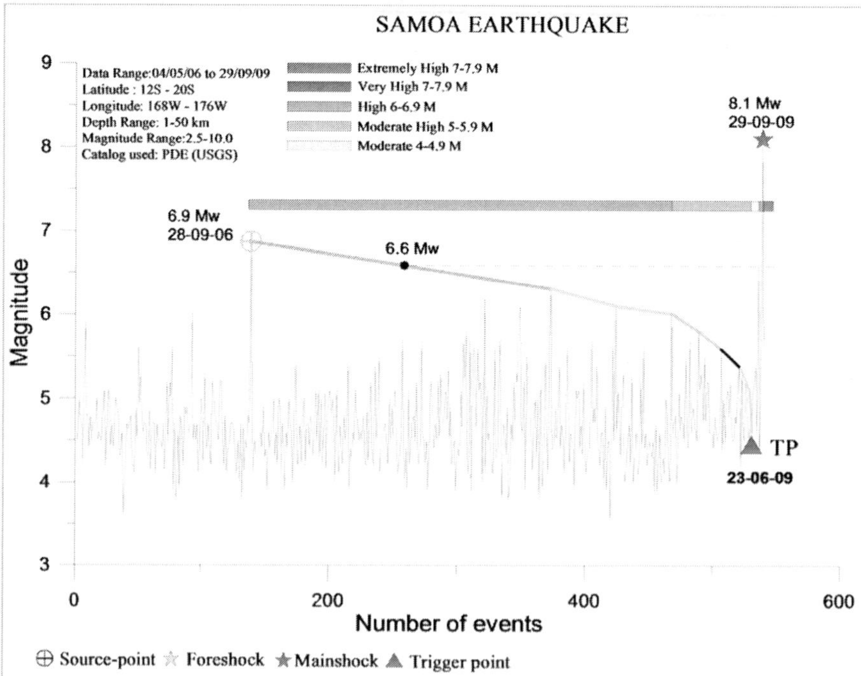

Figure 2. Samoan earthquake. Branching structure and hazard classes.

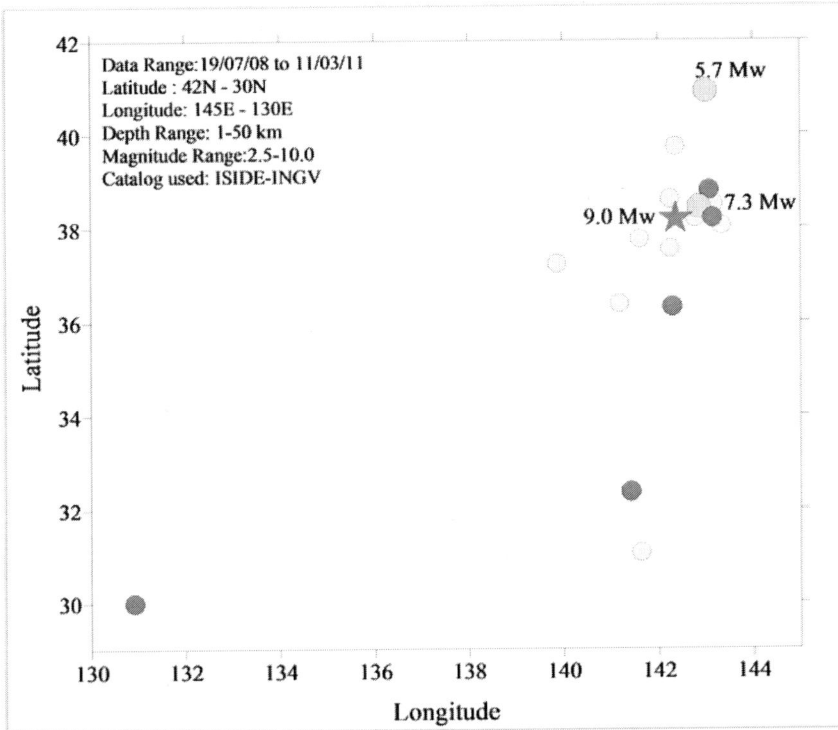

Figure 3. Japan's seismic sequence. Yellow circles: energy accumulation nodes' epicenters; red circles: TP epicenters; green circles: energy release stage epicenters; red star: mainshock epicenter.

Figure 4 shows the seismic sequence and the upper branching structure that preceded the earthquake on 11 March 2011.

2.4. Global Scale Branching Structures

Several earthquakes do not occur as independent events: just as faults' mutual interaction causes a change in the static stress, so too earthquakes can interfere through the dynamic variation of the stress field. The co-seismic stress change, caused by the wave passage, can lead to a seismicity increase even in some portions of the areas far from the mainshock, while in the neighbouring areas we can detect a seismicity quiescence resulting from the stress drop. A large number of studies describe the interference between earthquakes [3] - [5] .

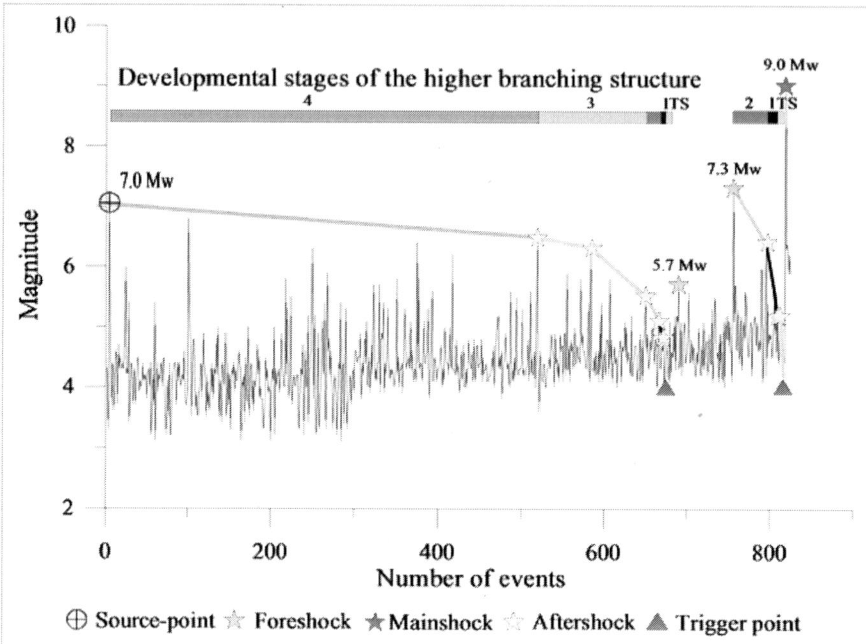

Figure 4. Earthquake's seismic sequence, Japan, 2011, and upper branching structure.

After a major seismic event, we observe the development of aftershocks, which generally may occur over a short period of time, due to the dynamic stresses associated with the passage of seismic waves produced by the main shock. Medium-long term aftershocks may develop, which are associated with changes in the static stress [1] [6] [7] .

In general, dynamic stresses can trigger long distance aftershocks, even more than 10,000 km, where the stress caused by the passage of the mainshock seismic wave, affects the crustal deformation by disrupting the crossed medium's elastic properties change [3] [8] [9] . Static stresses have a greater influence in the regions closest to the mainshock [3] [6] [10] .

The stress changes induced by static stresses, can anticipate or delay a series of aftershocks depending on the sign associated to the stress variation [6] [10] , while the trigger produced by dynamic stresses, linked to the medium's elastic variation, is capable of generating seismic events following the main one [11] .

The branching structure's long-term development [2] can provide important clues related to future events' occurrence. In addition, the analysis of the branching structure suggests that the triggering of other earthquakes can occur after a mainshock, in areas where faults may be in a state of critical stress.

Below we analyze on global scale the upper branching structure of two recent strong earthquakes in order to obtain information on the interaction between the most energetic earthquakes and the energy accumulation and release stage in the upper branching structure. From our studies we infer that the delays due to the co-seismic stress changes' application depend on the branching structure developmental stage.

We note a delays decrease in the final portion of the branching structure and during the energy release stage, while the different time-scales observed in the higher order branches most likely reflect different tectonic loading rates in the regions.

Figure 5 shows the branching structures of the Ecuador earthquake. The energy accumulation stage that developed after the 6.9 Mw earthquake occurred on 03 April 2016 (point source) consists of eight seismic events featuring a decreasing magnitude over time and delays between them ranging from a few hours to a maximum of 3 days (Table 3). The TP formed on 13/04/2016 with a delay of about 3 days compared to the mainshock. The energy release stage was "Progressive earthquakes"-type and consisted of 4 foreshocks of increasing magnitude and one mainshock [12] . The foreshocks straight interpolation line (red dashed line) has enabled us to make a preliminary estimate of the mainshock magnitude, while the minimum magnitude, calculated from the midpoint of the first seismic branch, is estimated to be 6.8 Mw (black dashed line). The trigger pattern preceding the mainshock was "Symmetrical triangle"-type [12] .

The hazard level of the upper branching structure shows a decreasing trend from high to moderate while after the TP it reaches very high values just before the mainshock. The achieved results make us assume a remote interaction between different faults stressed by the seismic waves generated by the most energetic branching structure's earthquakes.

Figure 6 shows the location of strongest earthquakes epicenters in the energy accumulation (yellow stars) and release (green and red stars) stage and of the three shocks the trigger point consists of (red-brown stars).

It appears that the epicenter of one of the three TP's shocks is located near the mainshock, while the energy accumulation and release stages have developed far from the epicenter.

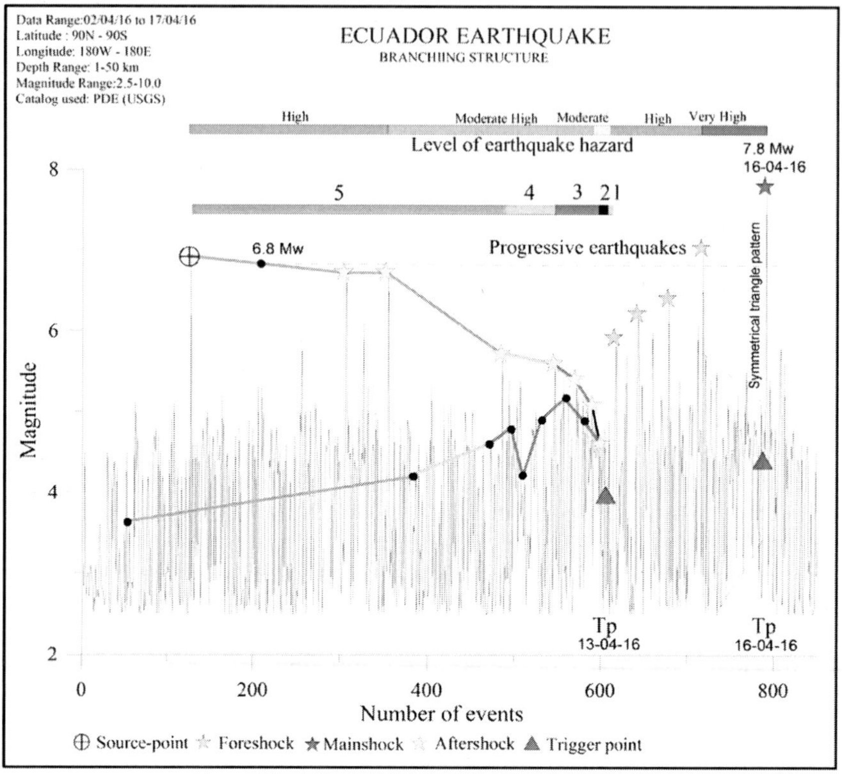

Figure 5. Ecuador earthquake. Branching structures and hazard classes.

Table 3. Coordinates and magnitude of the nodes and the warning signal in the 2016 Ecuador earthquake's upper branching structure.

No Node	Date	Time	Latitude	Longitude	Magnitude	Delay	Stage/Node
1	03/04/2016	08:23:52	−14.3235	166.8551	6.9 Mw	-	Source-point
2	06/04/2016	06:58:48	−14.0821	166.5981	6.7 Mw	3 g	Energy accumulation
3	07/04/2016	03:32:53	−13.9805	166.5943	6.7 Mw	1 g	Energy accumulation
4	10/04/2016	02:14:34	−4.1625	102.2024	5.7 Mw	3 g	Energy accumulation
5	11/04/2016	20:10:35	51.8428	176.6228	5.6 mb	1 g	Energy accumulation
6	12/04/2016	03:22:48	−3.7177	−103.9077	5.4 mb	1 g	Energy accumulation
7	12/04/2016	23:22:58	11.3235	140.8509	5.1 mb	0 g	Energy accumulation
8	12/04/2016	23:40:48	11.3791	140.9720	4.7 mb	0 g	Energy accumulation
9	13/04/2016	01:35:05	−14.4136	166.5495	4.6 mb	1 g	Energy accumulation
10	13/04/2016	06:34:35	63.0259	−151.7138	2.5 M_L	0	Trigger

A similar developmental pattern is shown by the Nepal earthquake's branching structures (Figure 7).

The energy accumulation stage that developed after the earthquake on 17 April 2015 (point source) consisted of six earthquakes with decreasing magnitude over time and with delays ranging from a few hours to a maximum of 3 days (Table 4). The TP formed on 23/04/2015 with a delay of about 2 days compared to the mainshock. The energy release stage, consisting of 4 foreshocks of increasing magnitude and one mainshock, was "Progressive earthquakes"-type [12] . The foreshock straight interpolation line (red dashed line) allowed us to calculate with sufficient accuracy mainshock's, while the minimum magnitude, which was estimated to be 6.45 Mw, was calculated from the midpoint of the first seismic branch (black dashed line). The breaking pattern preceding the mainshock was "Falling wedge"-type [12] .

The upper branching structure hazardousness shows a decreasing trend from high to moderate while after the TP it reaches very high values just before the mainshock.

In Figure 8 we can note how the epicenter of one of the three TP shocks (red-brown stars) is located close to the mainshock (red star), while the energy accumulation (yellow stars) and release (green stars) stages have developed away from the epicenter.

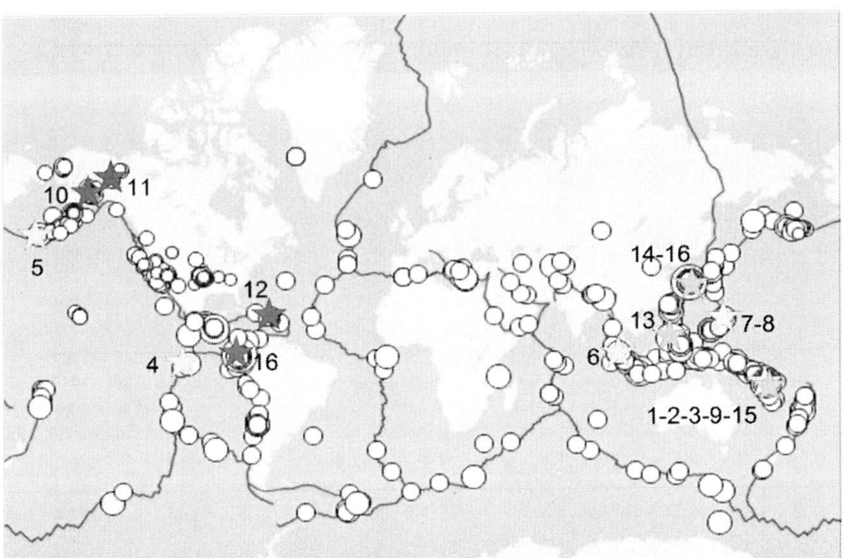

Figure 6. Ecuador earthquake. Earthquakes epicenters of the upper branching structure and TP.

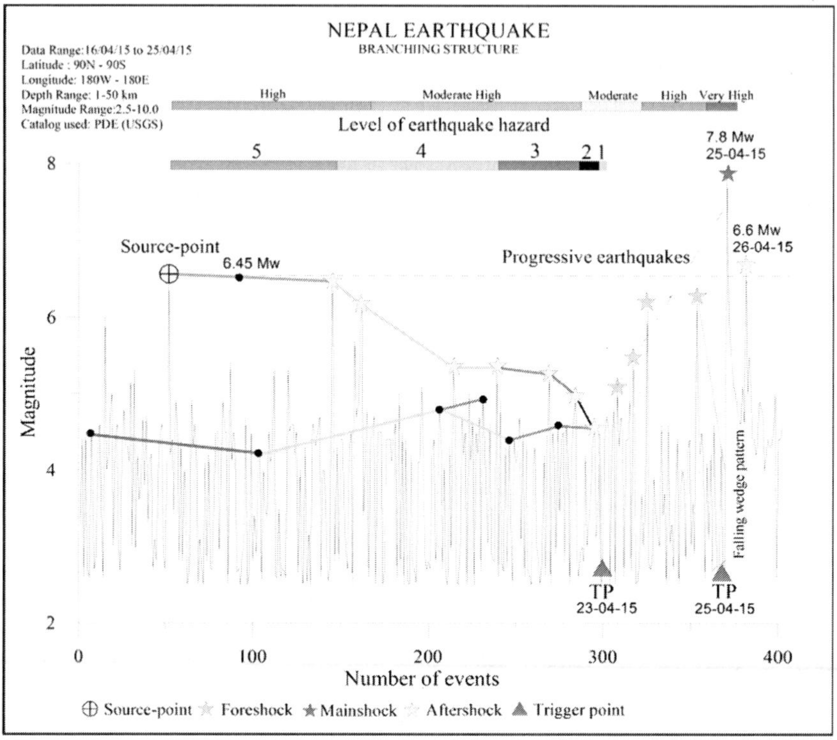

Figure 7. Earthquake in Nepal. Branching structures and hazard classes.

Table 4. Coordinates and magnitude of the nodes and the warning signal in the upper branching structure of Nepal 2015 earthquake.

No Node	Date	Time	Latitude	Longitude	Magnitude	Delay	Stage/Node
1	17/04/2015	15:52:51	−15.8815	−178.6005	6.5 Mw	-	Source-point
2	20/04/2015	01:42:58	24.2026	122.3158	6.4 Mw	3 g	Energy accumulation
3	21/04/2015	11:21:09	−6.0966	149.8353	5.3 Mw	1 g	Energy accumulation
4	21/04/2015	19:10:15	−6.3348	154.7629	5.3 Mw	0 g	Energy accumulation
5	22/04/2015	17:03:12	−30.6629	−178.3427	5.2 Mw	1 g	Energy accumulation
6	22/04/2015	23:07:30	−31.4858	76.7845	4.9 mb	0 g	Energy accumulation
7	23/04/2015	08:31:04	−33.2693	−179.2935	4.5 mb	1 g	Energy accumulation
8	23/04/2015	10:39:27	53.8161	−163.8865	2.5 ML	0 g	Trigger
9	23/04/2015	10:15:18	0.5147	92.2925	4.4 Mb	0 g	Trigger
10	23/04/2015	09:02:13	57.9079	−142.9024	3.1 ML	0 g	Trigger
11	23/04/2015	14:57:27	51.7501	176.3353	5.0 Mw	0 g	Foreshock
12	24/04/2015	01:34:55	40.4556	−127.1509	5.4 Mw	1 g	Foreshock
13	24/04/2015	03:36:42	−42.0602	173.0066	6.1 Mw	0 g	Foreshock
14	24/04/2015	13:56:15	51.6148	−130.7714	6.2 Mw	0 g	Foreshock
15	25/04/2015	06:11:25	28.2305	84.7314	7.8 Mw	1 g	Mainshock

2.5. Examples of Branching Structures

The examples shown in Figures 9-12 display the main branches in the upper and lower branching structures that have developed before some strong earthquakes as well as the branching structure's hazard classes developed over time.

The branches in the two branching structures, identified with the long-term seismic sequence hierarchization method, highlight how they tend to converge over time to generate the TP, which in some cases is followed by various order warning signs [13] . The TP may be followed by "Flash earthquake"-type (Figure 9 and Figure 10) [12] or "Progressive earthquakes"-type (Figure 11 and Figure 12) energy release stages.

The upper branching structure hazard level tends to decrease from the point source up to TP, then sharply increase up to the main event.

The alternation of the various seismic branches in the upper and lower branching structures show a repetitive development that can be used to obtain information on the seismic sequences' developmental state. In fact, it is clear that the formation of a certain order branch in the upper branching structure is subsequent to the formation of a branch of the same order in the lower one.

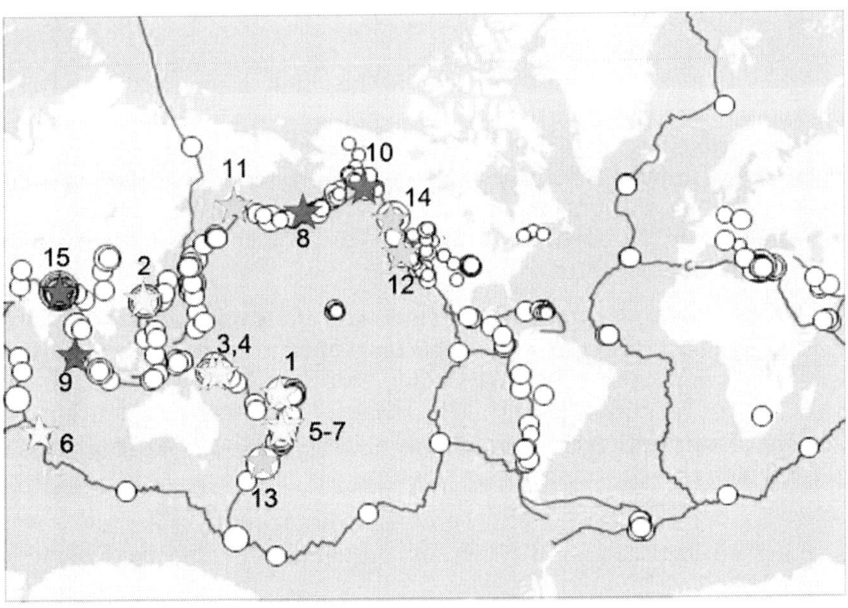

Figure 8. Earthquake in Nepal. Earthquakes' epicenters in the upper branching structure and TP.

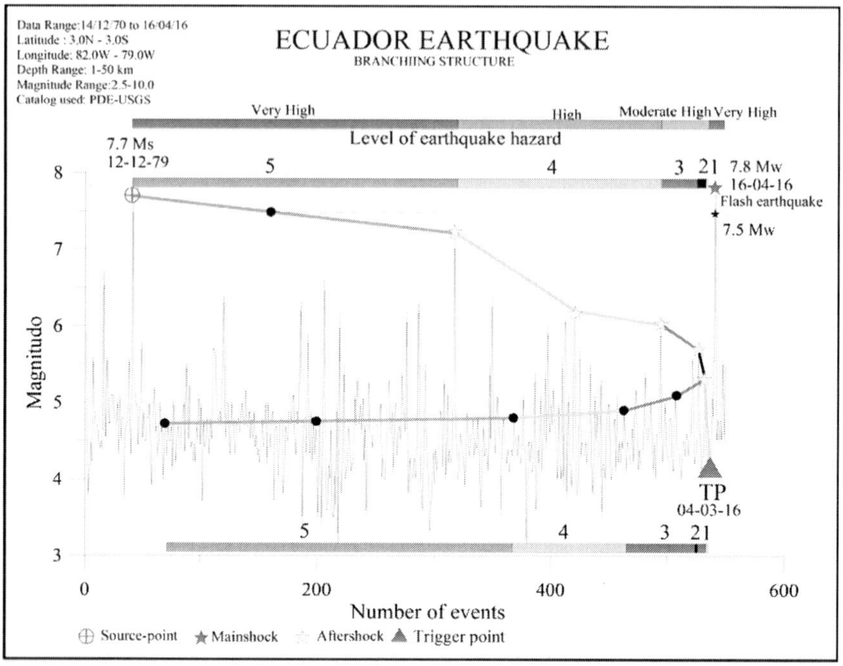

Figure 9. Ecuador earthquake. Branching structures and hazard classes.

For example, Figure 9, relative to the Ecuador earthquake's seismic sequence, shows how, after the formation of a fifth order seismic branch in the lower branching structure, a seismic branch of the same order has formed in the upper branching structure. Then a new fifth order seismic branch in the lower branching structure has formed, followed, in time, by the formation of a fourth order rod in the upper structure (seismic branches alternating process).

Usually, this developmental process ends with the convergence of a third order seismic branch in the lower branching structure with a second order seismic branch in the upper branching structure, followed by the TP formation. The repetition of seismic branches of the same order during the branching structures developmental process is suggestive of a seismic delay in the ongoing energy accumulation stage.

Table 5 shows the formation order of seismic branches in upper and lower branching structures and the seismic sequence's TP of the Ecuador's earthquake.

The upper branching structure's hazard level from the point source onward varies from very high to moderately high, while from the TP on, the hazard level returns to very high values. The energy release stage was "Flash earthquake"-type [12] , with a delay between the TP and the mainshock approximately of 44 days.

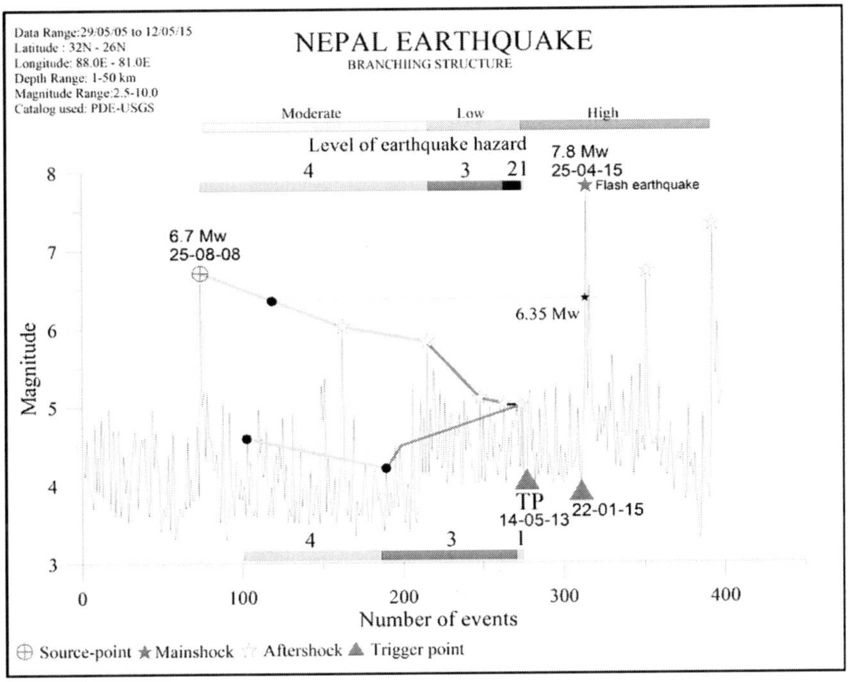

Figure 10. Earthquake in Nepal. Branching structures and hazard classes.

A similar trend is shown in the seismic sequence preceding the Nepal earthquake, on 25 April 2015 (Figure 10) where branching structures started from a fourth order branch and ended with the convergence of the third order branches with a second order one in the upper branching structure. The trigger point was formed on 14 May 2013 while the warning signs preceding the mainshock, generated by a DB-3SE microstructure, appeared on 22 January 2015 [13] . The delay between the TP and the mainshock was about 23 months, whereas that between the trigger point and the mainshock was approximately 3 months. The upper branching structure's hazard level after the TP decreased from moderate to low up, then returned to high values. The energy release stage was "Flash earthquake"-type.

Figure 11 shows the branching structure of the L'Aquila earthquake, occurred on 6 April 2009. We note a first upper branching structure with a point source that has formed on 15/08/1998, consisting of 5 developmental stages and an early convergence of the lower branching structure. The TP formed on 10 July 2007 was followed by a 4.0 M_L magnitude seismic event that generated a second, abnormal fourth order branching structure, with a TP formed on 15/02/2009. The energy release stage was "Progressive earthquakes"-type, with the last trigger point generated on 02/04/2009. The delays between the first and the second TP compared to the mainshock were approximately 21 months (duration of the whole energy release stage) and 46 days, respectively, while the delay between the last TP and the mainshock was around 4 days.

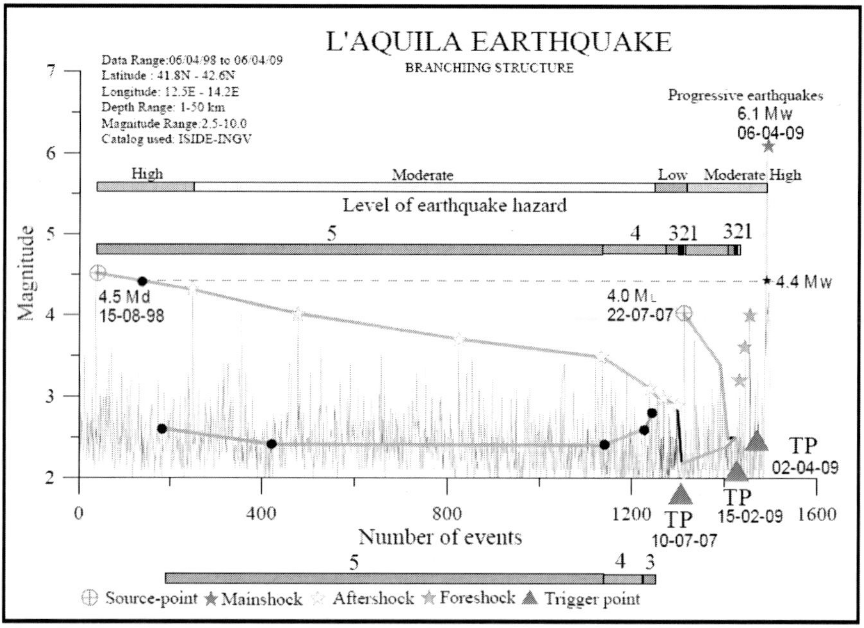

Figure 11. L'Aquila's earthquake. Branching structures and hazard classes.

The upper branching structure hazard level shows a decreasing trend from high to low up to the first TP, while during the energy release stage, it increased from moderate to potentially moderate-high if we consider the trendline joining the earthquakes the energy release stage.

Figure 12 shows the branching structures of the earthquake occurred in Emilia on 20 May 2012. The point source from which a fourth order upper branching structure has developed, has formed on 15/10/1996, while the TP was generated on 21/09/1997.

The energy release stage that followed was "Progressive earthquakes"-type with the last strongest earthquake, featuring a magnitude of 4.5 Md, recorded on 18/06/2000. This event is a second point source that generated a second third order upper branching structure and a TP that has formed on 22/02/2008.

The energy release stage consisted of two first order foreshocks of 4.5 Mw and 4.9 Mw magnitude respectively and a second order foreshock of 4.0 Md magnitude, recorded on 19/05/2012. The second order foreshock was preceded by a warning signal that was generated on 01/03/2012. The delay between the second TP and the mainshock was approximately 4 years and three months (duration of the whole energy release stage), the delay between the last TP on 01/03/2012 and the mainshock was about 80 days, while the delay between the second order foreshock and the mainshock was one day.

The upper branching structure hazard level remained moderate/moderate low until the mainshock and then became moderately high.

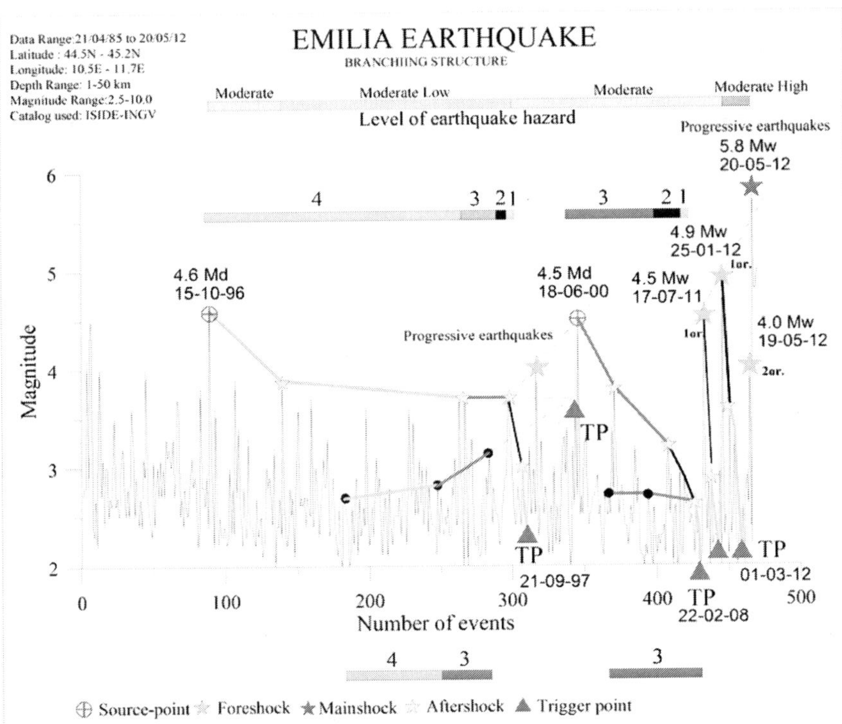

Figure 12. Earthquake in Emilia. Branching structures and hazard classes.

Table 5. Development of branching structures.

Branching structure Less	Branching structure Top
5 ⟶	5
5	⟵ 5
5 ⟶	4
4	⟵ 4
4 ⟶	4
3	⟵ 4
3 ⟶	3
3	⟵ 3
2 ⟶	⟵ 2
	1
	TP

3. CONCLUSIONS

By using the nodes of upper and lower branched structures' seismic branches, we have proposed new approaches to define a classification of the hazard levels reached by the upper branched structure during its development. This classification plays a crucial role in determining the hazardousness of the energy release phase that is triggered following the TP formation.

In fact, the upper and lower branched structures, in the last part of their development, show a convergence of branches of the same order with a lower order one, which is an important warning signal that usually precedes a strong earthquake and determines a warning sign of the sequence's hazardousness and its development.

Analyzing the upper branched structure's temporal and spatial changes both on global and local scale, it appears that they are adequate enough to assume a long-dis- tance interaction between the various earthquakes.

These interactions should, however, be linked both to the static stress changes and the passage of seismic waves from the most energetic earthquakes. These can generate transient stresses in the earth's crust, thus triggering earthquakes far from the original earthquake source.

We also believe that recurrence rates that can be observed in the branching structure's first order seismic branches most likely reflect the different tectonic loading rates of the seismic regions concerned.

Conversely, in the last stages, the decrease in the recurrence rates is probably due to a very high tectonic load, where small shocks can trigger the energy release stage.

CITE THIS PAPER

Riga, G. and Balocchi, P. (2016) Seismic Sequences' Bran- ching Structures: Long-Range Interactions and Hazard Levels. Open Journal of Earthquake Research, 5, 189-205.

REFERENCES

1. Pollitz, F.F. and Johnston, M.J.S. (2006) Direct Test of Static Stress versus Dynamic Stress Triggering of Aftershocks. Geophys?cal Research Letters, 33, L15318.
2. Riga, G. and Balocchi, P. (2016) Short-Term Earthquake Forecast with the Seismic Sequence Hierarchization Method. Open Journal of Earthquake Research, 5, 79-96.
3. Harris, R.A. (1998) Introduction to Special Section: Stress Triggers, Stress Shadows, and Implications for Seismic Hazard. Journal of Geophysical Research: Solid Earth, 103, 24347-24358.
4. Stein, R.S. (1999) The Role of Stress Trasfert in Earthquake Occurrence. Nature, 402, 605-609.
5. King, G.C.P. and Cocco, M. (2000) Fault Interaction by Elastic Stress Change: New Clues from Earthquake Sequences. Advances in Geophysics, 44, 1-38.
6. Stein, R.S., Barka, A.A. and Dieterich, J.H. (1997) Progressive Failure on the North Anatolian Fault Since 1939 by Earthquake Stress Triggering. Geophysical Journal International, 128, 594-604.
7. Brodsky, E.E. and van der Elst, N.J. (2014) The Uses of Dynamic Earthquake Triggering. Annual Review of Earth and Planetary Sciences, 42, 317-339.
8. Freed, A.M. (2005) Earthquake Triggering by Static, Dynamic, and Postseismic Stress Transfer. Annual Review of Earth and Planetary Sciences, 33, 335-367.
9. Delorey, A.A., Chao, K., Obara, K. and Johnson, P.A. (2015) Cascading Elastic Perturbation in Japan Due to the 2012 Mw 8.6 Indian Ocean Earthquake. Science Advances, 1, e1500468.
10. Lin, J. and Stein, R.S. (2004) Stress Triggering in Thrust and Subduction Earthquakes and Stress Interaction between the Southern San Andreas and Nearby Thrust and Strike-Slip Faults. Journal of Geophysical Research: Solid Earth, 109, B02303.
11. Belardinelli, M.E., Bizzarri, A. and Cocco, M. (2002) Earthquake Triggering by Static and Dynamic Stress Changes. Journal of Geophysical Research: Solid Earth, 108, 2135. http://www.bo.ingv.it/~bizzarri/Download/Productions/JGR_2003-1_Paper.pdf
12. Riga, G. and Balocchi, P. (2016) Seismic Sequence Structure and Earthquakes Triggering Patterns. Open Journal of Earthquake Research, 5, 20-34. http://dx.doi.org/10.4236/ojer.2016.51003 http://www.scirp.org/Journal/PaperInformation.aspx?PaperID=63538
13. Riga, G. and Balocchi, P. (2016) How to Predict Earthquakes with Microsequences and Reversed Phase Repetitive Patterns. Open Journal of Earthquake Research, 5, 153-164.

CHAPTER 9

Evaluation of Soil-Structure Interaction on the Seismic Response of Liquid Storage Tanks under Earthquake Ground Motions

Mostafa Farajian [1], Mohammad Iman Khodakarami [1,*] and Denise-Penelope N. Kontoni [2,*]

[1]*Faculty of Civil Engineering, Semnan University, Semnan 35131-19111, Iran*

[2]*Department of Civil Engineering, Technological Educational Institute of Western Greece, GR-26334 Patras, Greece*

ABSTRACT

Soil-structure interaction (SSI) could affect the seismic response of structures. Since liquid storage tanks are vital structures and must continue their operation under severe earthquakes, their seismic behavior should be studied. Accordingly, the seismic response of two types of steel liquid storage tanks (namely, broad and slender, with aspect ratios of height to radius equal to 0.6 and 1.85) founded on half-space soil is scrutinized under different earthquake ground motions. For a better comparison, the six considered ground motions are classified, based on their pulse-like characteristics, into two groups, named far and near fault ground motions. To model the liquid storage tanks, the simplified mass-spring model is used and the liquid is modeled as two lumped masses known as sloshing and impulsive, and the interaction of fluid and structure is considered using two coupled springs and dashpots. The SSI effect, also, is considered using a coupled spring and dashpot. Additionally, four types of soils are used to consider a wide variety of soil properties. To this end, after deriving the equations of motion, MATLAB programming is employed to obtain the time history responses. Results show that although the SSI effect leads to a decrease in the impulsive displacement, overturning moment, and normalized base shear, the sloshing (or convective) displacement is not affected by such effects due to its long period.

Keywords: liquid storage tanks; soil-structure interaction; seismic response; earthquake ground motions

1. INTRODUCTION

Liquid storage tanks are important structures which have a key role in human lives. The major existing codes, regulations, guidelines, and recommendations—such as API (650, 620), AWWA (D100, D103, D110, D115), Eurocode 8 Part 4 (and Eurocode 3), UNI EN 14015, Covenin, IITK, ACI (350.3, 371.R), NZSEE, AIJ—for the seismic design of tanks are widely used, although, in some cases, the codes were not correctly applied or there are imperfections in the code requirements. These structures have shown poor performance and failure modes during strong earthquakes—such as overturning, buckling, roof damage, sliding, uplift, different settlement, etc., which may also cause liquid leakage and fire after an earthquake—as described, for example, in [1,2,3,4].

It is clear that, in designing of such structures, all factors that affect seismic responses of these structures should be considered. One of these factors is soil-structure-interaction (SSI). The SSI affects earthquake ground motions, characteristics of structures, and also soil properties. Accordingly, depending on the period of structure, the seismic response of structure could either increase or decrease. Several studies showed that the SSI effect is more important for massive structures such as tall buildings, bridges, and liquid storage tanks which could cause to suspend their performance [5,6]. The most popular and relatively accurate model to represent the soil-structure-interaction is the substructure method which considers the soil as coupled springs and dashpots [7,8].

Liquid storage tanks behave differently from common structures, such as buildings, bridges, etc., due to fluid-structure-interaction. Housner's mass-spring model [7] was the first approximate model to obtain the seismic responses of rigid cylindrical liquid storage tanks. In Housner's model [7], the whole liquid is divided into two parts; a portion of liquid which excites independently of the tank wall near the free surface is named as "convective" and the other part of liquid exciting unison with the tank wall is named "impulsive". By increasing the tank's geometry (i.e., radius and height), Haroun and Housner [8] modified the Housner's model to consider the flexibility of the tank's wall; in their presented model, the liquid is divided into three portions; convective and impulsive masses which are attached to the tank's wall through springs and dashpots and the rigid mass which is attached to the tank's wall rigidly. Malhotra et al. [9] have proposed a simplified model by considering higher modes of impulsive mass with the first impulsive modal mass and higher mode of convective mass with the first convective modal mass. Bagheri et al. [10] studied the seismic responses of liquid storage tanks under near-fault ground motions; such earthquake ground motions have long-period components that may affect the long-period sloshing motion of liquid [10]. The effect of earthquake characteristics on seismic responses of base isolated liquid storage tanks is also studied by Bagheri and Farajian [11], who showed that the pulse-like earthquake-ground motions could cause excessive displacement in base isolation and, therefore, the impact could occur.

The SSI effect on the seismic response of liquid storage tanks has been studied by several researchers, mostly in the frequency domain. Veletsos and Tang [12] proposed a method to consider the SSI; they proposed to modify the impulsive mass frequency and damping to consider the SSI effect and their research has shown that the SSI has no special effect on convective mass displacement. Larkin [13] obtained the responses of steel and concrete liquid storage tanks considering SSI effect, and found that SSI affects the shear force and overturning moment, especially on soft soils. Foundation embedment effects on the behavior of elevated tanks were studied by Livaoglu and Dogangun [14], who concluded that embedment in soft soil significantly affects the tank roof's displacement. Livaoglu [15] shown that decreasing the stiffness of the soil leads to reduction of the base shear and impulsive displacement; on the other hand, sloshing displacement is not considerably affected due to SSI, embedment, and wall flexibility [15].

In this paper, the effect of SSI on the seismic response of liquid storage tanks is studied under earthquake ground motions in the time domain. Accordingly, after solving the equations of motion in the time domain, the peak responses are obtained and compared with the ones without considering SSI.

2. STRUCTURAL MODEL OF THE FLUID-TANK-SOIL SYSTEM

A simplified model is implemented here to model the fluid-tank-soil interaction. Figure 1 shows a cylindrical liquid storage tank rested on a half space soil. As modeling of the interaction effects is complicated, the theory of Malhotra et al. [9] is used to consider the fluid-structure-interaction (FSI), and the cone method [16] is employed to the simulation of soil-structure-interaction (SSI) effects. These models have been briefly described below.

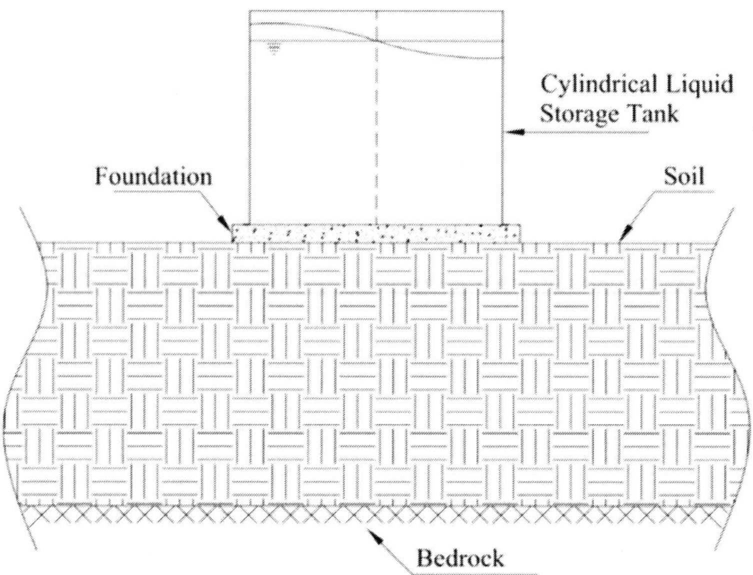

Figure 1. Liquid storage tank resting on a half space.

2.1. Fluid-Tank System

The 3D finite element model of a liquid storage tank is usually complicated due to hydrodynamic interaction effects. Accordingly, the simplified mass-spring model of Malhotra et al. [9] is used in the present study. The geometry of a cylindrical tank is the liquid height (H), the tank's radius (r), and the equivalent uniform thickness of the tank wall (t), as shown in Figure 2. According to Figure 3, the convective and impulsive masses (m_c and m_i) are connected to the tank's wall by springs and dashpots (k_c and c_c, k_i and c_i). The natural periods of the convective (T_c) and impulsive (T_i) of responses are [9]:

$$T_c = C_c\sqrt{r} \tag{1}$$

$$T_i = C_i\frac{H\sqrt{\rho_s}}{\sqrt{E.t/r}} \tag{2}$$

where, ρ_s and E are the mass density of liquid and modulus of elasticity of the tank's wall, respectively. The coefficients C_c and C_i, the relative convective and impulsive masses (m_c/m and m_i/m) and heights (h_c/H and h_i/H) are provided by Malhotra et al. [9]. The total liquid mass of the tank filling with water is equal to ($\pi r^2 H\rho_w$). The corresponding stiffness and damping ratio of springs and dashpots associated with convective and impulsive masses are equal to:

$$k_c = m_c \times \omega_c^2 \tag{3}$$

$$k_i = m_i \times \omega_i^2 \tag{4}$$

$$c_c = 2\xi_c m_c \times \omega_c \tag{5}$$

$$c_i = 2\xi_i m_i \times \omega_i \tag{6}$$

where ω_c and ω_i are the frequency of convective and impulsive responses. The damping ratio of the convective and impulsive modes (ξ_c and ξ_i) are 0.5% and 2%, respectively.

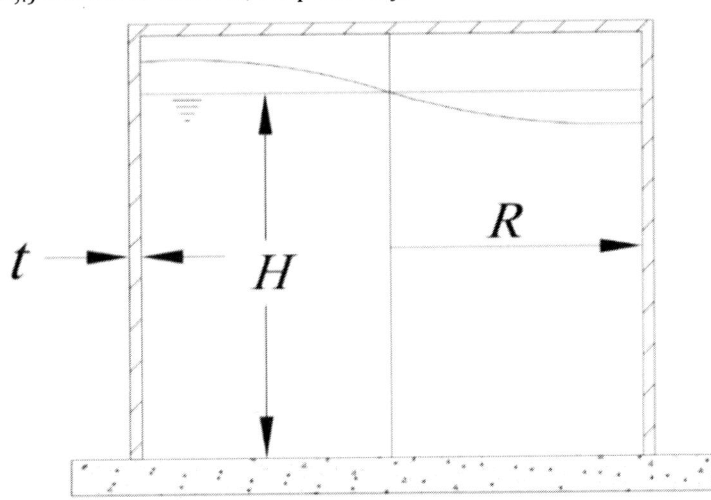

Figure 2. Geometry of the liquid storage tank.

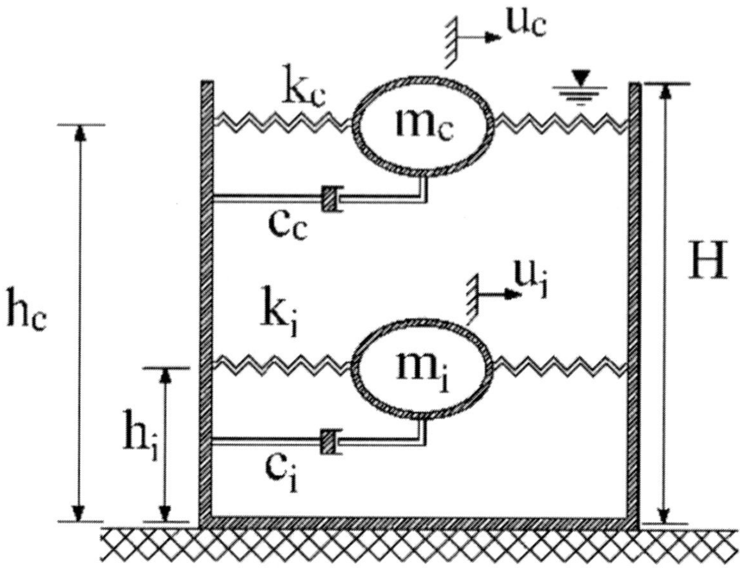

Figure 3. Simplified mass-spring model of Malhotra for fluid-structure-interaction.

2.2. Soil-Structure System

Considering real conditions, a liquid storage tank resting on soil consists of a tank (structure) and an adjacent bounded soil called near-field soil and unbounded soil called far-field. Both near-field and far-field soils affect the seismic response of the structure.

The soil-structure-interaction could be modeled using three proposed methods: (i) the direct method by employing numerical methods, such as the finite element method (FEM), boundary element method (BEM), and scaled boundary-finite element method (SBFEM); (ii) modifying the fixed base condition to take into account of SSI, where, in this model the effect of foundation embedment, layering, and material damping was ignored; and (iii) the substructure method which considers the soil by either dependent or independent frequency springs and dashpots which could be either used in the time or frequency domains. In order to obtain the corresponding stiffness and damping, three methods could be used: (i) the procedures presented in NIST GCR 12-917-21 [17]; (ii) the thin layer method which is used by SASSI software [18]; and (iii) the cone method proposed by Meek and Wolf [16]. Compared to other numerical methods, the cone model [16] has a simple numerical procedure and relative accurate response. In the cone method, the soil is modeled using springs and dashpots, and cones have translational, rotational, and torsional behavior. Only the translational motion is considered in this paper due to its simplicity, and remaining motions are ignored.

Based on the cone method theory, when a homogenous semi-infinite domain is subjected to a static load (P_0), the components of the displacement field will vary along the depth in the shape of a truncated cone, as shown in Figure 4, for the horizontal translational degree of freedom. The static stiffness of this truncated cone in a circular rigid foundation and equivalent circular foundation can be expressed by [19]:

$$K_{Static} = \frac{8\rho v_s^2 r_0}{2 - v} \tag{7}$$

where, ρ and v_s are mass density and shear velocity of the soil medium, r_0 is the radius of the equivalent circular foundation and v is the Poisson's ratio. For dynamic problems, the stiffness of the half-space in the cone model is frequency dependent and this static stiffness is used for calculating the dynamic stiffness $S(a_0)$ which is expressed by:

$$S(a_0) = K(k(a_0) + ia_0 c(a_0)) \tag{8}$$

where $k(a_0)$ is the dynamic spring coefficient, $c(a_0)$ is the dynamic damping coefficient, and a_0 is the dimensionless frequency equaled to $\omega r_0/v_s$ with implementing excitation frequency ω. In this study, these frequency dependent stiffness and damping coefficients are calculated using CONAN computer program.

In order to investigate the response of the structure due to various earthquakes on different soil conditions, four types of soil are considered, where, these soil properties are mentioned in Table 1; as it is clear from this table, soil S1 is known as a hard rock and by going to S4, the soils change to softer soil. As described, the cone method and CONAN program are used to evaluate the impedance functions of these four soil types by employing the soil characteristics.

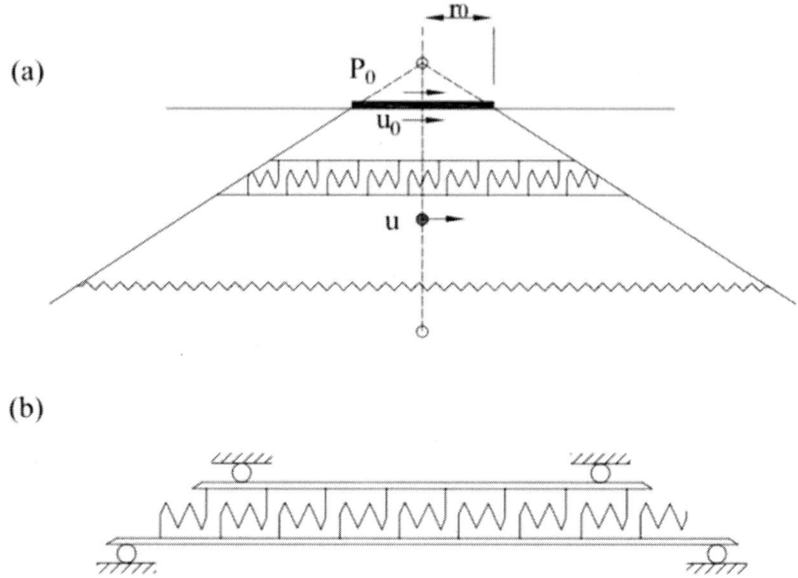

Figure 4. Translational truncated semi-infinite cone with horizontal motion, shear distortion and equilibrium of infinitesimal element, where rocking motion is prevented with: (**a**) infinite flexural rigidity; and (**b**) rollers for horizontal motion.

Table 1. Properties of considered soils types.

Soil Types	ζ_g	E (kN/m^2)	G (kN/m^2)	E_c (kN/m^3)	γ (kN/m^3)	υ	υ_s (m/s)	υ_p (m/s)
S1	0.05	7,000,000	2,692,310	9,423,077	20	0.30	1149.1	2149.89
S2	0.05	2,000,000	769,230	2,629,308	20	0.30	614.25	1149.16
S3	0.05	500,000	192,310	673,077	19	0.35	309.22	643.68
S4	0.05	35,000	12,500	75,000	18	0.40	82.54	202.18

2.3. Governing Equations of Motion

The equations of motion of the system of simplified model of liquid storage tank considering the SSI effect, as shown in Figure 5, can be written as:

$$m_c\ddot{u}_c + c_c(\dot{u}_c - \dot{u}_f) + k_c(u_c - u_f) = -m_c\ddot{u}_g \tag{9}$$

$$m_i\ddot{u}_i + c_i(\dot{u}_i - \dot{u}_f) + k_i(u_i - u_f) = -m_i\ddot{u}_g \tag{10}$$

$$m_f\ddot{u}_f + c_c(\dot{u}_f - \dot{u}_c) + c_i(\dot{u}_f - \dot{u}_i) + c_s\dot{u}_f + k_c(u_f - u_c) + k_i(u_f - u_c) + k_s u_f = -m_i\ddot{u}_g \quad (11)$$

where u_c, u_i, and u_f are convective, impulsive, and foundation displacements relative to the bedrock, respectively, and $\ddot{u}_g$ is the earthquake ground motion. The foundation mass is also represented by m_f. Other parameters are described in Section 2.1.

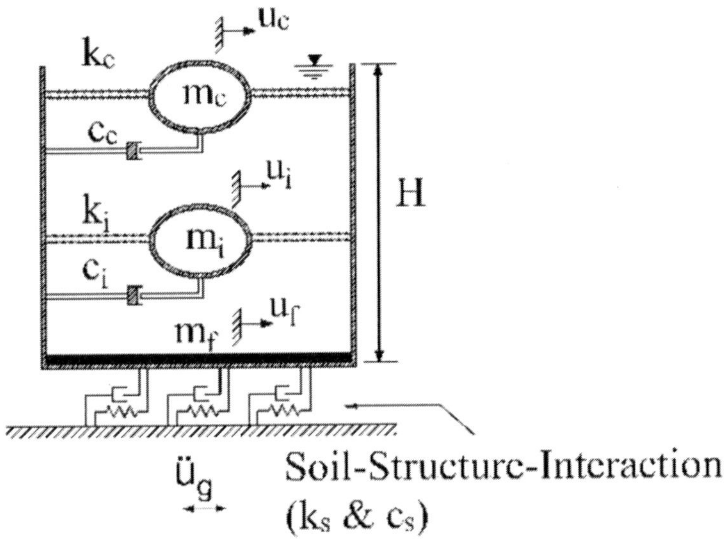

Figure 5. Simplified model of liquid storage tank considering soil-structure-fluid-interaction.

Using state-space method, a MATLAB (MathWorks Inc., Natick, MA, USA) routine is provided to solve the governing equations of motion. The numerical results will be mainly presented in terms of the convective and impulsive displacements relative to the foundation (x_c, x_i) according to Equations (12) and (13), free vertical surface displacement (d_x) according to Equation (14), and the overturning moment (OM) and structural base shear (F_s) according to Equations (15) and (16). The overturning moment and base shear are normalized by weight of the system:

$$x_c = u_c - u_f \quad (12)$$

$$x_i = u_i - u_f \quad (13)$$

$$d_x = 0.837R\frac{\omega_c^2(u_c - u_f)}{g} \tag{14}$$

$$OM = k_c \times x_c \times h_c + c_c \times v_c \times h_c + k_i \times x_i \times h_i + c_i \times v_i \times h_i \tag{15}$$

$$F_S = k_c \times x_c + c_c \times v_c + k_i \times x_i + c_i \times v_i \tag{16}$$

3. NUMERICAL STUDY

A parametric study has been done to evaluate the effect of SSI. For this purpose, a broad and a slender steel tank have both been considered as a numerical study. The resulting seismic responses of the tanks are compared with those of fixed ones. The geometric properties of the tank models are summarized in Table 2 and the resultant parameters of the equivalent mechanical models are listed in Table 3. The characteristics of selected earthquake ground motion records for time history analyses are tabulated in Table 4 and the six selected ground motions are classified, into two groups, namely far fault (no. 1, 3, 5) and near fault (no. 2, 4, 6) ground motions. The considered earthquake records are obtained from the web-based PEER ground motion database [20]. The selected near-fault ground motions have been recorded close to faults and have revealed near-fault pulses, while the selected far-fault ground motions have been recorded far from faults and have revealed far-fault pulses.

Table 2. Properties of the broad and slender tanks used in this study.

Tank Type	H (m)	R (m)	H/R	t (m)	E (GPa)	ρ (kg/m^3)
Broad	14.6	24.4	0.6	0.0203	200	1000
Slender	11.3	6.1	1.85	0.0058	200	1000

Table 3. Resultant parameters of the equivalent mechanical model for the broad and slender tanks.

Tank Type	m_c/m	m_i/m	h_c/H	h_i/H	C_c (s/m$^{0.5}$)	C_i	T_c (s)	T_i (s)
Broad	0.608	0.392	0.557	0.400	1.65	7.08	8.15	0.253
Slender	0.245	0.755	0.727	0.444	1.48	6.07	3.66	0.157

Table 4. Selected earthquake ground motions for time history analyses.

No.	Earthquake	Station	PGA (g)
1	Chichi, Taiwán, 1999	NST-E	0.309
2	Chichi, Taiwán, 1999	TCU075-W	0.333
3	Imperial Valley, 1979	6617 Cucapah	0.309
4	Imperial Valley, 1979	5155 EC Meloland	0.314
5	Northridge, 1994	90014 Beverly Hills	0.617
6	Northridge, 1994	24514 Sylmar	0.604

4. RESULTS

In this section, the effect of SSI on seismic responses of both broad and slender liquid storage tanks is studied. For example, Figure 6 depicts the time history of considered responses of two broad and slender tanks mounted on soil type 4 under Chichi-NST-E ground motion in both without SSI and with SSI, respectively.

The peak responses of broad and slender tanks rested on various soil types under different ground motions are tabulated in Table 5, Table 6, Table 7, Table 8, Table 9, Table 10, Table 11 and Table 12.

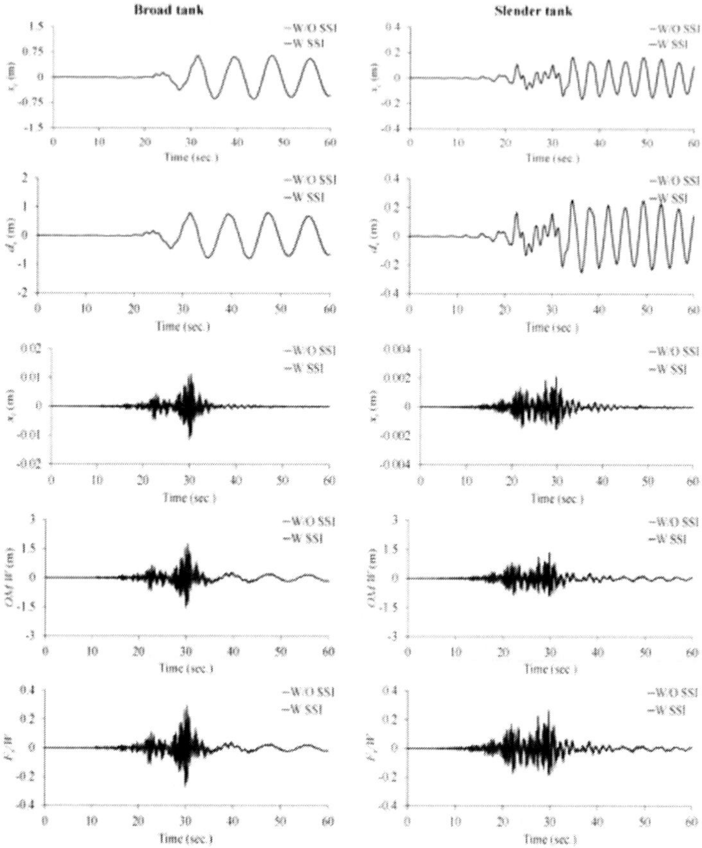

Figure 6. Time history of the broad and slender tanks under Chichi-NST-E ground motion resting on soil type 4.

Table 5. Peak responses of the broad tank on soil type 1.

Record No.	Tank Type	Condition	x_c (m)	d_x (m)	x_i (m)	OM/W (m)	F_s/W
1	Broad	w/o SSI	0.6259	0.7726	0.0102	1.544	0.2609
		w SSI	0.6259	0.7725	0.0102	1.526	0.2578
2	Broad	w/o SSI	1.4083	1.7382	0.0133	1.801	0.3145
		w SSI	1.4082	1.7381	0.0126	1.750	0.2958
3	Broad	w/o SSI	0.2738	0.3379	0.0070	1.033	0.1762
		w SSI	0.2738	0.3379	0.0072	1.037	0.1768
4	Broad	w/o SSI	0.3677	0.4539	0.0063	0.862	0.1491
		w SSI	0.3675	0.4536	0.0063	0.836	0.1463
5	Broad	w/o SSI	0.0951	0.1174	0.0260	3.766	0.6447
		w SSI	0.0951	0.1173	0.0257	3.711	0.6353
6	Broad	w/o SSI	0.1691	0.2087	0.0084	1.166	0.2021
		w SSI	0.1691	0.2087	0.0085	1.168	0.2023

Table 6. Peak responses of the slender tank on soil type 1.

Record No.	Tank Type	Condition	x_c (m)	d_x (m)	x_i (m)	OM/W (m)	F_s/W
1	Slender	w/o SSI	0.1320	0.2026	0.0041	2.572	0.5110
		w SSI	0.1320	0.2026	0.0041	2.572	0.5110
2	Slender	w/o SSI	1.4752	2.2636	0.0063	4.436	0.8281
		w SSI	1.4754	2.2638	0.0062	4.252	0.7913
3	Slender	w/o SSI	0.2612	0.4007	0.0096	5.866	1.1740
		w SSI	0.2612	0.4007	0.0094	5.821	1.1649
4	Slender	w/o SSI	0.5037	0.7729	0.0024	1.350	0.2748
		w SSI	0.5037	0.7729	0.0022	1.342	0.2732
5	Slender	w/o SSI	0.1303	0.1999	0.0092	5.735	1.1426
		w SSI	0.1304	0.2000	0.0090	5.609	1.1176
6	Slender	w/o SSI	0.4378	0.6718	0.0062	3.723	0.7522
		w SSI	0.4378	0.6718	0.0061	3.698	0.7473

Table 7. Peak responses of the broad tank on soil type 2.

Record No.	Tank Type	Condition	x_c (m)	d_x (m)	x_i (m)	OM/W (m)	F_s/W
1	Broad	w/o SSI	0.6264	0.7731	0.0225	3.304	0.5628
		w SSI	0.6264	0.7732	0.0213	3.127	0.5325
2	Broad	w/o SSI	1.4095	1.7396	0.0284	4.105	0.6959
		w SSI	1.4091	1.7391	0.0261	3.841	0.6539
3	Broad	w/o SSI	0.2739	0.3381	0.0175	2.522	0.4324
		w SSI	0.2739	0.3380	0.0176	2.526	0.4330
4	Broad	w/o SSI	0.3679	0.4540	0.0147	2.106	0.3573
		w SSI	0.3677	0.4539	0.0139	2.025	0.3434
5	Broad	w/o SSI	0.0961	0.1186	0.0648	9.319	1.5960
		w SSI	0.0959	0.1183	0.0604	8.717	1.4920
6	Broad	w/o SSI	0.1710	0.2110	0.0164	2.321	0.3999
		w SSI	0.1710	0.2110	0.0167	2.359	0.4059

Table 8. Peak responses of the slender tank on soil type 2.

Record No.	Tank Type	Condition	x_c (m)	d_x (m)	x_i (m)	OM/W (m)	F_s/W
1	Slender	w/o SSI	0.1322	0.2029	0.0074	4.551	0.9072
		w SSI	0.1323	0.2030	0.0072	4.547	0.9063
2	Slender	w/o SSI	1.4791	2.2696	0.0090	6.053	1.1480
		w SSI	1.4790	2.2694	0.0089	5.848	1.1071
3	Slender	w/o SSI	0.2619	0.4019	0.0167	10.245	2.0470
		w SSI	0.2619	0.4018	0.0165	10.190	2.0359
4	Slender	w/o SSI	0.5049	0.7747	0.0034	2.111	0.4238
		w SSI	0.5050	0.7748	0.0035	2.099	0.4215
5	Slender	w/o SSI	0.1307	0.2005	0.0158	9.781	1.9492
		w SSI	0.1308	0.2006	0.0156	9.613	1.9156
6	Slender	w/o SSI	0.4390	0.6736	0.0099	6.042	1.2016
		w SSI	0.4389	0.6735	0.0098	5.934	1.1831

Table 9. Peak responses of the broad tank on soil type 3.

Record No.	Tank Type	Condition	x_c (m)	d_x (m)	x_i (m)	OM/W (m)	F_s/W
1	Broad	w/o SSI	0.6273	0.7744	0.0153	2.165	0.3732
		w SSI	0.6260	0.7726	0.0149	2.100	0.3620
2	Broad	w/o SSI	1.4338	1.7696	0.0222	3.612	0.5985
		w SSI	1.4329	1.7686	0.0218	3.551	0.5880
3	Broad	w/o SSI	0.2755	0.3401	0.0132	1.909	0.3264
		w SSI	0.2753	0.3398	0.0130	1.901	0.3249
4	Broad	w/o SSI	0.3690	0.4555	0.0124	1.871	0.3171
		w SSI	0.3679	0.4541	0.0122	1.851	0.3136
5	Broad	w/o SSI	0.0953	0.1177	0.0430	6.204	1.0624
		w SSI	0.0954	0.1178	0.0411	5.928	1.0151
6	Broad	w/o SSI	0.1848	0.2281	0.0228	3.316	0.5662
		w SSI	0.1847	0.2280	0.0226	3.309	0.5650

Table 10. Peak responses of the slender tank on soil type 3.

Record No.	Tank Type	Condition	x_c (m)	d_x (m)	x_i (m)	OM/W (m)	F_s/W
1	Slender	w/o SSI	0.1334	0.2048	0.0041	2.469	0.4955
		w SSI	0.1335	0.2049	0.0040	2.443	0.4902
2	Slender	w/o SSI	1.5005	2.3023	0.0055	4.113	0.7617
		w SSI	1.5000	2.3015	0.0054	3.818	0.7028
3	Slender	w/o SSI	0.2636	0.4045	0.0071	4.379	0.8764
		w SSI	0.2635	0.4044	0.0068	4.140	0.8286
4	Slender	w/o SSI	0.5143	0.7892	0.0034	2.014	0.4073
		w SSI	0.5144	0.7893	0.0032	1.995	0.4036
5	Slender	w/o SSI	0.1334	0.2048	0.0108	6.671	1.3313
		w SSI	0.1336	0.2049	0.0105	6.474	1.2921
6	Slender	w/o SSI	0.4434	0.6804	0.0072	4.437	0.8905
		w SSI	0.4431	0.6800	0.0070	4.288	0.8606

Table 11. Peak responses of the broad tank on soil type 4.

Record No.	Tank Type	Condition	x_c (m)	d_x (m)	x_i (m)	OM/W (m)	F_s/W
1	Broad	w/o SSI	0.6568	0.8106	0.0114	1.765	0.2969
		w SSI	0.6475	0.7991	0.0099	1.406	0.2354
2	Broad	w/o SSI	1.5259	1.8833	0.0137	2.003	0.3286
		w SSI	1.5193	1.8751	0.0117	1.769	0.2990
3	Broad	w/o SSI	0.2755	0.3401	0.0097	1.356	0.2347
		w SSI	0.2734	0.3374	0.0093	1.318	0.2280
4	Broad	w/o SSI	0.4019	0.4961	0.0095	1.267	0.2219
		w SSI	0.4021	0.4963	0.0093	1.231	0.2157
5	Broad	w/o SSI	0.1417	0.1749	0.0273	3.933	0.6739
		w SSI	0.1418	0.1751	0.0235	3.395	0.5814
6	Broad	w/o SSI	0.3254	0.4017	0.0137	1.883	0.3271
		w SSI	0.3254	0.4017	0.0140	1.915	0.3325

Table 12. Peak responses of the slender tank on soil type 4.

Record No.	Tank Type	Condition	x_c (m)	d_s (m)	x_i (m)	OM/W (m)	F_s/W
1	Slender	w/o SSI	0.1633	0.2507	0.0021	1.351	0.2662
		w SSI	0.1632	0.2503	0.0021	1.351	0.2661
2	Slender	w/o SSI	1.8978	2.9119	0.0035	2.850	0.5001
		w SSI	1.8850	2.8923	0.0028	2.307	0.3875
3	Slender	w/o SSI	0.3377	0.5182	0.0054	3.323	0.6661
		w SSI	0.3365	0.5163	0.0047	2.870	0.5758
4	Slender	w/o SSI	0.6630	1.0174	0.0030	1.679	0.3413
		w SSI	0.6630	1.0174	0.0028	1.634	0.3321
5	Slender	w/o SSI	0.2115	0.3246	0.0056	3.496	0.6952
		w SSI	0.2115	0.3246	0.0048	3.014	0.5991
6	Slender	w/o SSI	0.7356	1.1286	0.0048	2.797	0.5756
		w SSI	0.7353	1.1282	0.0044	2.503	0.5142

It is observed that the impulsive mass displacement, normalized overturning moment and normalized base shear are reduced due to SSI effect. SSI causes a shift in the period of the structure; therefore, the responses get reduced. Such reduction will lead to better performance of these structures during earthquake events. On the other hand, the convective mass displacement is slightly affected. This phenomenon is related to the fact that the convective response has a relatively long period and, therefore, the SSI has no special effect on this response.

From Table 5, Table 6, Table 7, Table 8, Table 9, Table 10, Table 11 and Table 12, it is observed that the maximum values obtained for the reduction percentage of impulsive mass displacement, normalized overturning moment and normalized base shear are 13.2%, 20.3%, and 20.7% for broad tank under Chichi-NST-E ground motion, and 20%, 19.1%, and 22.5% for slender tank under Chichi-TCU075-W earthquake, when the liquid storage tank rested on soil type 4 (see Table 11 record no. 1 and Table 12 record no. 2). From all the data of Table 5, Table 6, Table 7, Table 8, Table 9, Table 10, Table 11 and Table 12 (for the six selected earthquake ground motions), Figure 7 shows the mean reduction percentages of peak responses in broad and slender tank due to SSI effect (where for the calculation of each mean reduction percentage, in those only very few specific cases in which there is amplification and not reduction because of the frequency content of the earthquake and the structure, those reduction percentages are taken as negative). Generally, as the shear velocity of medium soil decreases, the aforementioned responses reduce further. The mean reduction percentages of impulsive mass displacement, normalized overturning moment, and normalized base shear are 7.6%, 8.3%, and 7.9% for broad tank, and 10.4%, 9.9%, and 10.5% for slender tank, when the liquid storage tank rested on soil type 4. According to Figure 7, the convective mass displacement and also free vertical surface displacement is only slightly reduced. This is due to long period of convective mass. However, for soil type 4, the SSI causes to reduce the convective mass displacement and also free vertical surface displacement, compared to the without SSI condition. This phenomenon is observed in both broad and slender tanks.

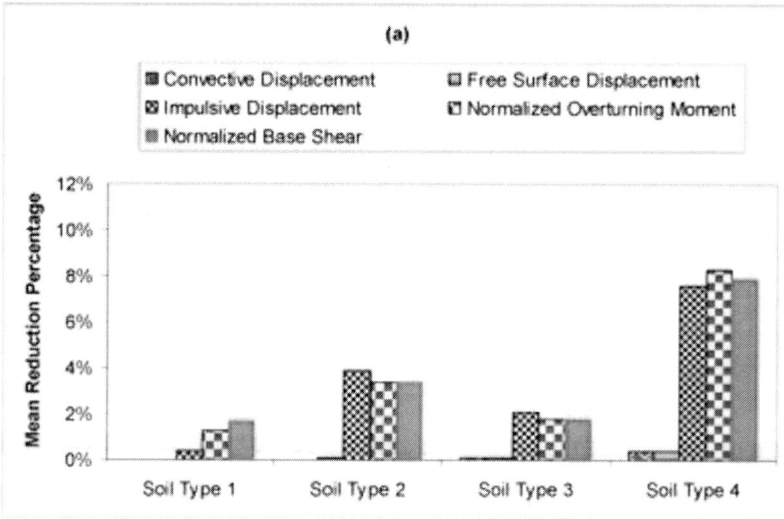

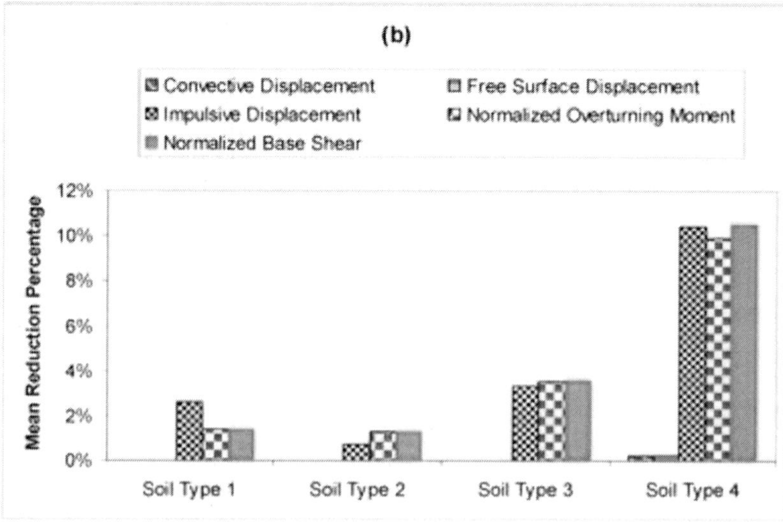

Figure 7. Mean reduction percentage of peak responses: (**a**) broad tank; and (**b**) slender tank.

5. CONCLUSIONS

The seismic behavior of liquid storage tanks, considering the SSI effect, is evaluated in this paper. The substructure method is used to consider the SSI effect, and dynamic stiffness and damping are obtained using the cone method. Two types of tanks rested on four soil types are considered as the case study. Then the peak responses of these tanks, in both with and without considering SSI, under six earthquake excitations are compared. According to obtained responses, the impulsive mass displacement, normalized overturning moment, and normalized base shear are reduced as the SSI effect is considered. However, for relatively stiff soil, this reduction is not considerable but, for soft soil, the SSI effect could shift the fundamental period of impulsive mass and, therefore, the impulsive displacement and other dependent responses reduce.

Since convective mass has a long period, the SSI did not considerably affect its seismic characteristics. Nevertheless, the transition from relatively stiff soil (S1) to softer soil (S4) could cause a shift in the fundamental period of the spectrum and, therefore, the convective displacement is also reduced.

AUTHOR CONTRIBUTIONS

Mostafa Farajian contributed to preparing the computer codes, performing the analysis, and preparing the first draft of the manuscript. Mohammad Iman Khodakarami contributed to checking the formulations and technical concepts of the research and also finalizing the results. Denise-Penelope N. Kontoni contributed to the writing and finalizing of this paper and also the reviewing process of this research.

CONFLICTS OF INTEREST

The authors declare no conflict of interest.

REFERENCES

1. Manos, G.C. Evaluation of the earthquake performance of anchored wine tanks during the San Juan, Argentina, 1977 earthquake. Earthq. Eng. Struct. Dyn. 1991, 20, 1099–1114.

2. Cooper, T.W. A Study of the Performance of Petroleum Storage Tanks during Earthquakes, 1933–1995; Report for US Department of Commerce: NIST GCR 97-720; US Department of Commerce: Gaithersburg, MD, USA, 1997.
3. Zareian, F.; Sampere, C.; Sandoval, V.; McCormick, D.L.; Moehle, J.; Leon, R. Reconnaissance of the Chilean Wine Industry Affected by the 2010 Chile Offshore Maule Earthquake. Earthq. Spectra 2012, 28, S503–S512.
4. Brunesi, E.; Nascimbene, R.; Pagani, M.; Beilic, D. Seismic performance of storage steel tanks during the May 2012 Emilia, Italy, Earthquakes. J. Perform. Constr. Facil. (ASCE) 2015, 29, 04014137.
5. Halabian, A.M.; El Naggar, M.H. Effect of non-linear soil–structure interaction on seismic response of tall slender structures. Soil Dyn. Earthq. Eng. 2002, 22, 639–658.
6. Stewart, J.P.; Fenves, G.L.; Seed, R.B. Seismic soil-structure interaction in buildings. I: Analytical methods. J. Geotech. Geoenviron. Eng. 1999, 125, 26–37.
7. Housner, G.W. Dynamic pressures on accelerated fluid containers. Bull. Seismol. Soc. Am. 1957, 47, 15–35.
8. Haroun, M.A.; Housner, G.W. Seismic design of liquid storage tanks. J. Tech. Counc. ASCE 1981, 107, 191–207.
9. Malhotra, P.K.; Wenk, T.; Wieland, M. Simple procedure for seismic analysis of liquid-storage tanks. Struct. Eng. Int.2000, 10, 197–201.
10. Bagheri, S.; Rofooei, F.; Bozorgnia, Y. Evaluation of the seismic response of liquid storage tanks. In Proceedings of the 10th International Conference on Civil, Structural and Environmental Engineering Computing, Rome, Italy, 30 August–2 September 2005; Topping, B.H.V., Ed.; Civil-Comp Press: Stirling, UK; Paper No. 216.
11. Bagheri, S.; Farajian, M. The effects of input earthquake characteristics on the nonlinear dynamic behavior of fps isolated liquid storage tanks. J. Vib. Control 2016.
12. Veletsos, A.S.; Tang, Y. Soil-structure interaction effects for laterally excited liquid storage tanks. Earthq. Eng. Struct. Dyn. 1990, 19, 473–496.
13. Larkin, T. Seismic response of liquid storage tanks incorporating soil structure interaction. J. Geotech. Geoenviron. Eng.2008, 134, 1804–1814.
14. Livaoglu, R.; Dogangun, A. Effect of foundation embedment on seismic behavior of elevated tanks considering fluid–structure-soil interaction. Soil Dyn. Earthq. Eng. 2007, 27, 855–863.
15. Livaoglu, R. Investigation of seismic behavior of fluid–rectangular tank–soil/foundation systems in frequency domain. Soil Dyn. Earthq. Eng. 2008, 28, 132–146.
16. Meek, J.W.; Wolf, J.P. Cone models for homogeneous soil. I. J. Geotech. Eng. 1992, 118, 667–685.
17. NEHRP Consultants Joint Venture, Soil-Structure Interaction for Building Structures; NIST GCR 12-917-21; NIST (National Institute of Standards and Technology), U.S. Department of Commerce: Gaithersburg, MD, USA, 2012.
18. Lysmer, J.; Tabatabaie-Raissi, M.; Tajirian, F.; Vahdani, S.; Ostadan, F. SASSI: A System for Analysis of Soil-Structure Interaction; Report No. UCB/GT/81/02; Department of Civil Engineering, University of California: Berkeley, CA, USA, 1981.
19. Wolf, J.P.; Deeks, A.J. Foundation Vibration Analysis: A Strength of Materials Approach; Butterworth-Heinemann: Oxford, UK, 2004.
20. PEER Ground Motion Database, Pacific Earthquake Engineering Research Centre. Available online: http://ngawest2.berkeley.edu (accessed on 2 January 2016).

CHAPTER 10

Estimating the Frequency, Magnitude and Recurrence of Extreme Earthquakes in Gulf of Aqaba, Northern Red Sea

Ali Hassan Baaqeel[1*], Said Ali El-Quliti[2], Yahya Ali Daghreri[3], Sultan Abdullah Bin Hajlaa[4], Hadi Hussain Al-Yami[5]

[1]*Health, Safety and Environment, Economic Cities Ports, King Abdullah Economic City, Saudi Arabia*

[2]*Department of Industrial Engineering, King Abdul-Aziz University, Jeddah, Saudi Arabia*

[3]*Crisis and Disasters Management Team, Jeddah, Saudi Arabia*

[4]*Tides and Notice to Mariners Department, General Commission for Survey, Jeddah, Saudi Arabia*

[5]*Health Affairs, Saudi National Guard, Riyadh, Saudi Arabia*

ABSTRACT

Humans coexist with extreme events all the time, only when the intensity of the event becomes greater than a certain level there is a resulting disaster. Small earthquakes occur all of the time with no adverse effects. Only large earthquakes cause disasters. Statistical analysis reveals that larger events occur less frequently than small events. In a year, we would have many values for the events; the annual maximum was the greatest of those values. Within an annual series, only the largest value per year is allowed, even if an additional significant peak occurs. As the magnitude of a hazardous increases, the frequency of occurrence (how often a given magnitude is equaled or exceeded) decreases. Thus, major disasters result from a small number of large events that rarely occur. A plot of recurrence intervals versus associated magnitudes produces a group of points that also approximates a straight line on semi-logarithmic paper. Therefore, past records of earthquakes at the Gulf of Aqaba, Northern Red Sea for months from May, 1999 to Feb, 2016 are used to predict future conditions concerning the annual frequency, the return period, the percentage probability for each event, and the probability of a certain-magnitude earthquake occurring in the region during any period.

Keywords: Earthquakes, Gulf of Aqaba, Northern Red Sea, Magnitude, Recurrence

1. INTRODUCTION

One of the most destructive disasters of nature is a severe earthquake and its destroying effects. If the earthquake occurs in a populated area, it may cause many deaths and injuries and extensive property damage regions. The ultimate goal of seismic hazard assessment and risk evaluation for a particular site or area is to condense seism-tectonic knowledge and experience into parameters used for predicting seismic parameters which in turn can be applied by engineers in design and subsequent earthquake resistant construction.

Statistical surveys support researches on the likelihood of future earthquakes. A primary goal of earthquake research is to increase the reliability of earthquake probability estimates. With a greater understanding of the hazard parameters of earthquakes, we may be able to reduce damage and loss of life from this destructive event. Statistics help us to predict the future events based on previous events.

1.1. Seismicity of the Western Region of Saudi Arabia

Recently, there has been an increasing concern about the seismic activity along the western coast of the Kingdom. Several studies were conducted to estimate the level of the seismic risk in the Kingdom [1] and [2] . Western Region of Saudi Arabia is considered to be a moderately active seismic zone as shown in Figure 1 [3] . Seismic events in the region that have been reported in literature include a significant earthquake, with a magnitude of 6.25 in Richter scale (a scale for expressing the magnitude of an earthquake in terms of the logarithm of the amplitude of the ground wave; values range from 0 to over 9), that occurred in 1941 at about 30 km to the east of Jizan city [4] . Seismic events also include a sequence of earthquakes which occurred in 1967 along the Red Sea rift system at a distance of about 150 km to the south west of Jeddah [5] . Recently, El-Isa et ale [6] reported that about 500 local earthquakes with magnitudes less than 4.85, occurred in the Gulf of Aqaba area during the period from January 21 to April 20, 1983. Merghelani [7] has also reported that a high level of micro-earthquake activity was detected near the border of the Red Sea and near the transition from oceanic to continental crust.

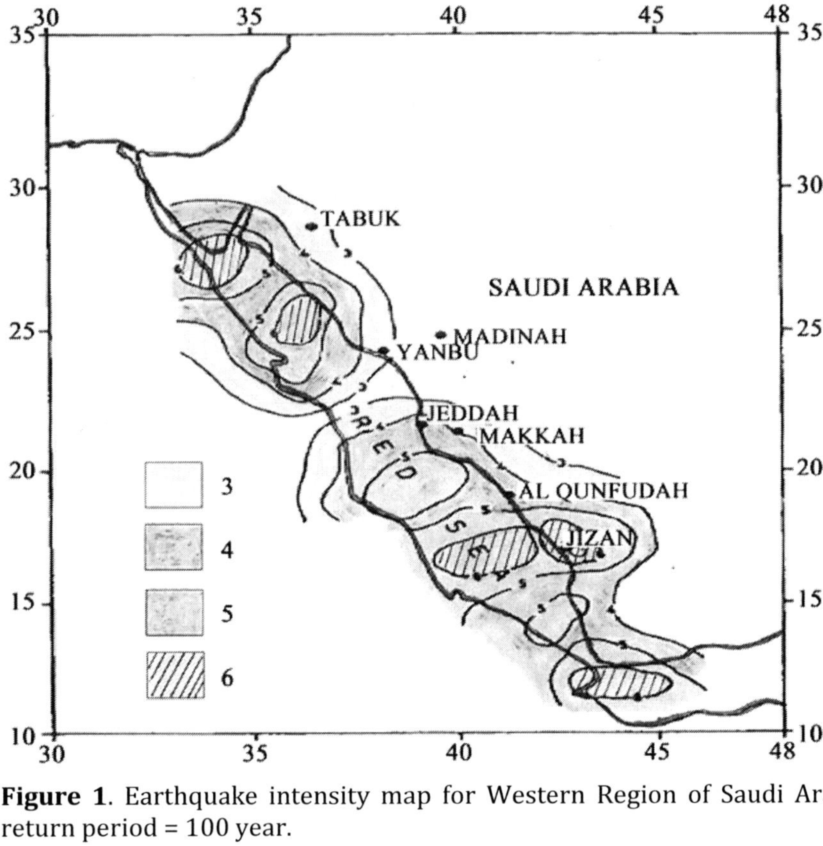

Figure 1. Earthquake intensity map for Western Region of Saudi Arabia, return period = 100 year.

1.2. Seismicity of Gulf of Aqaba

The Gulf of Aqaba is an elongated basin (~180 × 20 km) with depths reaching 1850 m. It represents the southern segment of the Dead Sea Transform (DST), which is one of the largest transform fault zones in the world. The opening of Gulf of Aqaba is thought to have originated from the relative displacement of the African and Arabian Plates. According to historical and recent earthquake records, it is seismically active [8] , as it is shown in Figure 2.

The Dead Sea transform fault borders the Arabian plate in the west and has experienced approximately 107 km of left-lateral displacement since the mid-Miocene, which translates into roughly 6 - 10 mm/yr of slip on the fault [10] , as it is shown in Figure 3.

The Gulf of Aqaba is east of the Sinai Peninsula and west of the Arabian Peninsula. With the Gulf of Suez to the west, it extends from the northern portion of the Red Sea. It reaches a maximum depth of 1850 m in its central area: The Gulf of Suez is significantly wider but less than 100 m deep. The gulf measures 24 kilometers (15 mi) at its widest point and stretches some 160 kilometers (99 mi) north from the Straits of Tiran. Like the coastal waters of the Red Sea, the gulf is one of the world's premier sites for diving. The area is especially rich in coral and other marine biodiversity and has accidental shipwrecks and vessels deliberately sunk in an effort to provide a habitat for marine organisms and bolster the local dive tourism industry. At this northern end of the gulf are important cities: Taba in Egypt and Aqaba in Jordan. They are strategically important commercial ports and popular resorts for tourists seeking to enjoy the warm climate. Further south, Haql is the largest Saudi Arabian city on the gulf. On Sinai, Sharm el-Sheikh and Dahab are the major centers. The largest population center is Aqaba, with a population of 108,000 (2009) [12] .

Figure 4 shows the position of Gulf of Aqaba and the surrounding area relative to the map of the Kingdom of Saudi Arabia.

2. LITERATURE REVIEW

When considering the earthquake, we have to answer the four questions: Where? How often? How big? And When? The goal of the earthquake prediction is to give warning of potentially damaging earthquakes early enough to allow appropriate response to the disaster, enabling people to minimize loss of life and property [14] . Many studies have been presented to develop reliable estimates, of probability, magnitude and recurrence relations given the large pattern of earthquake occurrence. The primary advantage of probabilistic seismic hazard analysis (PSHA) is that it integrates over all seismicity: temporal and spatial along with ground motions to calculate a combined probability of exceedance, which incorporates the relative frequencies of occurrence of different earthquakes and ground-motion characteristics. Practically in any earthquake catalogue the quality of different parts (periods and areas) varies significantly with

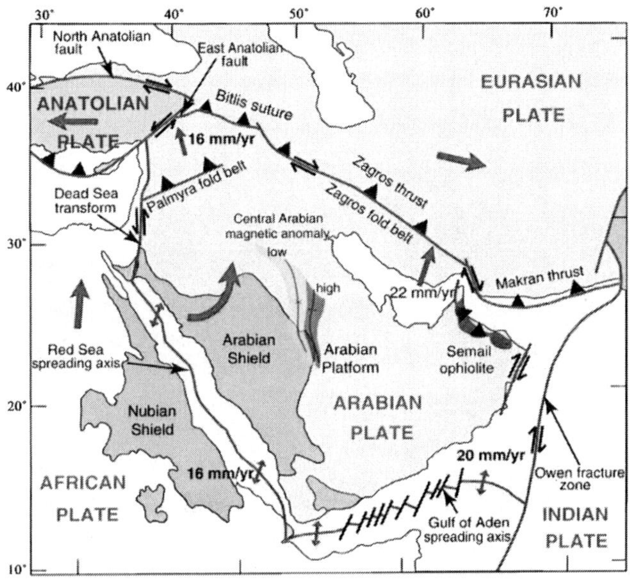

Figure 2. Arabian plate boundaries adapted from Johnson and Stern, 2010 [9] .

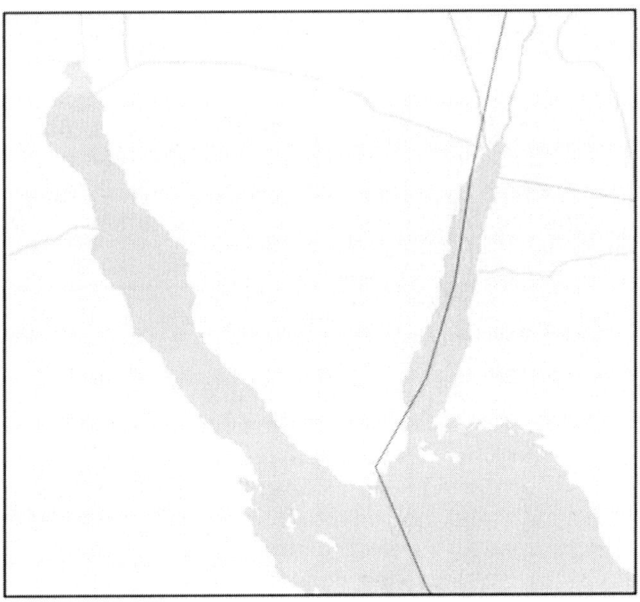

Figure 3. Gulf of Aqaba-Dead Sea transform fault [11] .

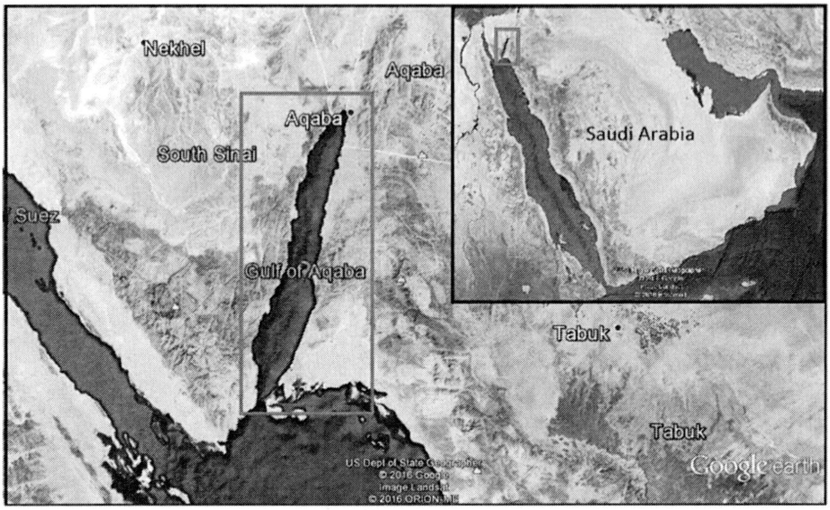

Figure 4. Gulf of Aqaba and surrounding area [13] .

respect to completeness, magnitude reliability, homogeneity and location accuracy [15] - [17] .

Statistical theory of extreme values has been used to analyze the observed extremes of any phenomena and to forecast the further extremes based on the appropriate distribution, Gumbel [18] . In earthquake engineering, this theory has been applied successfully by many researchers in the past few decades (Nordquist [19] ; Epstein [20] ; Yegulalp [21] ; Al-Abbasi [22] and Jaiswal [23]). This theory does not require analysis of the complete record of earthquake occurrence, but uses the sequence of earthquakes constructed from the largest values of the magnitude over a set of predetermined intervals.

Abe and Suzuki [24] analyzed the seismic data from the viewpoint of science of complexity, where one of the main goals of seismology is to predict when and where the next main shock will occur after an earlier main shock. The rate of recurrence of earthquakes on a seismic source can be represented with the Gutenberg-Richter relation [25] . Kasap and Gürlen [26] studied the return periods of earthquakes. Ogata [27] investigated the statistical models for earthquake occurrences. Utsu [28] applied gamma, log-normal, Weibull and exponential distributions to describe the probability distribution of inter-occurrence time of large earthquakes in Japan. Aktaş et al. [29] , used Poisson distribution to describe the recurrence times, and estimated the expected value and variance computed for the loss of life and damaged buildings after the change point using the compound Poisson process. Bayrak et al. [30] evaluated the seismicity and earthquake hazard parameters of Turkey based on maximum regional magnitude. Öztürk et al. [31] estimated the mean return periods, the most probable magnitude in a time period of t-years, and the probability of earthquake occurrence for a given magnitude during a time span of t-years for different regions in and around Turkey. They also showed that in the specific region, the most probable earthquake magnitude in the next 100 years would be over 7.5. Bayrak et al. [32] calculated the seismicity parameters for the 24 seismic regions of Turkey according to Gumbel and Gutenberg-Richter methods and concluded that b-values obtained from the maximum likelihood approach gives better results for the tectonics of the examined area. In a study [33] , 231 earthquake data of magnitude 5 and higher, between north (39.00° - 42.00°) and east (26.00° - 45.00°) coordinates in Turkey from July 12, 1900 to July 25, 2011 are analyzed. In this study, the probability distribution of magnitude is attempted and the statistical models are taken to interpret the observed frequency distribution. The earthquake catalog for Iraq covering an area between latitude 39 - 50 E and longitude 29 - 50 N and containing more than thousand events for the period 1905-2000 has been compiled The statistical parameters for Gumbel's have been estimated using both the least squares and maximum likelihood techniques. The goodness of fit is evaluated employing Kolmogorov-Smirnov test [34] .

3. EXTREME EARTHQUAKE ANALYSIS

Earthquake prediction can be considered into two types. First is the statistical prediction which is based on previous events; Data are collected from the records. Second is deterministic prediction which is made from the earthquake signs. The table at Appendix shows the data for earthquakes in Gulf of Aqaba and surrounding area representing the minimum magnitude and maximum magnitude.

Most extreme event analysis is concerned with the distribution of annual maximum or minimum values at a given site. These events are given a rank, m, starting with m = 1 for the highest value, m = 2 for the next highest and so on in descending order. Each earthquake magnitude is associated with a rank, m, with m = 1 given to the maximum magnitude over the years of record, m = 2 given to the second highest magnitude, m = 3 given to the third highest one, etc. The smallest earthquake magnitude will receive a rank equal to the number of years over which there is a record, n. Thus, the discharge with the smallest value will have m = n = 18.

There are several formulas for calculating the probability value. The Weibull formula will be used because of its ease of use. The US Geological Survey [35] , among others, also uses this formula.

According to the Weibull equation [36] , the return period or recurrence interval T (in years) is calculated using the following equation:

$$T(\text{years}) = (n+1)/m \tag{1}$$

where: m = event ranking (in a descending order), and n = number of events in the period of record.

The percentage probability the (annual exceedance probability) for each magnitude is calculated using the inverse of the Weibull equation as follows:

$$P(\text{percent}) = 100 \cdot m/(n+1) \tag{2}$$

From Equations ((1), (2)) it is clear that P = 100/T%. For example, an earthquake equal to that of a 10-year one would have an annual exceedance probability of 1/10 = 0.1 or 10%. This would say that in any given year, the probability that an earthquake with a magnitude equal to or greater than that of a 10-year earthquake would be 0.1 or 10%. Similarly, the probability of an earthquake with a magnitude exceeding the 50 year one in any given year would be 1/50 = 0.02, or 2%. Note that such probabilities are the same for every year, but in practice, such an earthquake could occur next year, or be exceeded several times in the next 50 years.

Table 1 shows the calculations of the rank m, the probability P and the return period T for the data of the yearly maximum magnitude given in the Appendix and Figure 5 shows the location of given data in the Appendix.

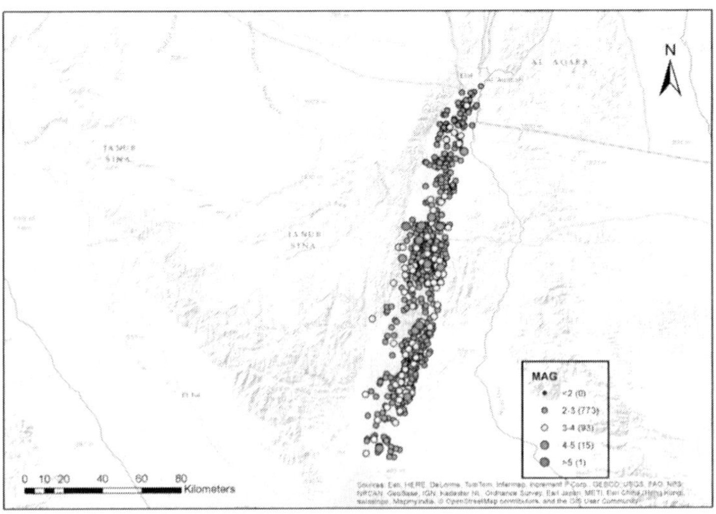

Figure 5. Location of give date in the Appendix [38] .

Table 1. The rank, probability and the return period results.

Rank (m)	Year	Maximum Magnitude	Probability (P) %	Return Period (T)
1	2015	5.08	5.26	19.00
2	1999	4.65	10.53	9.50
3	2000	4.44	15.79	6.33
4	2010	4.32	21.05	4.75
5	2007	4.25	26.32	3.80
6	2011	4.14	31.58	3.17
7	2002	4.08	36.84	2.71
8	2008	4.05	42.11	2.38
9	2012	4.02	47.37	2.11
10	2014	3.95	52.63	1.90
11	2001	3.82	57.89	1.73
12	2003	3.79	63.16	1.58
13	2004	3.56	68.42	1.46
14	2005	3.5	73.68	1.36
15	2009	3.4	78.95	1.27
16	2006	3.35	84.21	1.19
17	2013	3.32	89.47	1.12
18	2016	2.06	94.74	1.06

4. EARTHQUAKE PARAMETERS

4.1. Annual Exceedance Probability and Return Period

Return period or Recurrence interval is the average interval of time within which a flood of specified magnitude is expected to be equaled or exceeded at least once. 19-year earthquake is an earthquake that is expected to occur, on the average, once every 19 years, or has 5.26% chance of occurring each year.

Figure 6 is a plot of earthquake magnitude and annual exceedance probability relationship (linear scales) with the annual maximum magnitude per year on the Y axis versus the annual exceedance probability on the X axis. The X and Y axes both use linear scales.

A best-fit curve is drawn through the data points. From the best-fit curve, one can determine the earthquake magnitude associated with an earthquake with a recurrence interval of say 1.9 years, it is about 3.95 on Richter scale. This would be called the 2-year earthquake.

Similarly, the recurrence interval associated with an earthquake magnitude of magnitude of 5.08 on Richter scale is about 19 years.

The annual peak information may also be presented with a logarithmic rather than a linear scale. This is often done to make the curve appear as a straight line and also to avoid a graph that will suggest either a zero or a one-hundred percent exceedance probability. Moreover, a straight line curves are more easily allow extrapolation beyond the data extremes. Figure 7 represents the earthquake magnitude and the annual exceedance probability (log scale) relationship.

Percentage probability is determined by dividing one by the recurrence interval and multiplying by 100. For example, the probability that an earthquake magnitude will exceed the 19-year earthquake this year or any other year would be 5.26%.

Figure 8 shows the earthquake magnitude and return period relationship on linear scales. From the figure it can be noticed that the return period of an earthquake of magnitude 5.08 on Richter scale is about 19 years, and an earthquake of magnitude of 4.44 on Richter scale has a recurrence interval of about 6.33 years.

Sometimes it is suitable to add a second Y-axis to represent the return period to the first Y-axis representing the annual exceedance probability. Figure 9 shows the earthquake magnitude on the X-axis and the annual exceedance probability on the first Y-axis and the return period on the second Y-axis. Both the two Y-axes use avariable log scale so the relationship appears as a semi-parallel line, this will allow for easier findings.

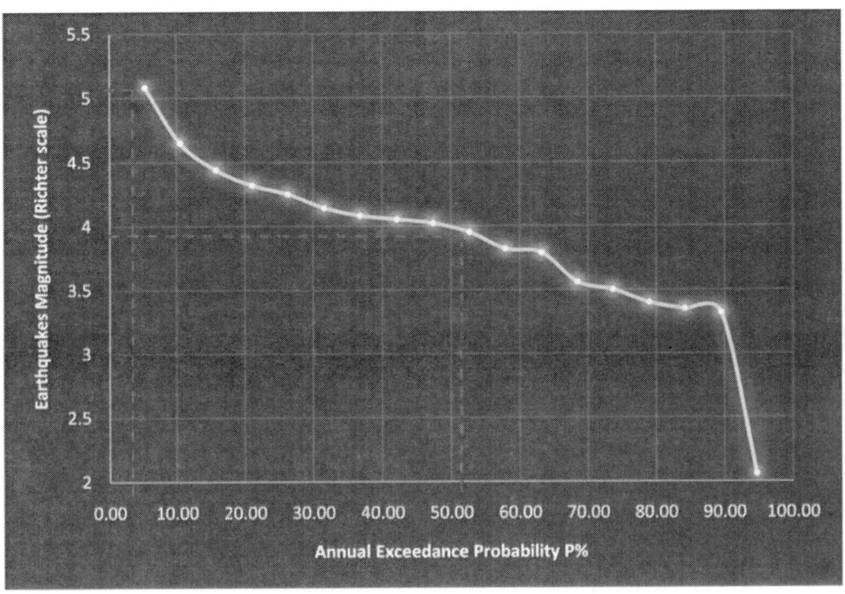

Figure 6. Earthquake magnitude and probability relationship (linear scales).

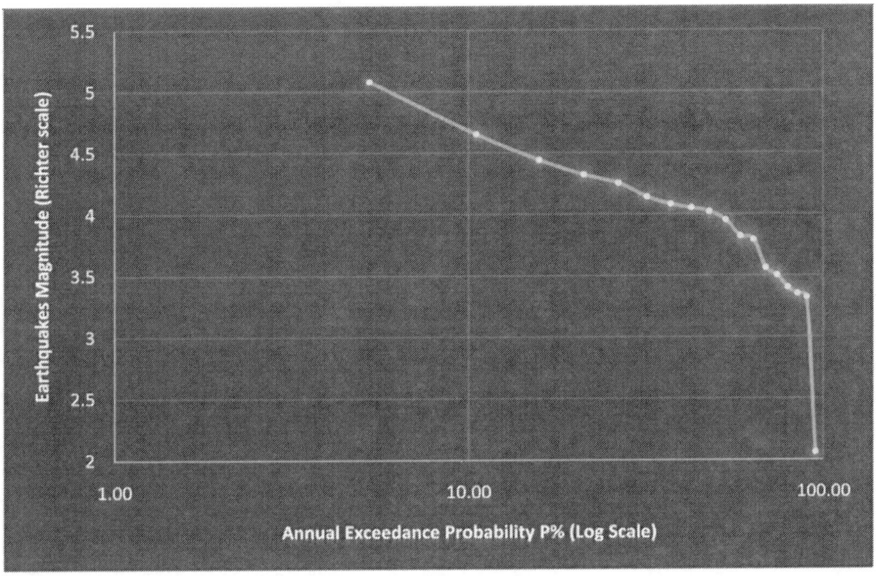

Figure 7. Earthquake magnitude and annual exceedance probability (log scale) relationship.

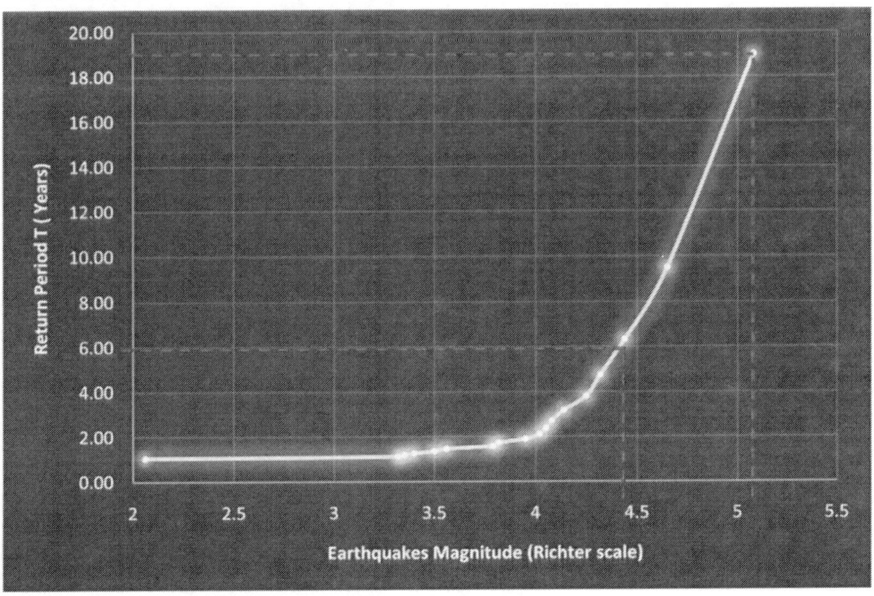

Figure 8. Earthquake magnitude and return period relationship (linear scales).

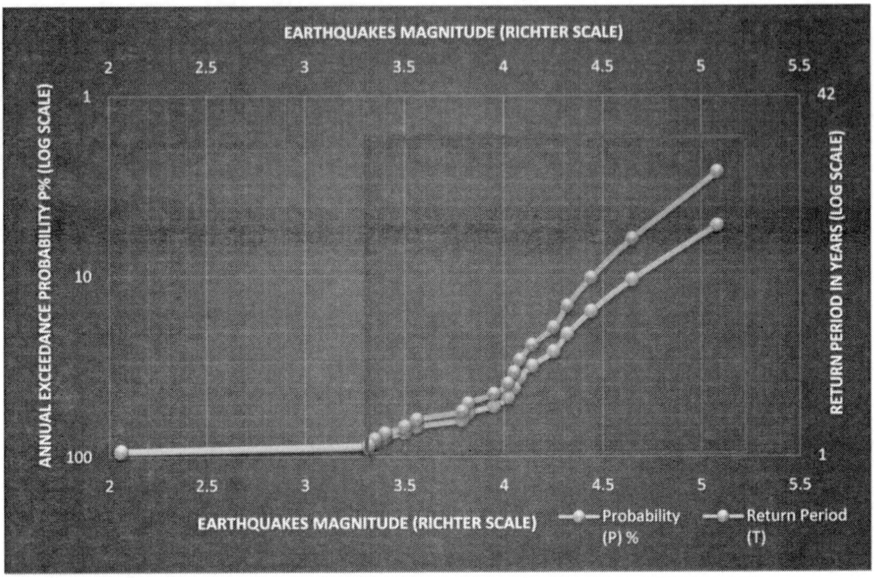

Figure 9. Earthquake magnitude, probability and return period relationship.

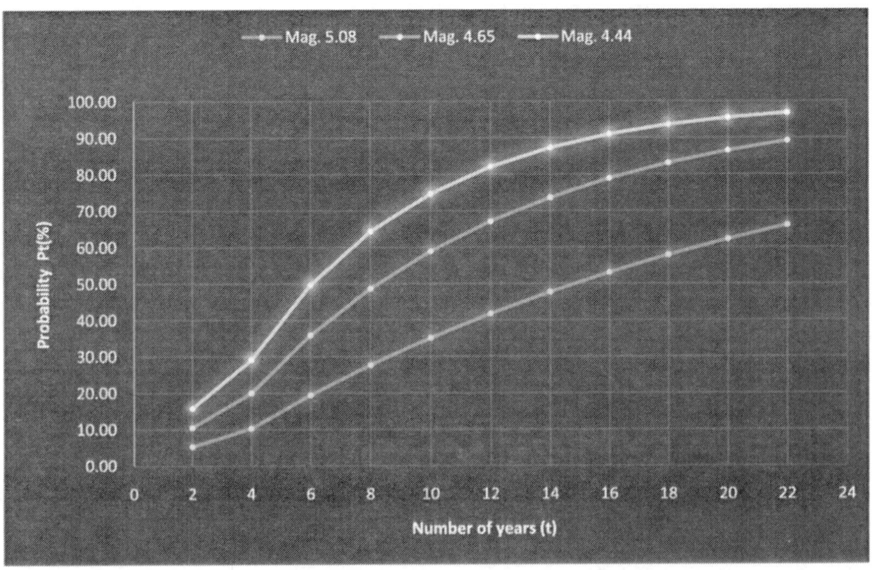

Figure 10. Earthquake probability for some earthquake magnitudes in a time span period.

From Figure 10, as the earthquake magnitude increased more than 3.3 on Richter scale the returned period will have increased and the probability will have decreased, this relation can be determined by the increment of the gap between the probability and return period lines.

4.2. The Probability during a Time Period

Theprobabilityofacertain-magnitudeearthquakeoccurringduringanyperiod t can be calculated using the following equation:

$$Pt = 1 - (1 - P)t \tag{3}$$

where P is the probability of occurrence over the entire time period, t, and P is the probability of occurrence in any year.

It is worth to apply equation (3) for earthquakes of highest magnitudes which represent the most dangerous events in the study. The equation is applied for earthquakes of magnitudes 5.08, 4.65 and 4.44 Richter scale of probabilities of 5.26%, 10.52% and 15.78% respectively. The result is depicted in Figure 11 for earthquakes of magnitudes: 5.08 Richter scale (P = 5.26%), 4.65 Richter scale (P = 10.52%) and 4.44 Richter scale (P = 15.78%).

A homeowner considering the costs of reinforcing a house against earthquakes will want to know how the risk varies during an average mortgage span of 5 years. Figure 11 shows the earthquake probability and earthquake magnitudes in a time span of 5 years. An earthquake of magnitude of 4.44 on Richter scale for example, has a 57.65% probability of occurrence but, if the earthquake of magnitude 5.08 on Richter scale is chosen, the probability drops to 23.69%.

In addition, from Figure 11, any earthquake of magnitude less than 3.56 on Richter scale has a 100% probability of occurrence.

5. CONCLUSIONS AND POINTS FOR FUTURE RESEARCHES

5.1. Conclusions

Earthquake is an unavoidable natural disaster for the region. Hence, to take precautions for the future by utilizing the past experiences is very substantial. This can be a kind of a proposition to the higher authorities to have an open eye to this particular region.

In this study, the statistical frequency analyses are applied to the recorded annual maximum earthquake magnitudes for Gulf of Aqaba since 1999.

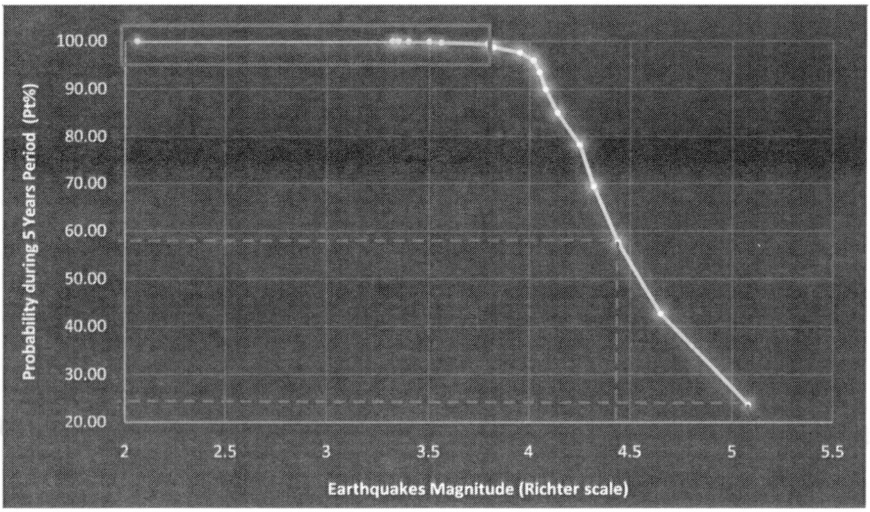

Figure 11. Earthquake probability and earthquake magnitudes in a time span of 5 years.

The earthquake hazard parameters are estimated, these are: the mean return periods (recurrence intervals), the frequency, the probability of earthquake occurrence (annual exceedance probability) for a given magnitude during any year, and the probability of earthquake occurrence for a given magnitude during a time span of t-years with a stress on a 18-year period. The Weibull equation is applied to estimate the return period, while the inverse of the Weibull equation is used to calculate the probability of occurrence.

The relation between magnitude and frequency and between magnitude and return period is represented as a curve in a linear scale graph and as a straight line on a logarithmic scale and variable scale graphs to facilitate the findings. The results lead to a general conclusion that Gulf of Aqaba is considered as a high seismic area and the region is exposed to earthquakes with strength ranging of 5 or more on the Richter scale with a high probability. The maximum magnitude is 5.08 with a return period of 19 years and probability of about 5.26%.

5.2. Points for Future Researches

Points for future researches can be summarized as follows:

- To study in details the period before 1999 where it is included a recorded earthquake with magnitude 7.3 Richter scale on 1995.
- To use other methods for evaluation of earthquake parameters and compare the obtained results.
- To estimate earthquake hazard parameters for regions around Gulf of Aqaba.
- To estimate hazard parameters for other events like: floods, subsidence, volcanic eruptions and severe storms in different regions around Gulf of Aqaba.
- To draw a seismic map for Gulf of Aqaba region and for other regions around it.

ACKNOWLEDGEMENTS

We would like to present our appreciation to Professor Amar Ameen, Dean of the Faculty of Earth Sciences, King Abdulaziz University, Saudi Arabia and Assistant Professor Abdul-elah Bahabri, Engineering and Environmental Geology Department, King Abdulaziz University for their appreciated help.

In addition, special tanks to Hani Zahran, Directing Manager, National Center for earthquakes and volcanoes, Geological Survey Authority, Saudi Arabia, for providing all seismic reading of Gulf of Aqaba, which used in this paper.

CITE THIS PAPER

Ali Hassan Baaqeel, Said Ali El-Quliti, Yahya Ali Daghreri, Sultan Abdullah Bin Hajlaa, Hadi Hussain Al-Yami (2016) Estimating the Frequency, Magnitude and Recurrence of Extreme Earthquakes in Gulf of Aqaba, Northern Red Sea. *Open Journal of Earthquake Research*,**05**,135-152.

REFERENCES

1. AL-Haddad, M., Siddiqi, G.S., Al-Zaid, R., Arafah, A., Necioglu, A. and Turkelli, N. (1992) A Study Leading to a Preliminary Seismic Design Criteria, for the Kindom. Final Report, KACST Project No. AR-9-31, Riyadh.
2. Al-Haddad, M., Siddiqi, G.S., Al-Zaid, R., Arafah, A., Necioglu, A., and Turkelli, N. (1994) A Basis for Evaluation of Seismic Hazard and Design Criteria for Saudi Arabia. Journal of Earthquake Engineering Research Institute, 10, No. 2.
3. Ashour, S.A. and Abdel-Rahman, H.H. (1994) Application of Seismic Risk Analysis and Earthquake Simulation Methods to the Western Region in Saudi Arabia. Journal of Kong Abdulaziz University, 6, 3-23.
4. Gutenberg, B. and Richter, C. (1965) Seismicity of the Earth and Associated Phenomena. Hafner Publishing Co., New York and London, 310.
5. Barazangi, M. (1981) Evaluation of Seismic Risk along the Western Part of the Arabian Plate: Discussion and Recommendations. Bulletin of Faculty of Earth Science, 4, 77-87.
6. El-Isa, Z.H., Merghelani, H.M. and Bazzani, M.A. (1984) The Oulf of Aqaba Earthquake Sworn of 1983, January-April. Geophysical Journal International, 78, 711-722.
7. Merghelani, H.M. (1979) Seismicity of the Tihamat-Asir Region. Kingdom of Saudi Arabia. US. Geological Survey Saudi Arabian Project Report 251, 20.
8. El Khrepy, S., Koulakov, I., Al-Arifi, N. and Petrunin, A. (2016) Seismic Structure Beneath the Gulf of Aqaba and Adjacent Areas Based on the Tomographic Inversion of Regional Earthquake Data. Solid Earth.
9. Stern, R.J. and Johnson, P. (2010) Continental Lithosphere of the Arabian Plate; a Geologic, Petrologic, and Geophysical Synthesis. Earth-Science Reviews, 101, 29-67.
10. Vita-Finzi, C. (2001) Neotectonics at the Arabian Plate Margins. Journal of Structural Geology, 23, 521-530.
11. USGS Website. http://earthquake.usgs.gov/
12. Wikipedia Website. https://en.wikipedia.org/wiki/Gulf_of_Aqaba
13. Google Maps Website. https://www.google.com/maps/@28.7151807,34.3586736,367486m/data=!3m1!1e3?hl=en
14. Brillinger, D.R. (1993) Earthquake Risk and Insurance. Environmetrics, 4, 1-21.
15. Ameer, S.A., Sharma, M.L. and Wason, H.R. (2002) Maximum Likelihood Estimation of Seismic Hazard for Iraq from Complete Data Files. Proceedings of the 12th Symposium on Earthquake Engineering, Roorkee, India, Vol. 1, 306- 312.
16. Ameer, S.A., Sharma, M.L. and Wason, H.R. (2002) Completeness of Earthquake Catalogue and Its Implications in Seismic Hazard Estimation—A Case Study for Iraq. Proceedings of the 12th Symposium on Earthquake Engineering, Roorkee, 16-18 December 2012, 342-349.
17. Ameer, S.A., Sharma, M.L., Wason, H.R. and Alsinawi, S.A. (2005) Probabilistic Seismic Hazard Assessment for Iraq Using Complete Earthquakes Catalogue Files. Pure & Applied Geophysics, 162, 951-966.
18. Gumbel, E. (1958) Statistics of Extremes. Columbia University Press, New York.
19. Nordquist, J.M. (1945) Theory of Largest Values Applied to Earthquake Magnitudes. Transactions—American Geophysical Union, 26, 29-31.
20. Epstein, B. and Lomnitz, C. (1966) A Model for the Occurrence of Large Earthquakes. Nature, 211, 954-956.
21. Yegulalp, T.M. and Kuo, J.T. (1974) Statistical Prediction of the Occurrence of Maximum Magnitude Earthquakes. Bulletin of the Seismological Society of America, 64, 393-414.
22. Fahmi, K.J. and Al Abbasi, J.N. (1991) GEMPAK: A FORTRAN-77 Program for Calculating Gumbel's First, Third and Mixture Upper Earthquake Magnitude Distributions Employing Maximum Likelihood Estimation. Computers and Geosciences, 17, 271-290.

23. Jaiswal, K., Gupta, S. and Sinha, R. (2002) Estimation of Maximum Magnitude Earthquakes in Peninsular India Using Extreme Value Statistics. Proceedings of the 12th Symposium on Earthquake Engineering, IIT Roorkee, 12-18 December 2002.
24. be, S. and Suzuki, N. (2005) Scale-Free Statistics of Time Interval between Successive Earthquakes. Physica A: Statistical Mechanics and Its Applications, 350, 588-596.
25. Gutenberg, R. and Richter, C.F. (1944) Frequencies of Earthquakes in California. Bulletin of the Seismological Society of America, 34, 185-188.
26. Kasap, R. and Gürlen, ü. (2003) Deprem Magnitüdleri Icin Tekrarlanma Yillarinin Elde Edilmesi: Marmara Bolgesi Ornegi. Dogus üniversitesi Dergisi, 4, 157-166.
27. Ogata, Y. (1988) Statistical Models for Earthquake Occurrences and Residual Analysis for Point Processes. Journal of the American Statistical Association, 83, 9-27.
28. Utsu, T. (1984) Estimation of Parameters for Recurrence Models of Earthquakes. Bulletin of the Earthquake Research Institute, 59, 53-66.
29. Aktas, S., Konsuk, H. and Yigiter, A. (2009) Estimation of Change Point and Compound Poisson Process Parameters for the Earthquake Data in Turkey. Environmetrics, 20, 416-427.
30. Bayrak, Y., Oztürk, S., Cinar, H., Kalafat, D., Tsapano, T.M., Koravos, G.C. and Leventakis, G.A. (2009) Estimating Earthquake Hazard Parameters from Instrumental Data for Different Regions in and around Turkey. Engineering Geology, 105, 200-210.
31. Oztürk, S., Bayrak, Y., Koravos, G.C. and Tsapanos, T.M. (2008) A Quantitative Appraisal of Earthquake Hazard Parameters Computed from Gumbel I Method for Different Regions in and around Turkey. Natural Hazards, 47, 471-495.
32. Bayrak, Y., Oztürk, S., Koravo, G.C., Leventakis, G.A. and Tsapanos, T.M. (2008) Seismicity Assessment for the Different Regions in and around Turkey Based on Instrumental Data: Gumbel First Asymptotic Distribution and Gutenberg-Richter Cumulative Frequency Law. Natural Hazards and Earth System Sciences, 8, 1090-1122.
33. Konsuk, H. and Aktas, S. (2013) Estimating the Recurrence Periods of Earthquake Data in Turkey. Open Journal of Earthquake Research, 2, 21-25.
34. Ameer, A.S., Sharma, M.L., Wason, H.R. and Alsinawi, S.A. (2004) Seismic Hazard Characterization and Risk Evaluation Using Gumbel's Method of Extremes (G1 and G3) and G-R Formula for Iraq. 13th World Conference on Earthquake Engineering, Vancouver, 1-6 August 2004, Paper No. 2898.
35. The US Geological Survey Website. Visited on 22 December, 2015. http: //www.usgs.gov/
36. Senocak, S., Düzgün, O. and Sengül, S. (2014) Frequency Analysis of Annual Maximum Earthquakes for Askale, Erzurum (Turkey) Province. Recent Advances in Civil Engineering and Mechanics, Mathematics and Computers in Science and Engineering Series 36, North Atlantic University Union. Proceedings of the 5th European Conference of Civil Engineering (ECCIE'14), Florence, 22-24 November 2014, 72-78. http://www.wseas.us/e-library/conferences/2014/Florence/ SEMOTEC /SEMOTEC-08.pdf
37. Data Provided from Saudi Survey Authority, Earthquakes and Volcanos National Center.
38. Map Prepared by Saudi Survey Authority, Earthquakes and Volcanos National Center.

CHAPTER 11

Seismic Sequence Structure and Earthquakes Triggering Patterns

Giulio Riga[1], Paolo Balocchi[2]

[1]*Geologist, Independent Researcher, Lamezia Terme, Italy* [2]*Geologist, Independent Researcher, Modena, Italy*

ABSTRACT

Within a time distribution of magnitude values, before any mainshock some earthquake triggering patterns with several features develop, under tectonic processes' influence, through which it is possible to early identify the preparation phase of big earthquakes. The purpose of this article was to identify and classify the warning patterns that develop before a big earthquake by considering space-time seismicity variations. The methodological approach adopted was of graphical type, based on procedures of technical analysis currently used to estimate the financial markets. In the initial phase of the study we have analyzed the seismic sequences types described in the bibliography (type 1: foreshocks-mainshock-aftershocks, type 2: mainshock-aftershock; type 3: swarm) and the main structure of the seismic cycle, within which maximum and minimum magnitude values characterize the pattern that it develops until the main event changes. Then, we assessed the position of foreshocks, mainshock and aftershocks within the seismic cycle in order to identify the warning pattern that characterized the exact time when the energy emission occurs. As to the evolution normally shown over time, we have grouped the warning patterns in 2 categories: 1) progressive earthquake pattern; 2) flash earthquake pattern. Finally, we have made a classification of the warning pattern related to the fluctuations of maximum and minimum magnitude values, compared its form with the mainshock's focal mechanism and suggested some graphic procedures in order to estimate the mainshock magnitude value associated with each warning pattern. The results we obtained unquestionably allow a better comprehension of preparation process of a large earthquake, improving the earthquakes forecasting probability in the next future.

Keywords: Foreshock, Classification, Sequence, Cycle, Pattern, Trigger, Earthquake

1. INTRODUCTION

The analysis of the seismic sequence structures underlines (points out) that magnitude values do not unpredictably change, but follow some trends that can be identified in a chart.

Starting from this principle, we can speculate that some behaviors cyclically recur, making it possible, if identified on time, for the formulation of hypothesis about the future evolution of the seismic sequence.

The checks we have performed on several big earthquakes occurred in the last 40 years, have highlighted that it is possible to identify some earthquakes triggering patterns in the short and medium term, whose space-time complexity is strictly linked to the following items: a) the seismic region heterogeneity [1] ; b) the stress state inside and around the source of a big earthquake [2] [3] ; c) the long-range interactions among different seismic and heterogenic regions.

The identification, in advance, of the warning pattern within the seismic sequence, provides several useful insights to understand what is happening in the source region area, which can be used to improve earthquakes predictions.

Below, we will explain the known seismic sequences features, which include warning patterns we have classified and discussed in relation to the magnitude values' trend.

We performed the analysis of the seismic sequence using: a) spatial coverage of datasets ranging from portions of a region to several countries (from a few tens of kilometers to thousands of kilometers); b) magnitude range of 2.0 - 10 M; c) depths range of 1 - 50 km.

2. METHOD OF ANALYSIS

We analyzed and classified the time series of magnitude values using the results of some researches carried out in the field of financial markets studies.

The analysis provides some characteristics of the values' trend in a time series which are useful for assessing whether the phenomenon occurs according to the predicted scheme.

In particular, for the study of the movements of the seismic cycle we used the Elliott's wave theory [4] - [6] , where the concept of pattern is one of the most significant aspects.

The methodological approach adopted for the classification of earthquake triggering patterns and the assessment of the main earthquake magnitude is based on the technical analysis [7] [8] , which consists in a methodology for the evaluation of financial markets used to identify trends and assess future trends possibility. Similarly, it may be used as a safe tool to analyze the most significant magnitude values' oscillations and to predict likely future events, mainly through graphical methods and statistics.

3. STRUCTURE OF SEISMIC SEQUENCE

3.1. Types of Seismıc Sequences

Earthquakes are generally part of a sequence with variable and sharp features. Mogi [9] divided these sequences in three types: 1) mainshock-aftershock (Figure 1(a)); 2) foreshock-mainshock-aftershock (Figure 1(b)); 3) swarm (Figure 1(c)). The foreshocks and aftershocks sequences are strictly associated to a bigger event called mainshock, while the swarm is a sequence of earthquakes and is not associated with a dominant earthquake. The subsequent events of mainshock-aftershock sequences of similar dimension (Figure 1(d)) are defined as swarm, but they clearly differ from an ordinary swarm (Figure 1(c)).

Moreover, numerical results [9] show that different homogeneity index will produce different patterns of seismic sequences. We have three types of sequences: a) swarm; b) foreshock-mainshock-aftershock; c) mainshock-aftershock that happen specifically in these cases: a) extremely heterogeneous; b) mildly heterogeneous and c) homogeneous [1] .

Isolated earthquakes occur in presence of a great pre-stress and/or low stress at break. Foreshocks-main- shock-aftershock seismic sequences happen in presence of a mild pre-stress and stress at break. Swarms happen in case of low pre-stress and/or high resistance to break. Very low values of pre-stress and/or very high break strength are the best conditions for the aseismic creep event [10] to occur. Other studies underline that in areas where the time variation of the strain-rate (measured by geological surveys GPS) is low, higher magnitude earthquakes are expected compared with areas with high strain-rate [11] .

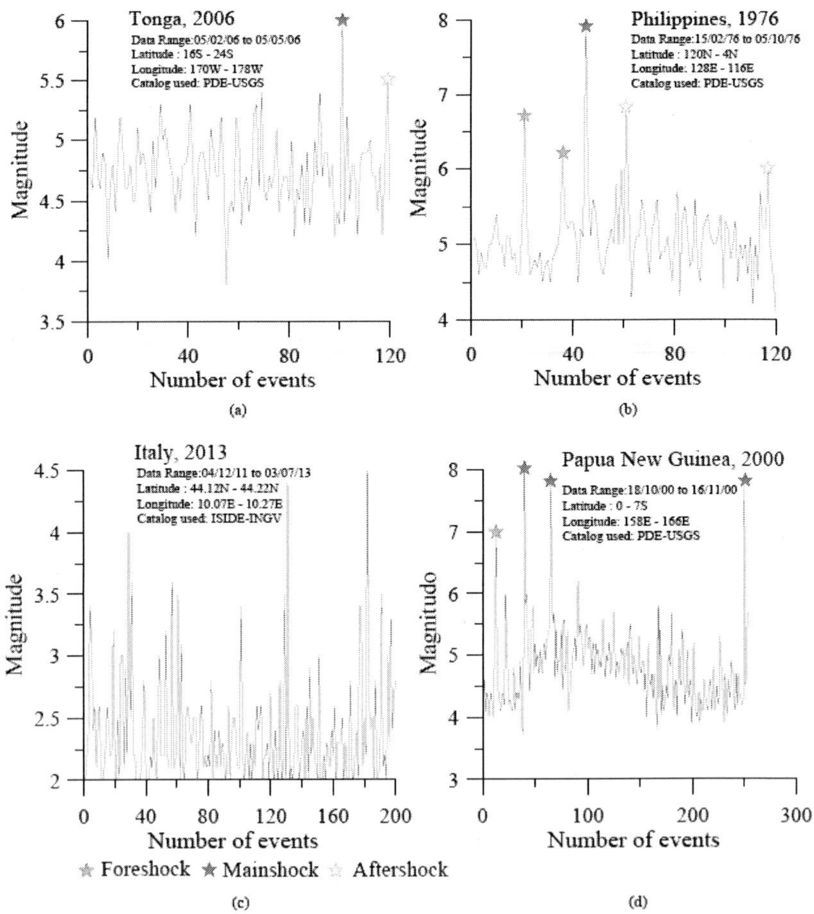

Figure 1. Examples of earthquake sequences: (a) mainshock-aftershock sequence; (b) foreshock-mainshock-after- shock sequence; (c) earthquake swarm; (d) subsequent occurrence of mainshock-aftershock sequence.

Simulations [10] suggest that an isolated earthquake can be interpreted as the extension of a break in a non-uniform field, while a foreshocks-mainshock-aftershocks sequence results from a break extension to a non-uniform field having modest fluctuations and swarms are a consequence of a break extension to a very inhomogeneous sector.

Swarms are a sequence of earthquakes that often start and finish gradually, where there is no evidence of any earthquake shock with higher magnitude [12] . However, shocks that make up swarms are actually included in seismic cycles in which the main shock, usually of mild magnitude, occurs always at the end of the nucleation process.

Swarms are divided in two typologies. The first one is the ordinary swarm in which the activity changes fairly irregularly (Figure 1(c)), and has a duration between some hours and little more than a year. Most of the swarms happening in volcanic zones show this model, which reflects the magmatic activity in depth. The second typology features a succession of main shocks with similar magnitude (Figure 1(d)) whose duration is between several days and months.

3.2. Seismic Cycle's Basic Structure

The long-term seismic sequence consists of several cycles that are repeated according to the representation scale and whose duration varies from a few minutes to several years.

The basic seismic cycle is a collection of processes that develop before, during and after an earthquake, which are highlighted by the variations in the magnitude values that develop in short and medium term time windows. The term "cycle" suggests that big earthquakes can be either periodic or regular events over time. The analysis of the succession of the seismic cycles included in a seismic sequence shows that the periodicity can be prevalent among big earthquakes recorded in areas even with different extension.

According to the magnitude values fluctuations, seismic cycles can be divided in four stages: 1) trigger; 2) energy release; 3) breaking; 4) adjustment (Figure 2).

Trigger stage. In this phase, the magnitude values vary within a very close range in which a small shock, after some fluctuations, triggers the next one.

Energy release stage. This step follows the trigger phase, where the magnitude values fluctuations produce increasing maximum values along with an increase in the number of shocks. In type 1 sequences, (mainshock-aftershock) these characteristics are often not found.

Breaking stage. This is the energy release phase's top. It corresponds to the main breaking point i.e. when the mainshock occurs, generally marked by the highest number of recorded shocks. If no additional main breaking (multiple events) follows, the magnitude values fall quickly under the peak magnitude value recorded.

Adjustment stage. The magnitude values and the number of shocks decrease, until it gets back to the trigger phase once again.

The transition from the energy release phase to the adjustment one can be confirmed by the values decrease before and after the mainshock, called "transition line" (Figure 3(a)).

We should bear in mind, however, that usually after being dropped below the "transition line" (point 3), the magnitude values temporarily return to the same line level or even slightly above, before definitely starting to decrease up to the next minimum value.

In case of "transition line" crossing during the increase phase, a second mainshock may occur (subsequent occurrence of mainshock-aftershock sequences).

In the case of minimum "1" and "2" descending values, the step from energy release to the adjustment one is always confirmed by the minimum "2" value formation (Figure 3(b)).

3.3. The Seismic Cycle Movements

In many long-term seismic sequences, the analysis of magnitude values trends during the energy release phase shows a cyclical fluctuation, which is characterized by one or more waves (foreshocks-mainshock). In the adjustment phase, we observe several aftershocks that partially rebalance all the impulsive movement in the previous step.

Based on Elliot's theory [4] - [6] , a seismic sequence can be considered as a succession of cyclic variations of magnitude values whose duration may vary over time (Figure 4 and Table 1), where the most energetic seismic events in the energy release phase occur according to a pattern consisting of several waves. Most frequently the pattern consists of five waves (1, 2, 3, 4, 5) whose values in 1-3-5 show an increasing magnitude trend (foreshocks-mainshock), while in the adjustment stage it consists of five waves (a, b, c, d, e) whose magnitude values in "b" and "d" are increasing compared to the values in "a" and "c" (significant aftershocks).

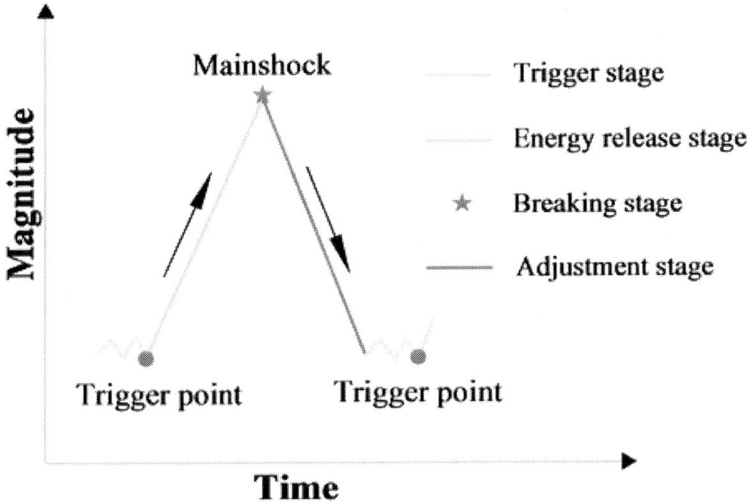

Figure 2. Schematic representation of the seismic cycle's four stages.

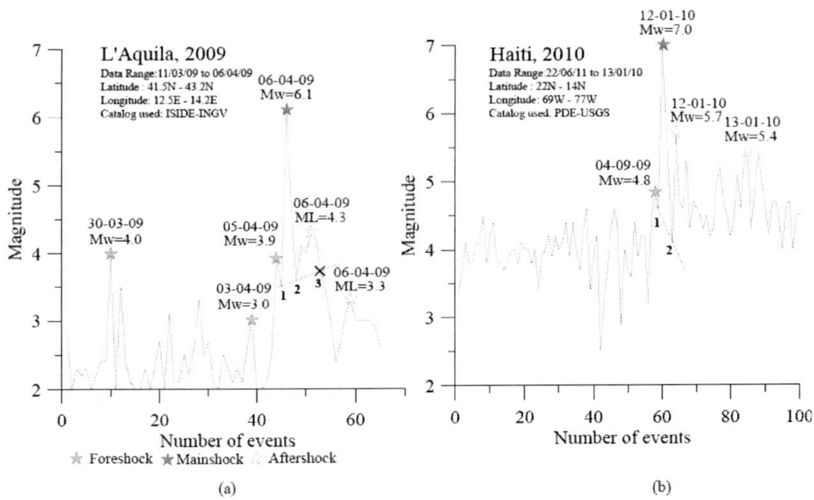

Figure 3. Sequence of earthquake imminent period: (a) L'Aquila earthquake, 06-04-2009 (Italy); (b) Haiti earthquake, 12-01-2010.

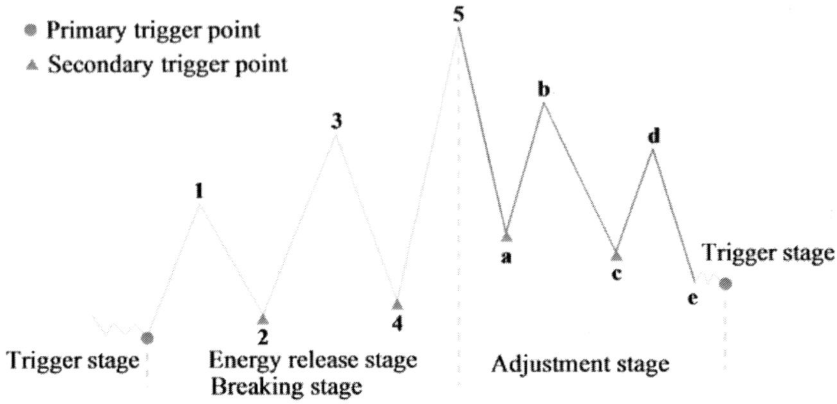

Figure 4. Schematic representation the seismic cycle's waves.

Table 1. Seismic cycle-energy release and adjustment phases' waves.

Point	Waves	Description
1	Energy release	Formation of a first foreshock
2	Adjustment	Formation of a relative minimum
3	Energy release	Formation of a second foreshock
4	Adjustment	Formation of a relative minimum
5	Energy release	Formation of a mainshock
a	Adjustment	Formation of a relative minimum
b	Energy release	Formation of a first aftershock
c	Adjustment	Formation of a relative minimum
d	Energy release	Formation of a second aftershock
e	Adjustment	Formation of a relative minimum

The energy release stage starts after the formation of the primary trigger point, while the transition from a decreasing movement to an increasing one and vice versa is marked by a secondary trigger point.

Table 1 describes the fluctuations of the magnitude values over time and reports the description of the maximum values that are formed at the end of each wave is given as well.

The basic chart is characterized by a series of several seismic cycles, showing a nonlinear structure with an up and down progression (Figure 5) in which the different phases of the seismic process before a big earthquake are clearly visible.

Typically, a big earthquake may be preceded by two types of energy release phase: 1) progressive earthquakes and 2) flash earthquakes.

In the progressive earthquakes-type energy release phase, a big earthquake is preceded by one or more foreshocks of various kinds, while in the flash earthquakes type, a big earthquake is not preceded by any foreshock at all.

The adjustment phase consists of two or more primary aftershocks with a lower magnitude compared to the mainshock. They usually occur days, months, sometimes even years after the main event to rebalance the elastic deformation caused by the mainshock.

Figure 6(a) shows the chart of the magnitude values recorded in Japan from 1973 to 2015. Here, we can observe two similar medium-term seismic cycles. The first consists of three energy release phases and one adjustment phase. The first ended on May 8th, 1974 with a 6.7 Mw foreshock, the second on June 12th, 1978 with a 7.7 Mw foreshock and the third on May 26th, 1983 with the 7.8 Mw mainshock, followed by the adjustment phase which ended on November 11th, 1989 with the aftershock.

The second cycle, not completed yet, consists of the energy release phases, represented by the foreshocks occurred on 28 December 1984 (7.8 Mw) and the 25 September 2003 (8.3 Mw), while the last one, with the high energy earthquake occurred on 11 March 2011 which has not been followed by a significant aftershock yet.

Figure 6(b) shows the chart of the magnitude values recorded in Greece in 2008: here, we can observe a traditional cycle consisting of one energy release phase with two short-term foreshocks, the mainshock and one adjustment phase with two main aftershocks.

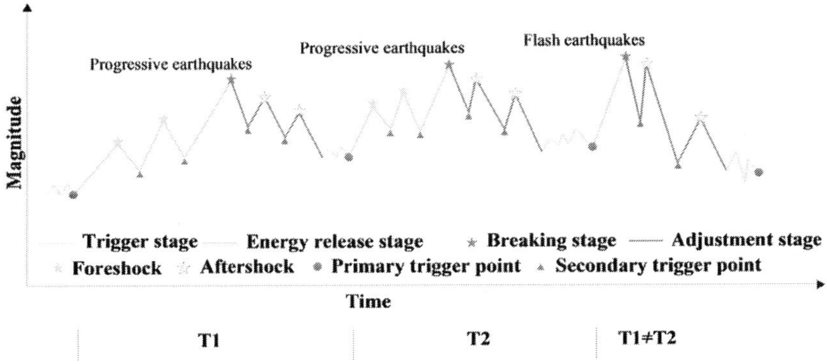

Figure 5. Schematic representation of the seismics cycle. The green, red and yellow-colored stars indicate the foreshocks, the mainshocks and the aftershocks, respectively. The red-colored circle indicates the energy release phase's trigger point.

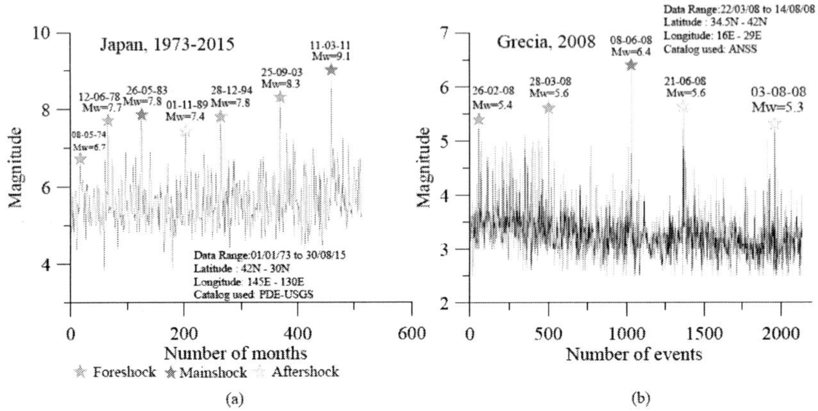

Figure 6. Examples of seismic cycles: (a) long-term sequence; (b) short-term sequence.

3.4. Progressive Earthquakes Type

Typically, the energy release phase that precedes a large earthquake consists of a series of warning shocks of different order (foreshocks) featuring an increasing magnitude characterizing the breaking process that ends with a big earthquake.

It begins to develop at the end of an adjustment phase with an opposite motion compared to the preceding one.

Based on the magnitude values' fluctuations that characterize the energy release phase, it becomes clear that the greater the amplitude between two foreshocks, the greater the magnitude of the mainshock expected.

The form varies according to the temporal position and the foreshocks' magnitude values that precede the mainshock.

Depending on the foreshocks position, the pattern may consist of:

a) One or more first-order or medium-short-term foreshocks and manshock (Figure 7(a));

b) One second-order or impending term foreshock and mainshock (Figure 7(b));

c) One or more first-order or medium-short-term foreshocks, one second order or impending-term foreshock and mainshock (Figure 7(c));

d) One or more various order foreshocks and several mainshocks (Figure 7(d)).

The adjustment phase consists of two or more main aftershocks with a lower magnitude compared to the mainshock. Usually the aftershocks occur days, months, sometimes even years after the main event to rebalance the elastic deformation the mainshock caused.

Figure 8(a) shows the pattern that anticipated the earthquake occurred in Nicaragua on September 2nd, 1992. This pattern consists of two first-order foreshocks and one second-order foreshock (M 5.8) that occurred shortly before the mainshock on 2 September 1992 in the proximity of its hypocenter.

The second-order foreshock recorded on 10 August 1992 with 5.8 M, most likely indicates the beginning of the breaking phase that ended with the mainshock.

Figure 8(b) shows the pattern of the earthquake occurred in Nicaragua on September 2nd, 1992, which was processed by using a seismic area extended to the state of Guatemala. We can note how the pattern retains the same foreshocks, mainshocks and aftershocks distribution of Figure 13(a), but includes seismic events that occurred in different times and different areas.

Figure 9(a) shows the pattern that preceded the earthquake that took place in Vanuatu in 2009. Usually, only one mainshock occurs, but here the strongest earthquake was preceded on the same day (October 7th, 2009) by a high magnitude event classifiable as mainshock.

The second-order foreshock on 10 August 2009, with 6.6 Mw magnitude, marks the beginning of the breaking phase which ended with the three strong mainshocks.

Figure 9(b) shows the pattern which preceded the earthquake in Loma Prieta (California, USA). We can see the two first-order foreshocks that preceded the mainshock on 18 October 1989 and the two main aftershocks that followed it.

3.5. Flash-Type Earthquake

This type of energy release phase is quite uncommon. Here, the big earthquake develops suddenly (with no foreshocks) after a long adjustment phase following the breaking stage and consists of aftershocks whose magnitude slowly decreases over time.

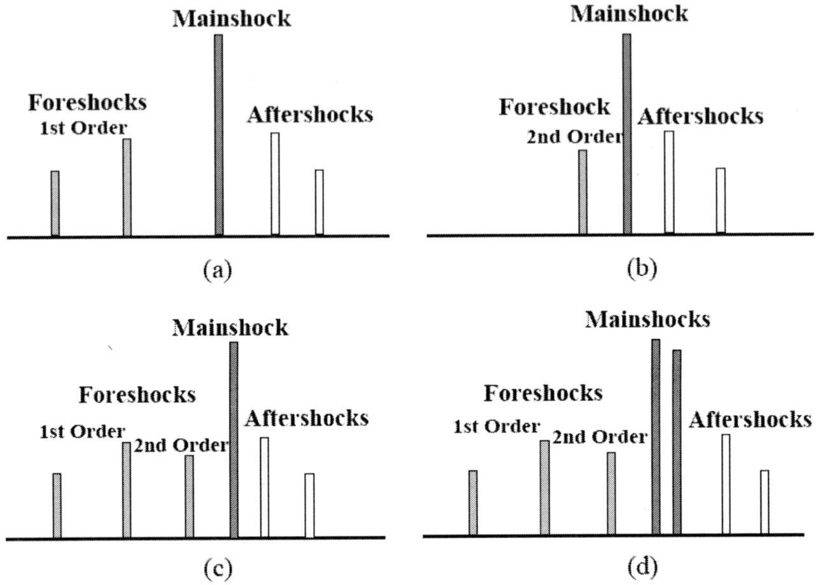

Figure 7. Schematic representation of progressive earthquakes patterns.

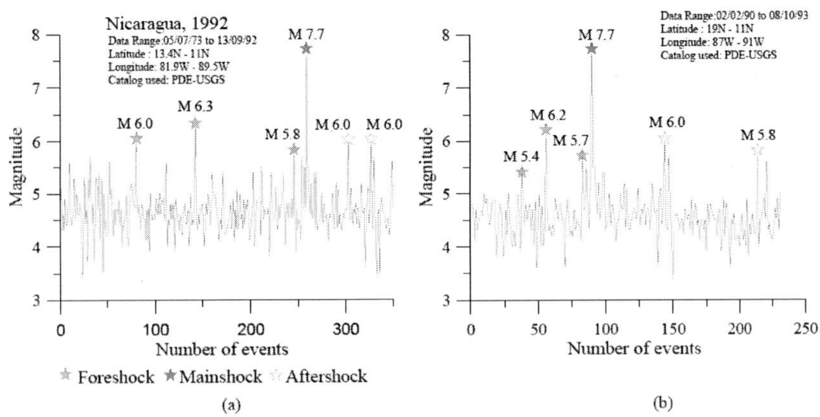

Figure 8. Progressive earthquakes patterns of Nicaragua earthquake, 1992: (a) restricted area; (b) wide area.

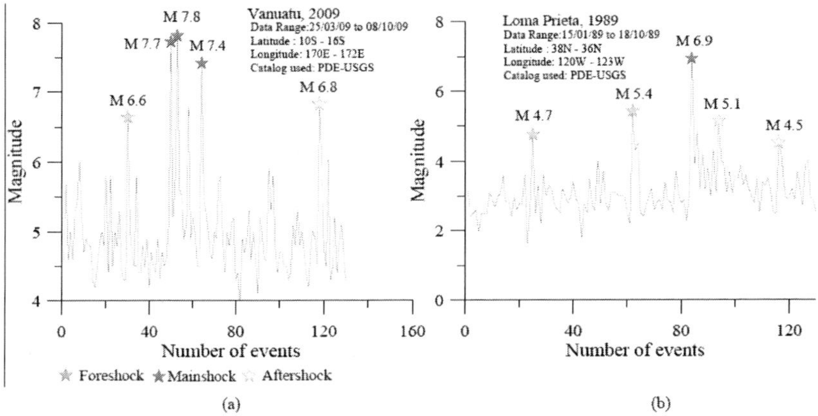

Figure 9. Progressive earthquakes patterns: (a) Vanuatu earthquake, 2009; (b) Loma Prieta earthquake (USA).

The magnitude of the expected mainshock depends on the adjustment stage time amplitude.

Based on the adjustment stage duration, it is clear that the more extended over time, the stronger the mainshock.

Depending on the aftershocks magnitude the pattern may consist of:

a) One or more significant aftershocks with decreasing magnitude (Figure 10(a));

b) One or more significant aftershocks with a similar magnitude (Figure 10(b)).

Figure 11(a) shows the pattern that preceded the earthquake in Taiwan on 20 September 1999. The maishock was followed by a first main aftershock whose magnitude was slightly lower than the mainshock's, while Figure 11(b) shows the pattern that preceded the earthquake in Tonga on 3 May 2006. Here we observe how a high magnitude mainshock was preceded by a sequence, whose magnitude values are consistent with the rest of the sequence in the post-mainshock period.

4. EARTHQUAKE-TRIGGERING PATTERNS

The sequence of aftershocks that develops after a mainshock or a foreshock event may have a significant impact on the next earthquake triggering pattern that anticipates another mainshock formation.

The effect is more apparent in those with the longest duration that consist of aftershocks associated with a large earthquake. Some earthquakes triggering patterns demonstrate how a sequence of earthquakes can trigger other earthquakes or how small earthquakes significantly contribute to the triggering of a big earthquake.

Statistical studies performed on seismic sequences worldwide showed that pre-earthquake seismic patterns, in relation to the evolution usually occurring over time, can be broadly divided in two types:

- -Flash earthquakes patterns;
- -Progressive earthquakes patterns.

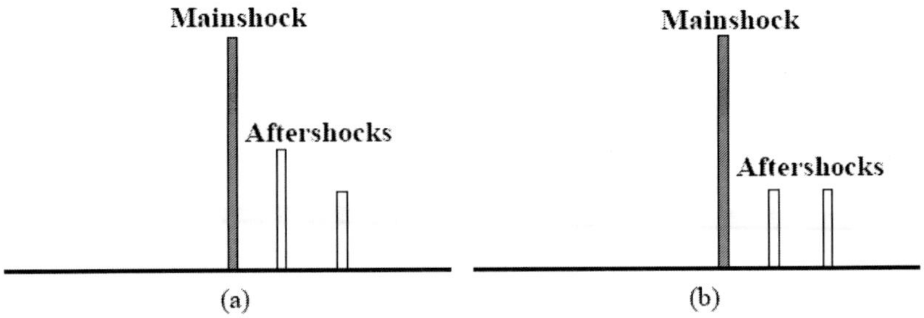

Figure 10. Flash type earthquakes' pattern illustration.

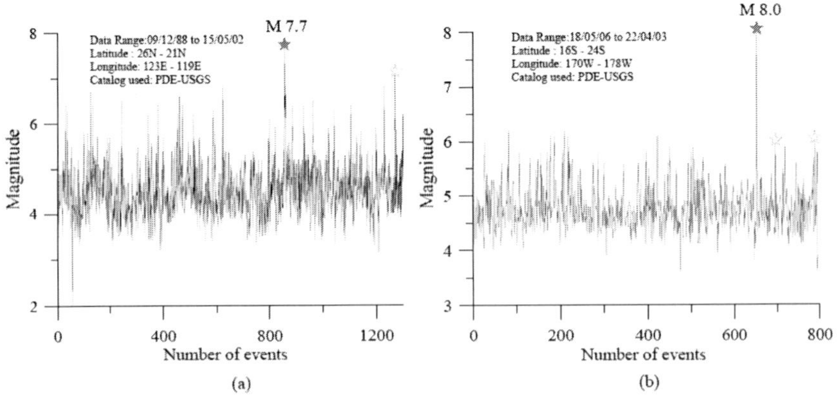

Figure 11. Examples of flash earthquales types: (a) Taiwan earthquake, 1999; (b) Tonga earthquake, 2006.

In flash earthquakes patterns, the magnitude values vary within two trend lines - inclined or horizontal - until the pattern reaches a critical state (breaking conditions) imposed by the rocks' strength properties and by the fault surface, where the small shock triggers the energy release phase that occurs all of sudden.

The upper trend line crossing determines the beginning of the breaking stage (breaking stage).

A small scale decrease in the crustal heterogeneity along with an increase in its depth, suggests that an increase in the normal stress inhibits the foreshocks occurrence [13] .

Usually, the mainshock is followed by aftershocks that are smaller compared to the mainshock, although occasionally an aftershock can be strong enough to assume the appearance of a subsequent mainshock.

Overall, the adjustment stage appears to be extended due to the effect of a large number of events that occur after the mainshock, but first and foremost it is characterized by events with decreasing magnitude.

In particular, the number of shocks recorded from the trigger point (first minimum value) up to the mainshock is 50% smaller compared to the total number of shocks recorded up to the second minimum value formed after the first significant aftershock (Figure 12(a)).

As to the chart, the final minimum value will be placed on a lower level compared to the trigger point: this makes the pattern take an apparently sinusoidal shape with maximum decreasing values (mainshock and aftershocks).

The time this pattern needs to complete varies from a few weeks up to a number of years.

In progressive earthquakes patterns, the magnitude values oscillate within two trend lines which diverge until the main shock.

The mainshock, therefore, is preceded by foreshocks that may occur at different times both inside and outside of the source region.

The number of shocks usually used from the trigger point up to the mainshock is greater than 50% of the total number of registered shocks to reach the second minimum value that is formed after the mainshock (Figure 12(b)).

This means that, until the mainshock, the energy release phase usually is more extended than the adjustment one due to a higher number of events with increasing magnitude.

As regards the chart, the second minimum value will be higher than the Trigger point's: this makes the cycle take the apparently classical sinusoidal shape with increasing maximum and minimum values that tend to return to the initial level.

Earthquake Triggering Patterns Classification

The proposed classification is the result of a thorough graphical analysis of seismic sequences, which is the basis for the first part of the short-term, deterministic forecasting model currently in the testing phase.

The characteristics taken into consideration in the classification process are: a) the fluctuations of the magnitude minimum and maximum values; b) the foreshocks observation before the mainshock.

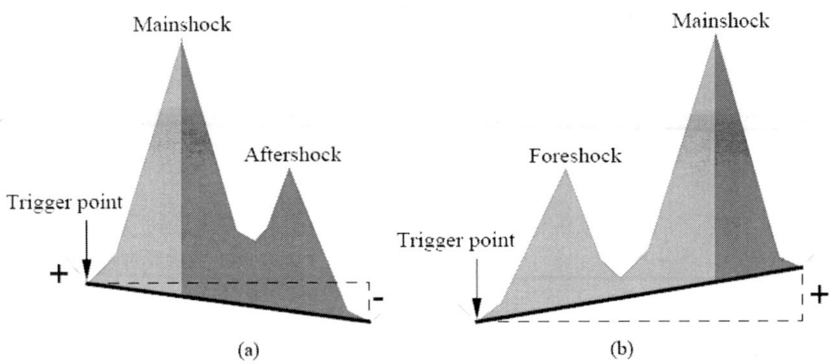

Figure 12. Schematic representation of pre-earthquake seismic types (a) flash earthquakes patterns; (b) progressive earthquakes parterns.

The patterns identified and classified using the technical graphic analysis [7] [8] , reflect the energy release and breaking phases and attempt to provide useful information to create connections between the pattern and the preparation process for a big earthquake.

(A) Symmetrical triangle pattern

The preparation process for a big earthquake starts with a first medium-strong shock, a short-medium term foreshock, followed by minor aftershocks, which are increasingly weaker and less frequent, included in a phase in which the deformation of the rock is probably minimal.

Observing this diagram in details (Figure 13), we note how the magnitude values oscillate between two trend lines that define maximum decreasing and minimum increasing values, which converge towards a triangle intersection point (Apex). The fluctuations' amplitude, which is formed inside the triangle, gradually decreases and the earthquake occurs when the magnitude values are nearing the apex. Therefore, the more extended over time the magnitude and amplitude values fluctuations, the stronger the earthquake associated with the pattern.

Usually, the critical phase begins when the difference in magnitude between points 4 and 5 is 50% less compared to the one between points 2 and 3 (triangle base). The crossing of the upper trend line currently occurs between 50% and 75% of its length.

The energy release phase may consist of the mainshock preceded by an impending term foreshock (progressive earthquakes), or by the mainshock preceded by no foreshock (flash earthquakes).

The pattern is considered as an "energy adjustment or accumulation", because the magnitude maximum values and the number of shocks decrease during the development, while increase upon the pattern breaking.

The symmetrical triangle pattern is typical of areas with little fractured, homogeneous rocks and it is most likely associated with planes of fault featuring little ruggedness.

The magnitude target (TM) of this pattern, once completed, is obtained by projecting from the breaking point (point 5), the width of the triangle base.

(B) Falling wedge pattern

The wedge is a pattern very similar to the triangle's, being the magnitude values limited by two converging trend lines, but differs from it because both sides have a different downward inclination and often the magnitude values remain locked up in the pattern up to the apex (Figure 14).

The upper trend line crossing usually occurs between 50% and 75% of the base length.

Usually, the energy release phase consists of the mainshock preceded by an impending term foreschock (progressive earthquakes), or by the mainshock preceded by no foreshock (flash earthquakes).

Just like in the triangle pattern, the number of shock decreases during its formation.

The TM of this pattern is calculated by projecting the width of the wedge base from the breaking point (point 3).

(C) Rectangle pattern

The rectangular pattern (Figure 15 and Figure 16) is one of the easiest patterns to identify because it is characterized by fluctuations in magnitude values within a more or less extensive strip. Most likely, it represents a stage of energy accumulation before the medium-strong event.

This pattern does not allow us to early identify the exact time when the upper trend line will be crossed.

The most frequent energy release phase consists of the mainshock preceded by no foreshock (flash earthquakes).

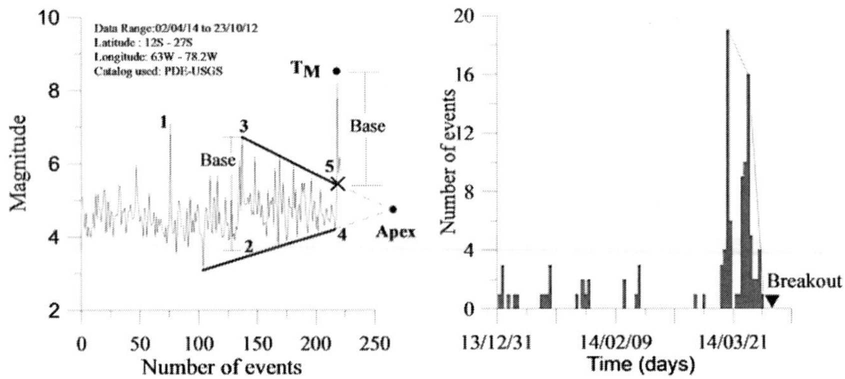

Figure 13. Symmetrical triangle pattern (Chile earthquake, 2014).

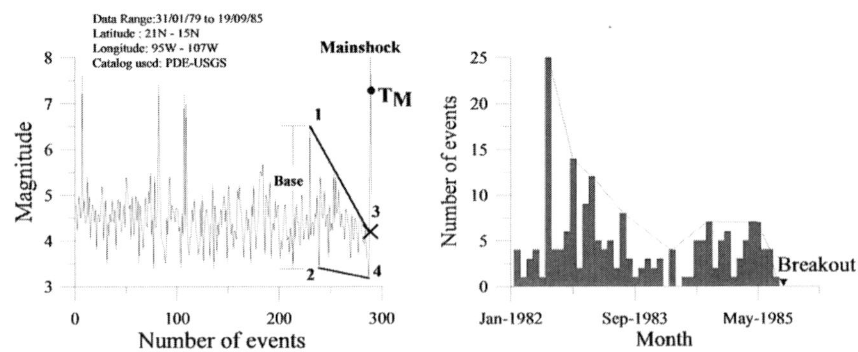

Figure 14. Falling wedge pattern (Mexico earthquake, 1985).

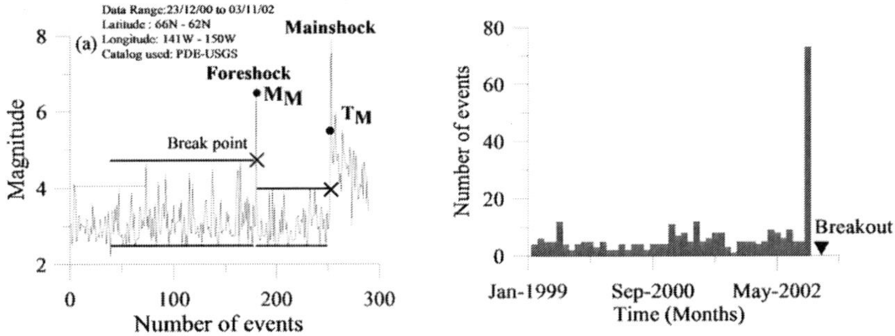

Figure 15.Horizontal rectangle pattern (Denali earthquake , 2002).

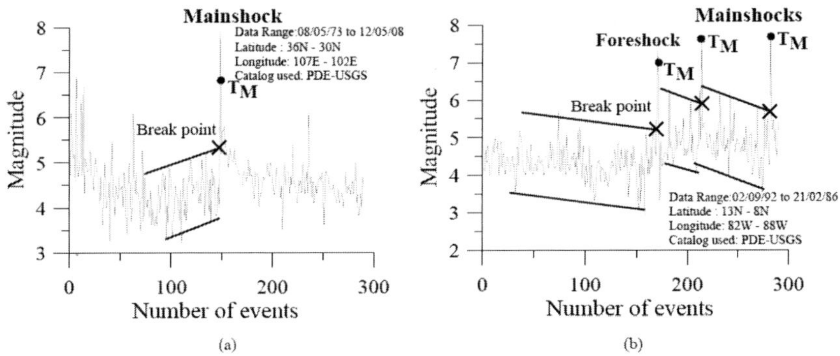

Figure 16. Rectangle Pattern: (a) ascending (China earthquake, 2008); (b) descending (Costa Rica-Nicaragua earth- quakes, 1990-1992).

Sometimes it is characterized by an increase in the number of shocks before the upper trend line crossing, as it happened, for example, during the earthquake in Denali (Alaska) coinciding with the foreshock on 23 October 2002 that preceded the earthquake on November 3rd, 2002 whose magnitude amounted to 7.9 Mw (Figure 15).

It includes three types: a) horizontal, b) ascending, c) descending, according to the arrangement of the minimum and the maximum values that are formed.

The TM of this pattern is calculated by projecting the rectangle height from the breaking point.

(D) Expansion triangle pattern

This is an uncommon pattern consisting of a series of oscillations of the magnitude minimum and maximum values that expand over time (Figure

17). The completion of the model does not coincide with the upper trend line crossing.

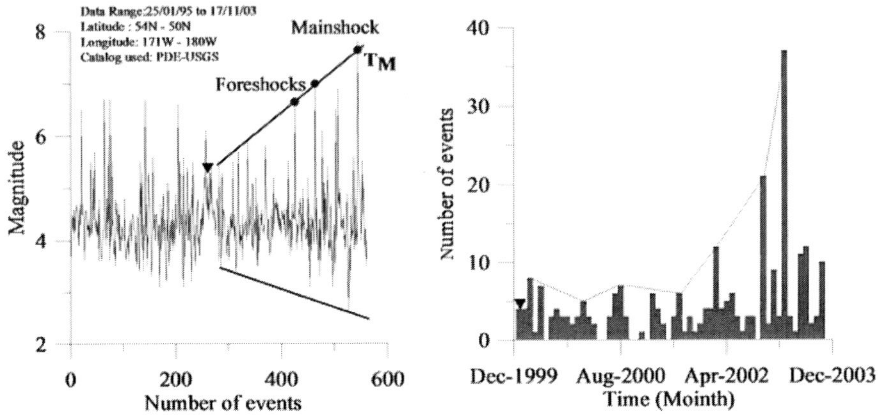

Figure 17. Expansion triangle pattern (Aleutian Islands Earthquake of 17 November 2003).

This pattern can be interpreted as alternate small cycles of energy accumulation and gradual release that continue until they reaches the critical stress level and, therefore, the mainshock.

The number of shock decreases during the first part of the pattern development and increases when it gets close to the breaking phase.

Pattern's special features considered, it is possible to calculate the TM at the intersection point of the line joining the maximum points with the vertical line passing from the last recorded event plus one (n + 1).

5. CORRELATION BETWEEN THE SEISMIC PATTERN AND THE FOCAL MECHANISM

The seismic patterns observed in many sequences can be correlated with the major earthquakes' focal mechanisms [14] . Table 2 shows some patterns that we could identify within the short-medium term time windows. As we see, the flash earthquakes patterns are generally associated with reverse or strike-slip faults, while progressive earthquakes patterns are related to normal faults.

Table 2. Earthquake pattern type and focal mechanism correlations.

Earthquake	Date	Earthquake triggering patterns	Focal mechanism
L'aquila-Italy	06/04/2009	Progressive earthquakes pattern	Normal
Colfiorito-Italy	26/09/1997	Progressive earthquakes pattern	Normal
Irpinia-Italy	23/11/1980	Progressive earthquakes pattern	Normal
Emilia-Italy	20/05/2012	Progressive earthquakes pattern	Thrust
Loma Prieta-CA, USA	19/10/1989	Progressive earthquakes pattern	Transpression
Nicaragua	02/09/1992	Progressive earthquakes pattern	Thrust
Vanuatu	07/10/2009	Progressive earthquakes pattern	Thrust
Aleutian Islands	17/11/2003	Progressive earthquakes pattern	Thrust
Haiti	12/01/2010	Progressive earthquakes pattern	Srike-slip
Turkey	23/10/2011	Progressive earthquakes pattern	Reverse strike-slip
Taiwan	14/11/1986	Progressive earthquakes pattern	Reverse
Chile	01/04/2014	Progressive earthquakes pattern	Low-angle reverse strike-slip
Japan	11/03/2011	Progressive earthquakes pattern	Thrust
New Zealand	03/09/2010	Flash earthquakes pattern	Dextral strike-slip
Landers-CA, USA	28/06/1992	Flash earthquakes pattern	Srike-slip
China	14/04/2010	Flash earthquakes pattern	Srike-slip
Mexico	19/09/1985	Flash earthquakes pattern	Reverse
Pakistan	18/11/2011	Flash earthquakes pattern	Normal strike-slip
Pakistan	24/09/2013	Flash earthquakes pattern	Normal strike-slip
Nicaragua	02/09/1992	Flash earthquakes pattern	Reverse
Nicaragua	04/02/1976	Flash earthquakes pattern	Srike-slip
Uzbekistan	19/03/1984	Flash earthquakes pattern	Reverse
Uzbekistan	17/05/1976	Flash earthquakes pattern	Reverse
Japan	26/05/1983	Flash earthquakes pattern	Reverse
Alaska Gulf	30/11/1987	Flash earthquakes pattern	Srike-slip
Philippines	16/08/1976	Flash earthquakes pattern	Reverse
Australia	22/01/1988	Flash earthquakes pattern	Reverse
Colombia	15/11/2004	Flash earthquakes pattern	Reverse
Colombia	18/10/1992	Flash earthquakes pattern	Reverse strike-slip
Colombia	13/07/1974	Flash earthquakes pattern	Low-angle reverse
Indonesia	11/04/2012	Flash earthquakes pattern	Low-angle reverse
Indonesia	06/04/2010	Flash earthquakes pattern	Srike-slip
Sumatra	26/12/2004	Flash earthquakes pattern	Thrust
Nepal	25/04/2015	Flash earthquakes pattern	Reverse
Emilia-Italy	29/05/2012	Flash earthquakes pattern	Reverse
Friuli-Italy	06/05/1976	Flash earthquakes pattern	Reverse strike-slip
Denali (Alaska)	03/11/2002	Flash earthquakes pattern	Reverse strike-slip
Tonga	03/05/2006	Flash earthquakes pattern	Srike-slip (transform)
Russia	05/12/1997	Flash earthquakes pattern	Srike-slip (transform)
Australia	22/01/1988	Flash earthquakes pattern	Reverse
Uganda	05/12/2005	Flash earthquakes pattern	Transtensive
Taiwan	20/09/1999	Flash earthquakes pattern	Reverse

Reverse faults tend to feature a higher percentage of big earthquakes not preceded by foreshocks compared to normal faults, while we have an average percentage with the strike-slip ones.

Previous studies on sequences with foreshocks showed that it would be probably very difficult to observe foreshocks before a big earthquake associated with reverse faults because normal stress would increase significantly from reverse faulting to normal [15] .

Some patterns with foreshocks recorded in an environment with reverse faults seem to be associated with a greater degree of small-scale heterogeneity in crust [16] .

6. CONCLUSIONS

With this study, we wanted to define the earthquakes triggering patterns that develop in seismic sequences before an energy earthquake that contains a series of information including the seismic history of the area under investigation.

Following a thorough graphical analysis performed over several seismic sequences, we classified the earthquakes triggering patterns according to the fluctuations of the minimum and maximum magnitude values, which in some cases may be considered as true earthquakes precursors.

The statistical studies performed on seismic sequences worldwide demonstrated that the earthquakes triggering patterns, in relation to the evolution usually shown over time, tend to group into two types: 1) flash earthquakes patterns; 2) progressive earthquakes patterns.

By considering the focal mainshock mechanisms, we saw how the earthquakes triggering patterns, associated with reverse and strike-slip faults tend to have a higher percentage of big earthquakes of flash earthquakes patterns type compared to normal faults that develop earthquakes of progressive earthquakes patterns type.

Some patterns featuring foreshocks recorded in an environment with reverse faults seem to be associated with a greater degree of small scale crustal heterogeneity and maybe even to a reduced normal stress on the fault plane.

The aforementioned approach unquestionably has some advantages in terms of prediction. In fact, the early detection of the earthquakes triggering patterns, allows us not only to predict the likely future changes in magnitude values but also to graphically estimate the magnitude of the strongest event.

CITE THIS PAPER

Giulio Riga, Paolo Balocchi (2016) Seismic Sequence Structure and Earthquakes Triggering Patterns. *Open Journal of Earthquake Research*,**05**,20-34.

REFERENCES

1. Jiao, M., Tang, C., Zhang, G., Shi, Y. and Hou, W. (2003) A Numerical Test on Influence of Mesoscopic Heterogeneity on Macroscopic Behavior of Rock Failure and Seismic Sequance Types. Chinese Journal of Geophysics, 46, 943-953.
2. Matsu'ura, R.S. (1986) Precursory Quiescence and Recovery of Aftershock Activities before Some Large Aftershocks. Bulletin of the Earthquake Research Institute, 61, 1-65.
3. Matsu'ura, R.S. (1991) Case 1: Precursory Quiescence and Recovery of Aftershock Activity before Some Large Aftershocks. In: Wyss, M., Ed., Evaluation of Proposed Earthquake Precursors, American Geophysical Union, Washington DC, 8-11. http://dx.doi.org/10.1029/SP032p0008
4. Phetking, C., Sap, M.N.M. and Selamat, A. (2008) Time Series Representation for Elliott Wave Identification in Stock Market Analysis. http://comp.utm.my /pars/files/2013/04/Time-Series-Representation-for-Elliott-Wave-Identification-in-Stock-Market-Analysis.pdf
5. Frost, A.J. and Prechter, R.R. (1985) Elliott Wave Principle. New Classics Library, Georgia.
6. Prechter Jr., R.R. (1995-2004) The Basics of the Elliott Wave Principle. New Classics Library, Gainesville. http://www.yinyangforex.com/download/trading_ books_ and_documents/EW%20Basics%20Booklet.pdf
7. Bulkowski, T.N. (2002) Trading Classic Chart Patterns. John Wiley & Sons, Inc., Hoboken. http://fx-arabia.com/vb/uploaded/178_01275676445.pdf
8. Matras, K. (2010) Chart Patterns Trader Supplement Understanding and Trading Classic Chart Patterns. Zacks Investment Research, Inc. http://www.sju.edu/ int/academics/hsb/finance/wstr/pdf/CPGuide1.pdf
9. Mogi, K. (1963) Some Discusions on Aftershocks, Foreshocks and Earthquake Swarms: The Fracture of a Semi-Infinite Body Caused by Inner Stress Origin and Its Relation to the Earthquake Phenomena (3rd Paper). Bulletin of the Earthquake Research Institute, 41, 615-658.
10. Chen, Y.T. and Knopoff, L. (1987) Simulation of Earthquake Sequences. Geophysical Journal International, 91, 693-709.
11. Riguzzi, F., Crespi, M., Devoti, R., Doglioni, C., Pietrantonio, G. and Pisani, A.R. (2012) Geodetic Strain Rate and Eartquake Size: New Clues for Seismic Hazard Studies. Physics of the Earth and Planetary Interiors, 206-207, 67-75.
12. Scholz, C.H. (2002) The Mechanics of Earthquakes and Faulting. 2nd Edition, Cambridge University Press, Cambridge, UK.
13. Abercrombie, R.E. and Mori, J. (1996) Occurrence Patterns of Foreshocks to Large Earthquakes in the Western United States. Nature, 381, 303-307.
14. Global CMT Catalog Search. http://www.globalcmt.org/CMTsearch.html
15. Sibson, R.H. (1974) Frictional Constraints on Thrust, Wrench and Normal Faults. Nature, 249, 542-544.
16. Lin, C.-H. (2004) Repeated Foreshock Sequences in the Thrust Faulting Environment of Eastern Taiwan. Geophysical Research Letters, 31, L13601.

CHAPTER 12

Seismic Hazard Assessment of Western Coastal Province of Saudi Arabia: Deterministic Approach

Faisal Rehman[1],Sherif M. El-Hady[2,3] Ali H. Atef[2], Hussein M. Harbi[2]

[1.] *Department of Earth SciencesUniversity of SargodhaSargodhaPakistan*

[2.] *Geophysics DepartmentKing Abdulaziz UniversityJeddahSaudi Arabia*

[3.] *Earthquake DepartmentNational Research Institute of Astronomy and GeophysicsHelwanEgypt*

ABSTRACT

Seismic hazard assessment is carried out by utilizing deterministic approach to evaluate the maximum expected earthquake ground motions along the Western Coastal Province of Saudi Arabia. The analysis is accomplished by incorporating seismotectonic source model, determination of earthquake magnitude (M_{max}), set of appropriate ground motion predictive equations (GMPE), and logic tree sequence. The logic tree sequence is built up to assign weight to ground motion scaling relationships. Contour maps of ground acceleration are generated at different spectral periods. These maps show that the largest ground motion values are emerged in northern and southern regions of the western coastal province in Saudi Arabia in comparison with the central region.

Keywords: Earthquakes Seismic hazard Logic tree Peak ground acceleration

1 INTRODUCTION

A major reason for anxiety arises from the increase of seismic vulnerability of most urban structures especially in developing countries (Re 2000). The most important element in the seismic design of the structure and its vulnerability assessment against the earthquake damage is ground motion vibrations. These parameters are evaluated through probabilistic seismic hazard assessment (PSHA) or deterministic seismic hazard assessment (DSHA) approaches (Waseem et al. 2013). The DSHA depends on selected earthquake scenario, specific ground motion probability, and closest distance to the site of interest. However, PSHA deals with numerous scenarios for earthquake and ground motion probabilities (Lin and Baker 2011).

In DSHA investigation, a maximum credible earthquake is supposed to occur at the closest distance to the site of interest. Meanwhile, the occurrence of earthquake probability during the specific period of exposure is neglected (Tavakoli et al. 2013). It must be noted that there is no commonly accepted deterministic seismic hazard analysis approach that is applicable to all parts of the world as it is experienced in assorted ways in different parts of the world. The traditional DSHA methodology delineates the seismic source or sources that might affect the site of concern and then computes the maximum possible earthquake magnitude for each of these sources. By considering each of these maximum earthquakes placed at a location that put the earthquake at the minimum possible distance to the site, the ground motion is expected, mostly, applying an empirical attenuation relation.

Previous studies, about seismic hazard assessment along Western Coastal Province of Saudi Arabia are limited and focus only on PHSA [e.g., Al-Amri (2013) and Al-Arifi et al. (2013)]. Recently, there are few published researches related to DSHA that are applied to some local cities located within Western Coastal Province of Saudi Arabia [e.g. Almadani et al. (2015)]. Yet there is no publication which accounts for the mapping of DSHA that could cover whole Western Coastal regions of Saudi Arabia.

The deterministic approach is subject to the epistemic and aleatory uncertainties (Stepp et al. 2001). In this research, the epistemic uncertainties are treated by taking alternatives for the ground motion prediction equations, which in concern link several different assessments of the ground motion. The aleatory uncertainties are related to various input parameters used to describe the seismicity and the ground motion prediction equation.

In the present study, DSHA is carried out along the whole Western Coastal Province of Saudi Arabia (Fig. 1). The grid spacing for regional studies is usually 50 km × 50 km or half degree (latitude or longitude) e.g., (Deif et al. 2013; Mohamed et al. 2012). A grid of 50 by 50 km (about 155 points) is selected along the eastern coast of Red Sea. For each point, the DSHA is evaluated with associated uncertainty and adopted by a variable logic tree (essentially due to change in distance between points of the grid). In current study, the used methodology is proposed by Campbell (2005) and applied by Deif et al. (2013, 2009). This methodology deals with the uncertainty problem and provides ground motion at different percentile levels at various spectral periods. This provides great flexibility to the engineer to select the appropriate input ground motions for selected site.

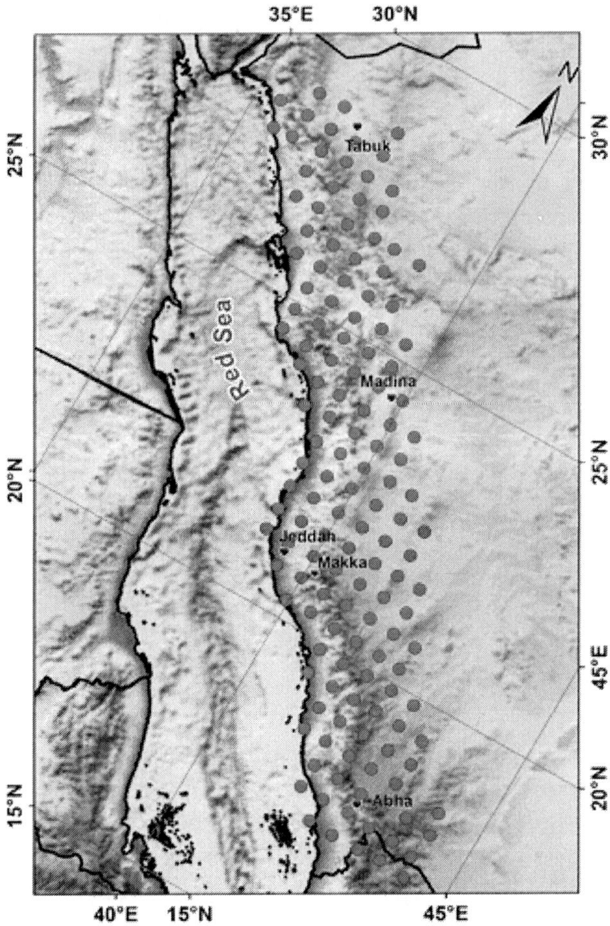

Figure 1. Location map for the hazard assessment grid point along the western coast of Saudi Arabia

2 CRUSTAL STRUCTURE AND REGIONAL SEISMICITY

Lithospheric structure and composition of the Arabian Plate are constructed using a broad choice of geophysical data (Stern and Johnson 2010). The Arabian Plate has a varying crustal thicknesses range. It varies from 22 to 53 km across the plate from west to east (Fnais et al. 2013). The variation is at modest amount in central Arabia ranging from 32 to 46 km in west to 35–50 km in east (Al-Damegh et al. 2005; Rodgers et al. 1999). The crustal thickness approaches 43 km at east of the Central Arabian Magnetic Anomaly (CAMA) and 38 km thick in western part of CAMA. There is a considerable crustal discontinuity in lateral direction beneath the CAMA, causing velocity increment around 0.2 km/s in northeast (Gettings et al. 1986).

The dominant mafic composition of the Arabian Shield is the reason for high crustal velocities in lower crust. The studies of entire shield by surface wave designated the Moho's depth to be at 41–46 km (Mokhtar 2004; Mokhtar et al. 2001). The crust composition, which is inferred from velocity structure, indicates lower mafic crust overlying by upper felsic crust, having low velocities (Stern and Johnson 2008).

The Arabian plate seismicity is clustered and controlled along its major tectonic borders, Red Sea rifting, Dead Sea transform fault, Zagros fold and thrust belt, Biltis thrust, and Makran subduction zone. Seismic activities in the vicinity of the study area are mainly controlled and narrowed along the Red Sea rifting, Gulf of Aqba, and Gulf of Aden (Fig. 2). The seismic activity is concentrated along Red Sea axial trough (Al-Malki and Al-Amri 2013).

3 METHODOLOGY

The estimation of ground motion parameters is carried out through seismic hazard assessment (SHA) at a site of interest for seismic design (Hashemi et al. 2013). SHA analysis depends on seismic activity and attenuation relations (Dowrick 2009). However, DSHA is restricted to a specific earthquake scenario.

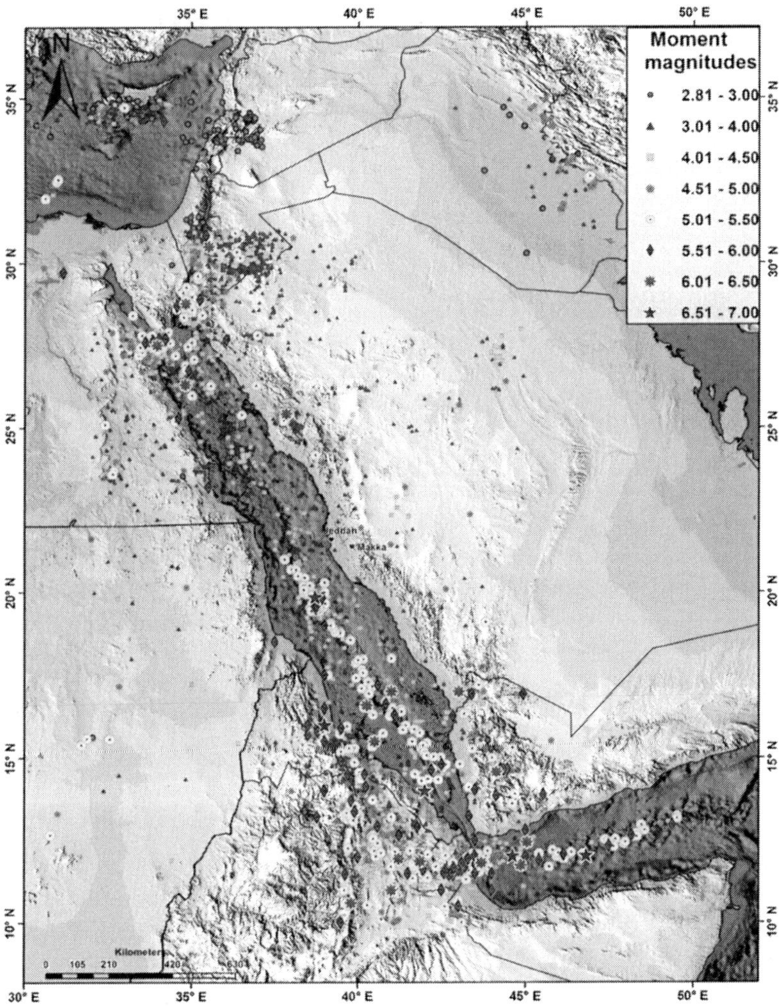

Figure 2. Instrumental seismicity map of the Arabian Shield and its surrounding for the period from 1900 to 2014 for magnitude 2.81 and above

The methodology for the current study followed Campbell (2005) procedure for the DSHA which is described in context to the study area as below:

1. *Identification of the seismogenic sources* having possible impact on Western Coastal Provinces of Saudi Arabia.

2. *Calculation of maximum earthquake* (M_{max}) for each seismic source zone using *earthquake catalogue and Gutenberg and Richter relationship*.
3. Selection of a specific set of earthquake magnitude and distance scenarios.
4. Selection of appropriate set of attenuation laws (*ground motion scaling relationships*).
5. Incorporation of uncertainties using *logic tree*.
6. Calculation of the median ground motion for each scenario.
7. Delineate the largest median value among calculated median ground motions.
8. Calculate fractile (percentile) largest median value.

In the current study, the logic tree is utilized to capture uncertainty. Thus the following structure of current methodology for DSHA is build up; identification of seismic sources, calculation of maximum credible earthquake magnitude M_{max}, selection of appropriate ground motion attenuation equations, and finally building logic tree to assign weight to GMPE.

4 IDENTIFICATION OF SEISMOGENIC SOURCES

The seismic source delineation is a major key parameter in hazard assessment (Vipin and Sitharam 2013). The seismogenic source characterizations follow historical and recent seismicity, seismicity pattern, seismogenic potential of active faults (Meletti et al. 2008). The seismogenic source model is identified and delineated by consideration from ancient times and current seismicity, tectonic and geological setting, crustal tomographic studies, crustal heat flow measurements, and current hazard studies (Al-Amri 2013; Al-Arifi et al. 2013; Al-Damegh et al. 2005; Al-Malki and Al-Amri 2013; Ares 2010; Burkhard and Grünthal 2009; Pailoplee et al. 2010). In current study, seismogenic source model by Rehman (2016) is utilized to carry out hazard assessment process (Fig. 3).

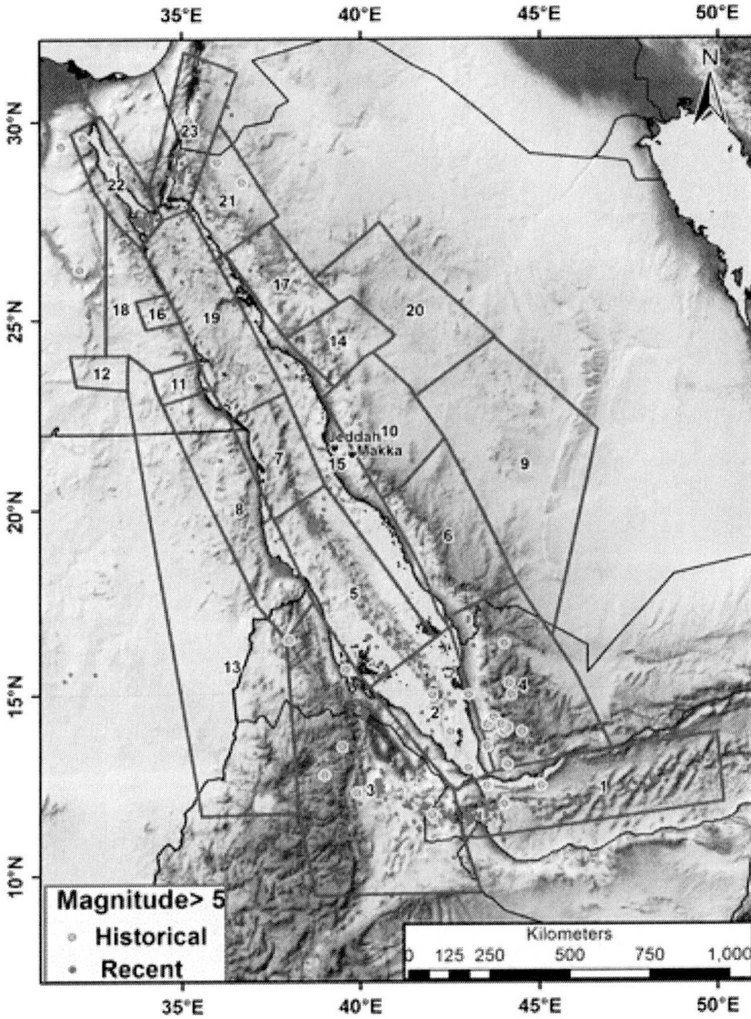

Figure 3. Seismogenic source model (Rehman 2016)

5 EARTHQUAKE CATALOGUE

One of the most important products of seismology is earthquake catalogue which provides a broad dataset in earthquake events. This can be used in various analyses associated with seismicity and seismotectonic, hazard assessment, and physics of earthquake. The hazard parameters are determined well if the catalogue has longer time coverage (Woessner and Wiemer 2005). A catalogue is a basic requirement for seismicity analysis in space-time volume, and in SHA (Gupta et al. 2012; Leonard et al. 2011).

In current study, the earthquake catalogue is compiled for spatial region extending from 15° to 35°N and 29° to 47°E and include events from magnitude 3 and above from 827 to 2013 AD. This catalogue is used to characterize seismicity of study area. It is compiled by combining information from different sources which include

- An earthquake catalogue provided by Dr. Abdullah M.S. Al-Amri (Personal Communication).
- The International Seismological Center online bulletin (http://earthquake.isc.ac.uk/).
- Preliminary determination of epicenters (PDE), online bulletin provided by the National Earthquake Information Center (NEIC) (http://earthquake.usgs.gov/earthquakes/).
- Incorporated research institutes for seismology online bulletin (http://www.iris.edu/SeismiQuery/sq-events.htm).
- European Mediterranean Seismological Center (EMSC) (http://www.emsc-csem.org/Earthquake/).
- National Research Institute of Astronomy and Geophysics (NRIAG) bulletins (http://www.nriag.sci.eg/).

Poirier and Tahir (1980), Badawy (1999), Ambraseys et al. (2005a, b), Ambraseys et al. (1995, 2009), Guidoboni et al. (1994), and Guidoboni and Comastri (2005).

6 COMPLETENESS MAGNITUDE (M_C)

The completeness magnitude (M_c) is defined theoretically as the lowest magnitude at which 100 % of earthquakes are detected in space-time volume. The precise estimation of M_c is critical because higher values of M_c lead to under-sampling, and too low values are erroneous. Mainly catalogue-based and network-based techniques are applied for M_c estimation (Mignan and Woessner 2012). Magnitude of completeness is a basic requirement to model seismicity in an area. The maximum curvature technique is mainly used (Wiemer and Wyss 2000) for the completeness magnitude.

In current study, M_c for each seismogenic source zone is calculated using maximum curvature technique. The Fig. 4 is an example of M_c calculation for one zone.

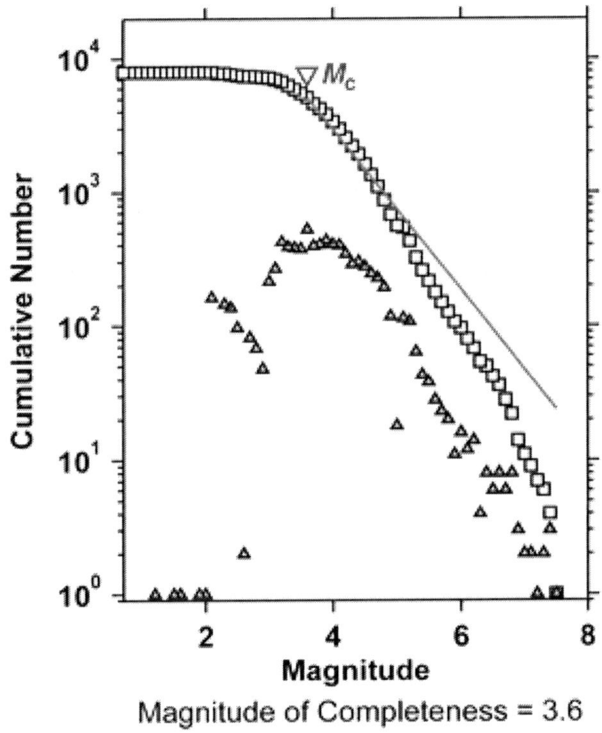

Figure 4 Maximum curvature technique for completeness magnitude determination

7 GUTENBERG AND RICHTER RELATIONSHIP (G-R)

The seismicity of the seismic source zone is described by the means of famous recurrence relationship termed as G-R relationship.

LogN=a−bM,LogN=a−bM,

N represents the earthquakes of specific magnitudes (*M*) or larger per year, *a* is activity rate and defines the intercept of the above equation at $M = 0$. The factor *b* is the slope which depicts the comparative proportion of small and large magnitudes.

The Gutenberg and Richter (1944) relationship introduces an impractical supposition in which the largest size possible earthquake in any zone being studied, is unrestrained and unconnected toward seismotectonic setting. Kijko and Sellevoll (1989, 1992) extended the Gutenberg-Richter equation from data that contain large historical events and recent observation with that of different quality and heterogeneity. In current study, the *b* values are calculated using Kijko and Sellevoll (1992) approach for each zone (Table 1).

Table 1.M_{max} and *b* values for each seismogenic source zone

Zone no.	M_{max}			*b* values	
	Kijko	Error	Obs	Kijko	Error
1	6.62	0.50	6.60	0.94	0.02
2	6.66	0.50	6.60	1.06	0.03
3	6.95	0.52	6.80	1.13	0.02
4	6.45	0.50	6.40	0.99	0.03
5	6.79	0.22	6.70	0.75	0.03
6	6.00	0.54	5.50	0 .94	0.07
7	5.47	0.26	5.30	1.01	0.05
8	6.30	0.54	5.80	0.85	0.07
9	5.30	0.54	4.80	0.98	0.07
10	5.53	0.50	5.00	0.99	0.07
11	4.72	0.23	4.60	1.11	0.06
12	5.35	0.32	5.10	0.97	0.05
13	4.99	0.53	4.90	0.99	0.07
14	5.84	0.52	5.70	0.98	0.07
15	4.40	0.22	4.30	1.04	0.07
16	4.86	0.26	4.70	1.13	0.06
17	6.20	0.54	5.70	1.00	0.06
18	5.19	0.35	4.90	1.11	0.06
19	5.87	0.26	5.70	1.27	0.03
20	5.52	0.50	5.00	1.06	0.07
21	7.32	0.59	7.00	1.18	0.04
22	6.69	0.51	6.60	1.19	0.03
23	7.33	0.50	7.20	1.13	0.03

8 MAXIMUM MAGNITUDE (M_{MAX})

Another most important parameter beside recurrence parameters is maximum expected magnitude (M_{max}). The SHA is strongly influenced by the choice of M_{max}. The M_{max} can be estimated using either deterministic or probabilistic approach. Deterministic approach comprises empirical regression relationships between the magnitude and various tectonic and fault rupture parameters, however the probabilistic approach involves extreme value statistics (Gupta 2002).

In current study, M_{max} for each seismogenic source zone is calculated using methodology proposed by Kijko (2004) based on observed seismicity. This method is designed for the calculation of the maximum earthquake magnitude, M_{max}, for a seismogenic source zone. This technique is based upon broad equation for the estimation of M_{max}, which is very flexible. This equation is applicable to the following three different scenarios:

- Distribution of earthquake magnitude follows doubly truncated Gutenberg-Richter relation.
- Moderate deviations occur from the Gutenberg-Richter relation for empirical magnitude distribution.
- There is no specific form of magnitude distribution assumed.

The first two solutions of the generic equation are parametric and the third solution non-parametric. The M_{max} values calculated for each seismogenic source zone are described in Table 1.

9 GROUND MOTION SCALING RELATIONSHIPS

The ground motion scaling relationships are functions of seismological parameters which include earthquake magnitude, source to site distance, and site conditions (Atkinson and Boore 1995; Peruš and Fajfar 2009; Yazdani and Kowsari 2013). Joyner and Boore (1981) postulated functional form to derive these empirical relationships. The geometrical spreading for all distances is accounted in this functional form (Ambraseys et al. 2005a; Ambraseys et al. 1996; Boore et al. 1997; Sabetta and Pugliese 1987; Smith et al. 1996). The selection of GMPE is accomplished preliminarily from a list of available equations.

The selection and rejection criteria for a GMPE in this study are based upon Cotton et al. (2006) guidelines. The key consideration for GMPE selection/rejection is as described below:

- Irrelevant tectonic regime,
- Insufficient dataset,
- Model is not yet published in peer-reviewed journal,
- Unsuitable frequency range for engineering applications, and
- Inappropriately regression analysis or regression coefficients misjudged.

Six different ground motion scaling relations are selected in accordance to Cotton et al. (2006) guidelines, which fulfill the tectonic conditions of the study area. These models include Abrahamson and Silva (1997), Campbell and Bozorgnia (2003), Sadigh et al. (1997), Atkinson and Boore (1995), (Boore et al. 1997) and one next generation attenuation equation by Campbell and Bozorgnia (2008).

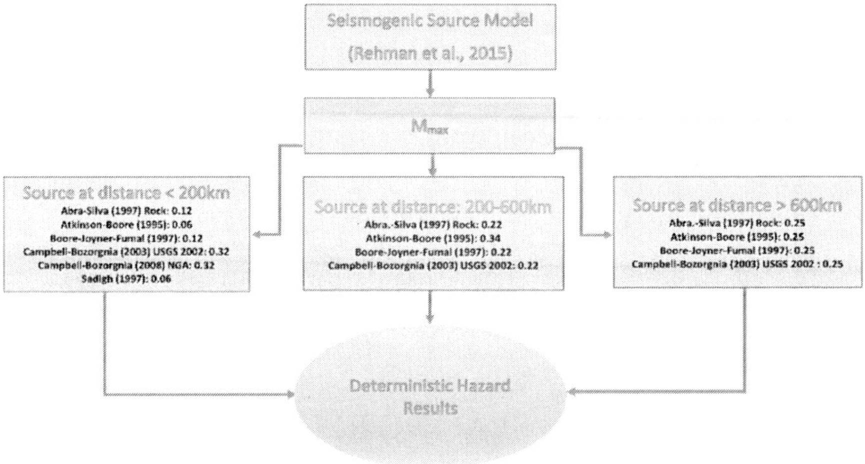

Figure 5. Logic tree to incorporate various GMPE

10 LOGIC TREE

The logic tree comprised several branches which portray confidence on different input parameters. Weight is assigned to each parameter according to the confidence level on each input (Abrahamson and Bommer 2005; Bommer 2002; Bommer and Scherbaum 2008). In current study, logic tree framework is utilized to incorporate several GMPE. The GMPE are assigned different weights in accordance with applicability of each model with respect to source-site distance. The source to site distance is divided into three ranges 0–200, 200–600, and 600–1000 km. Maximum weight is assigned to next generation attenuation model (Fig. 5).

11 RESULTS AND DISCUSSION

The DSHA is carried out with EzFrisk 7.52™ program and ground motion acceleration values are calculated in cm/s^2 at bedrock. The deterministic response spectrum is generated for MEAN and 0.84 fractile. The maps for MEAN and 0.84 fractile at peak ground acceleration (PGA), 0.1, 1, and 2 s spectral periods are generated (Figs. 6, 7, 8, 9, 10, 11, 12, 13)

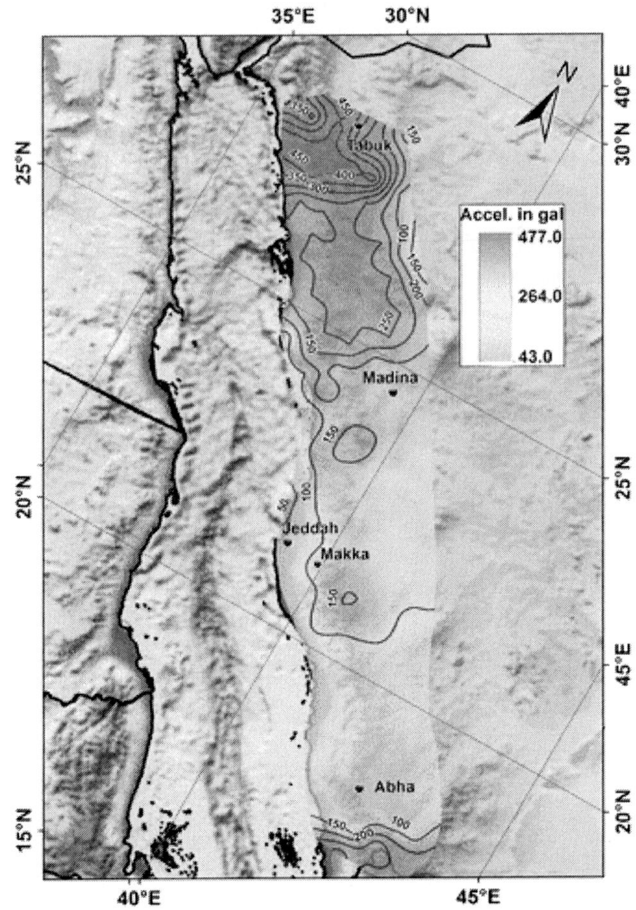

Figure 6. MEAN peak ground acceleration map

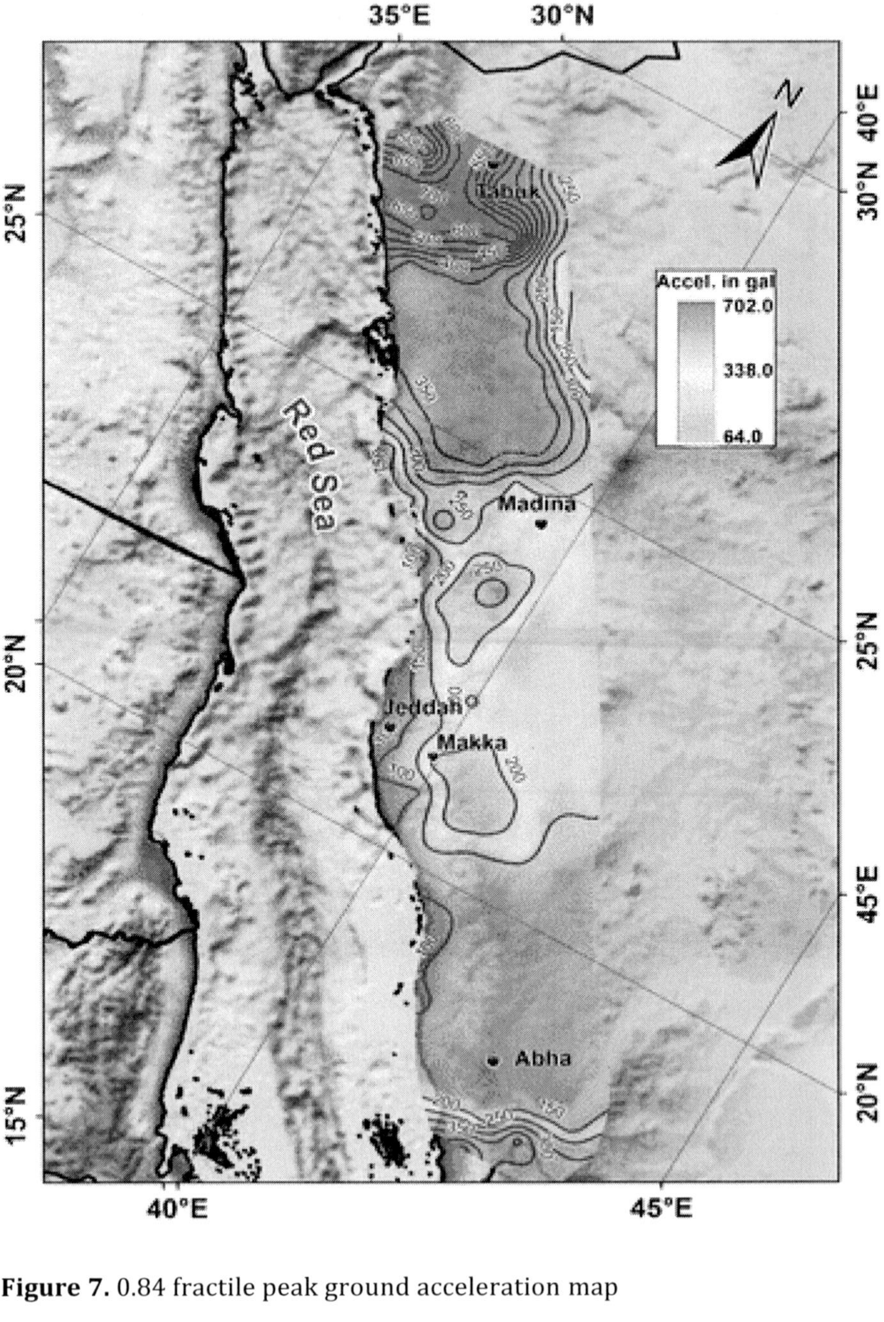

Figure 7. 0.84 fractile peak ground acceleration map

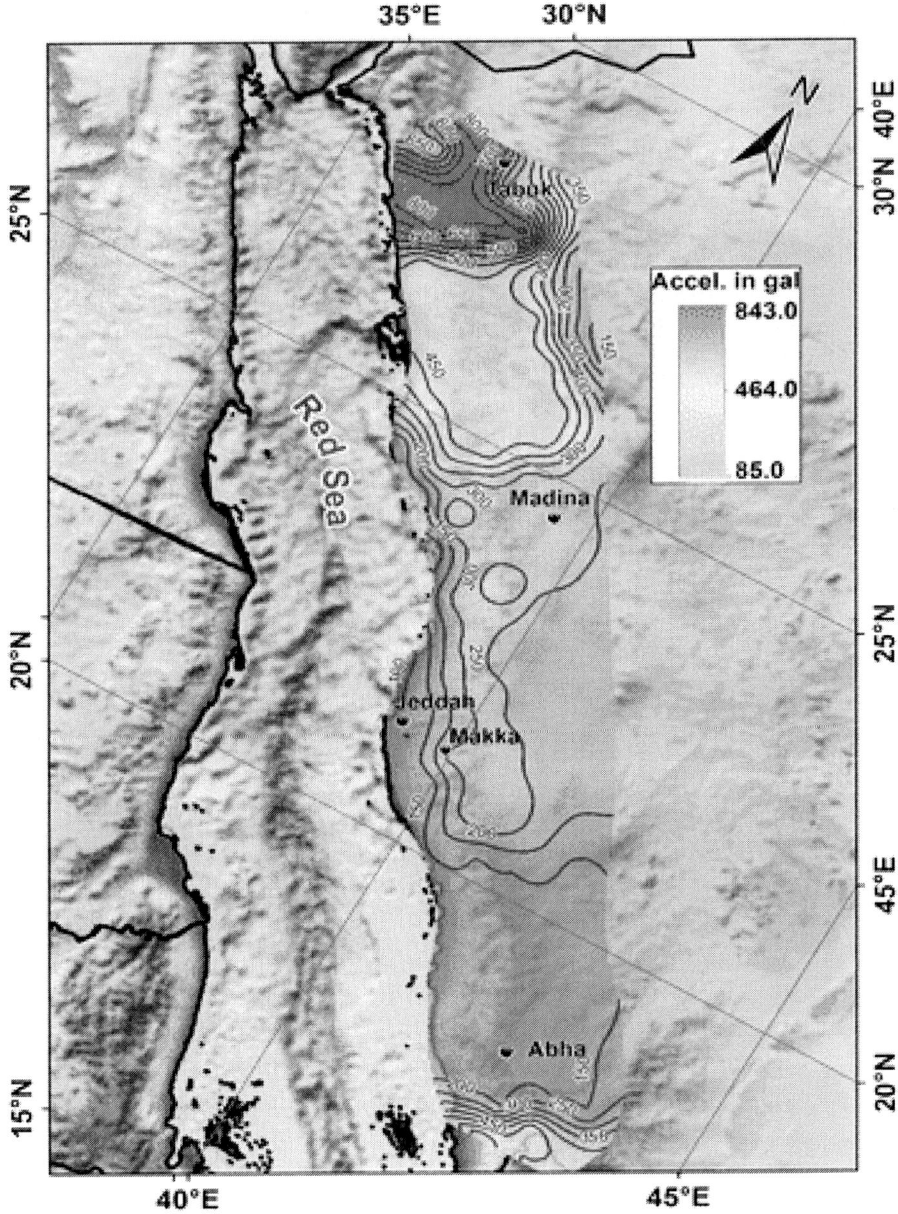

Figure 8. MEAN acceleration map at 0.1 s

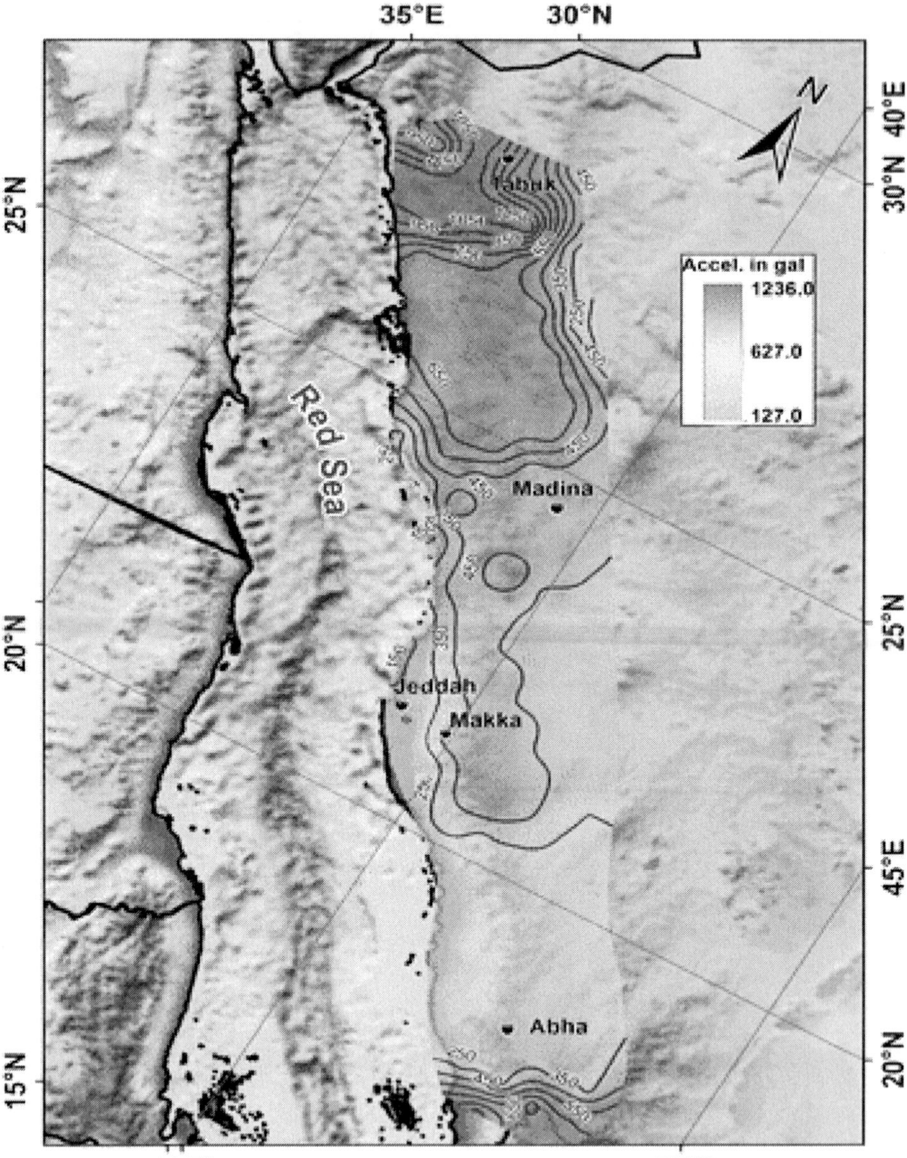

Figure 9. 0.84 fractile acceleration map at 0.1 s

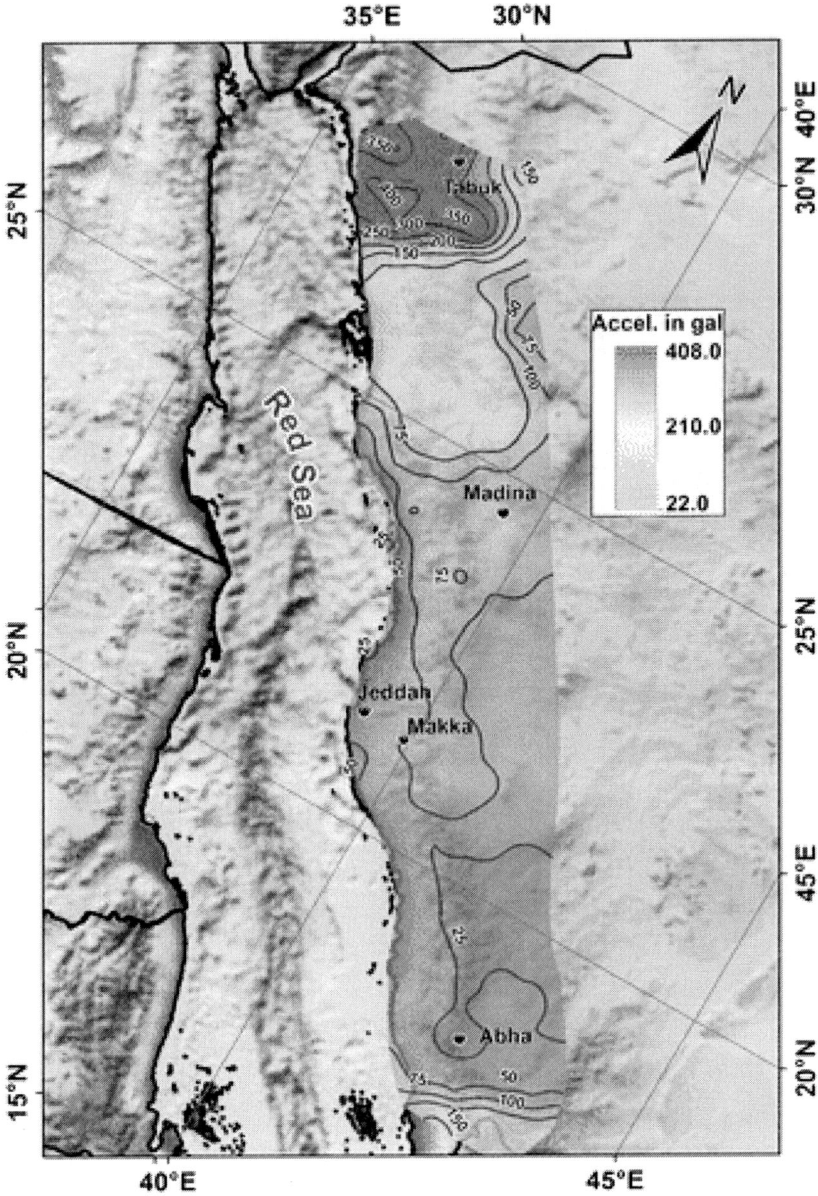

Figure 10. MEAN acceleration map at 1 s

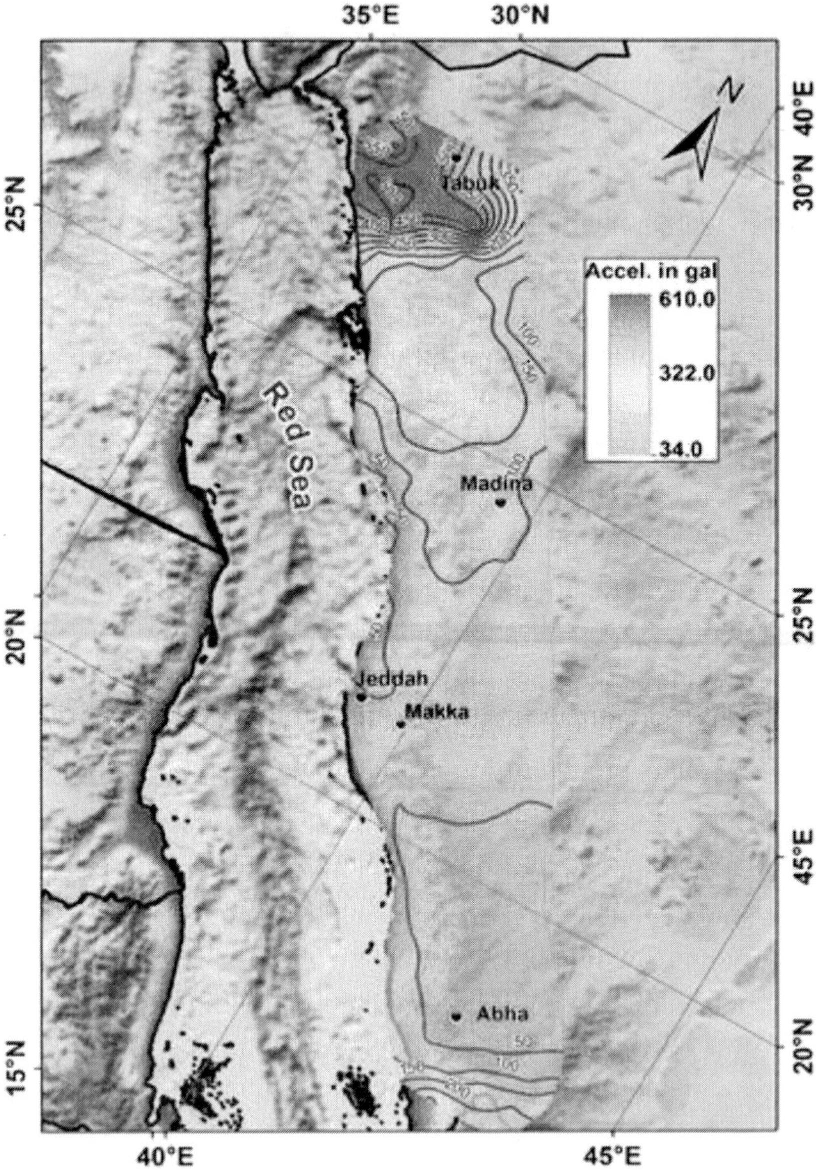

Figure 11. 0.84 fractile acceleration map at 1 s

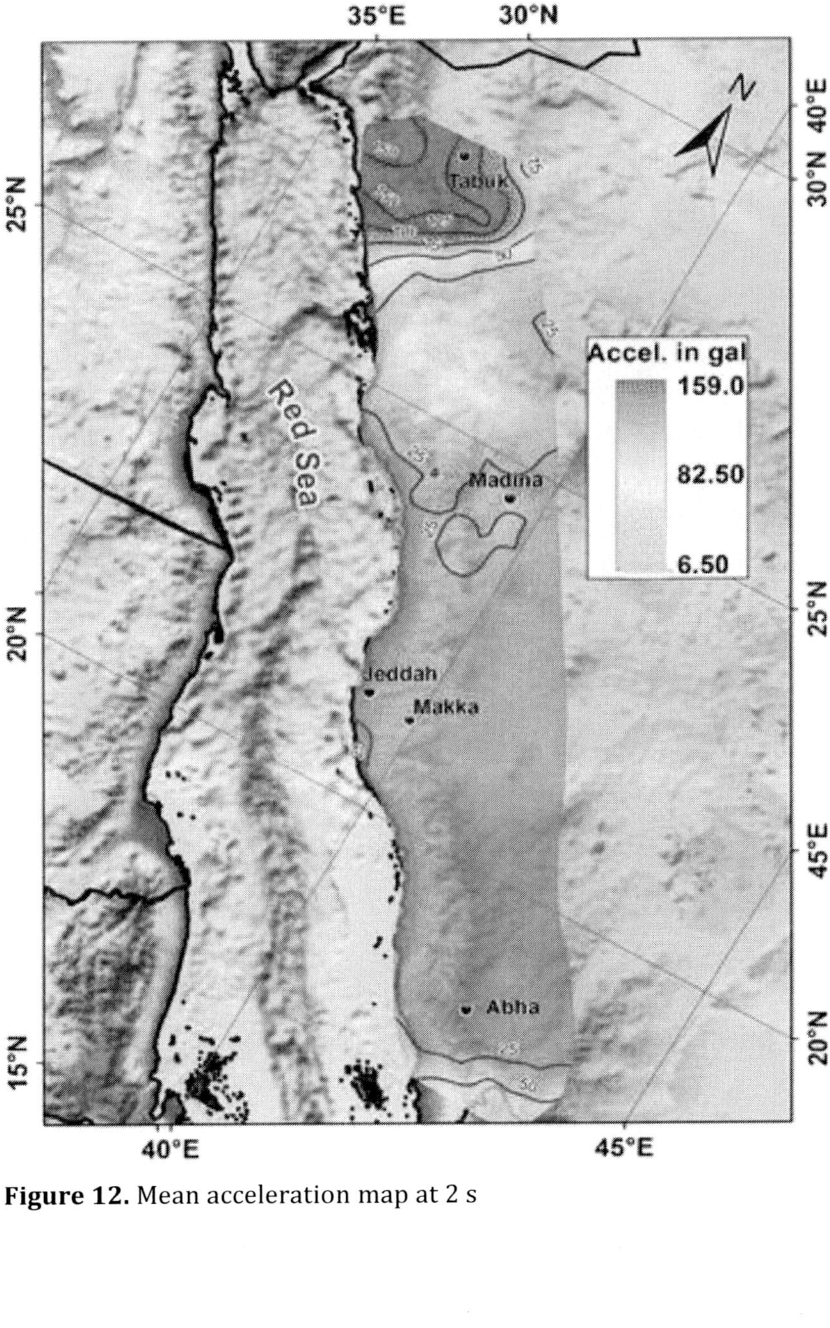

Figure 12. Mean acceleration map at 2 s

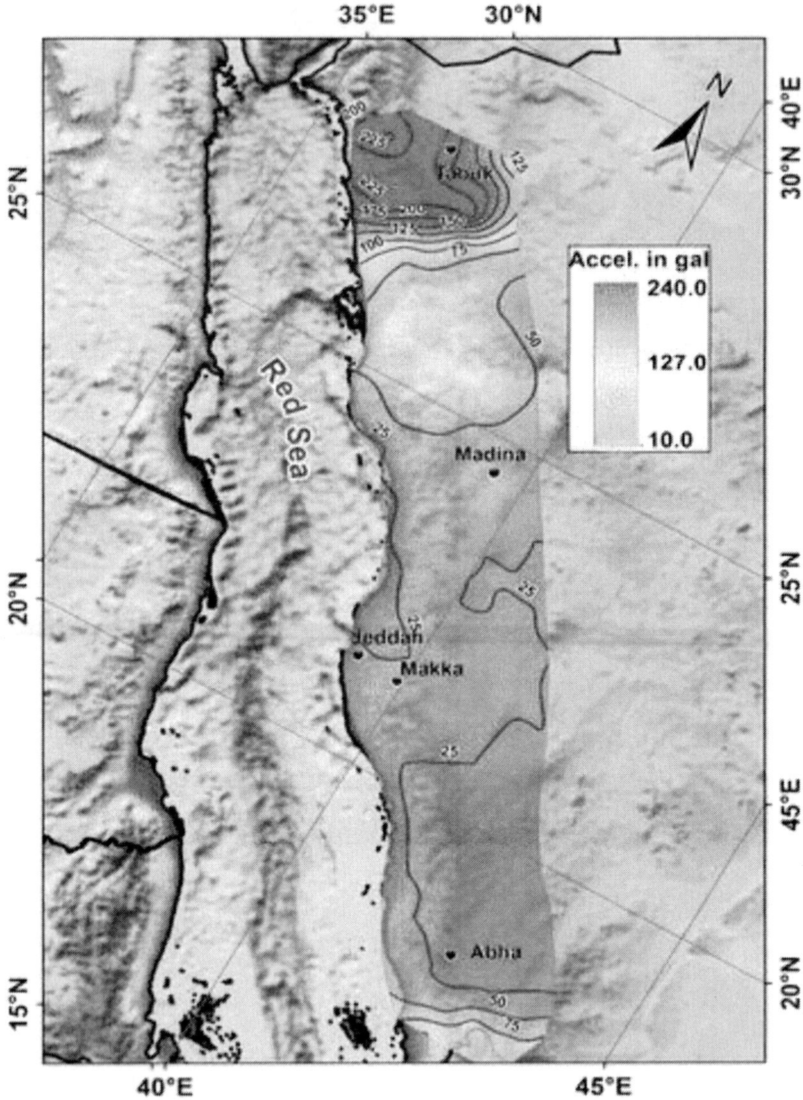

Figure 13. 84 % fractile acceleration map at 2 s

The ground acceleration values are extremely high in northern part of the study area. The possible explanation for such acceleration is because of close proximity of northern part of study area to seismogenic sources 19, 21 22, and 23 (Fig. 3). The ground acceleration decreases toward central part where moderate acceleration values are present. However, the moderate values can be categorized into two distinct behaviors for accelerations (Figs. 5, 6, 7, 8). The values for ground acceleration are comparably low for shoulder area along the Red Sea coast as compared to its eastern part. The controlling source for this area is Zone 7 and Zone 15, which have low seismicity. However, the eastern parts of central area are controlled by higher seismicity zones than western part (Zones 6, 10, and 14). At southern part of the study area high ground acceleration behavior is observed. The controlling sources for this part of the study area are Zones 1, 2, 4, and 5.

12 CONCLUSIONS

The deterministic approach actually considers the worst case ground motion. The worst case ground motion affects the design and cost of building. These ground motion spectra generated are used as essential input parameters for the upgradation of design parameters for local regulatory requirements in accordance with international standards. The design parameters are dependent on the local geology and local site soil conditions. In the current study we calculated ground motion at bedrock. However, the effect of soil must be taken into consideration for local site-based design parameters.

The DSHA technique is utilized to determine maximum expected ground acceleration for western coastal regions of Saudi Arabia. This research concludes that northern and southern part of western Saudi Arabia are more prone to higher seismic risks compared to central coastal areas. A detailed seismic risk assessment based on site response analysis is recommended for whole coastal regions. The top priority shall be given to northern and southern coastal regimes. This detailed risk assessment will facilitate qualitative decision making for redundant industrial systems and post-earthquake recovery plans.

ACKNOWLEDGMENTS

The authors gratefully appreciate the support by Deanship of graduate studies and Geophysics Department, at King Abdulaziz University. The authors would also like to thank Dr. Abduallah M. S. Al-Amri for his assistance and cooperation in providing scientific data. Finally the authors would profoundly like to thank Earthquake Monitoring Center, Sultan Qaboos University, Oman, who graciously granted us permission to run EZ-FRISK software for academic purposes and scientific cooperation.

REFERENCES

1. Abrahamson NA, Bommer JJ (2005) Probability and uncertainty in seismic hazard analysis. Earthq Spectra 21(2):603–607
2. Abrahamson N, Silva W (1997) Empirical response spectral attenuation relations for shallow crustal earthquakes. Seismol Res Lett 68(1):94–127
3. Al-Amri AM (2013) Seismotectonics and seismogenic source zones of the Arabian Platform. In: Lithosphere dynamics and sedimentary basins: the Arabian plate and analogues. Springer, Berlin pp 295–31
4. Al-Arifi NS, Fat-Helbary R, Khalil AR, Lashin AA (2013) A new evaluation of seismic hazard for the northwestern part of Saudi Arabia. Nat Hazards 69(3):1435–1457
5. Al-Damegh K, Sandvol E, Barazangi M (2005) Crustal structure of the Arabian plate: new constraints from the analysis of teleseismic receiver functions. Earth Planet Sci Lett 231(3):177–196CrossRef
6. Almadani S, Al-Amri A, Fnais M, Abdelrahman K, Ibrahim E, Abdelmoneim E (2015) Seismic hazard assessment for Yanbu metropolitan area, western Saudi Arabia. Arab J Geosci 8(11):9945–9958
7. Al-Malki M, Al-Amri A (2013) Seismic zones regionalization and hazard assessment of SW Arabian Shield and Southern Red Sea Region. In: Lithosphere dynamics and sedimentary basins: the Arabian Plate and analogues. Springer, pp 317–331
8. Ambraseys N (2009) Earthquakes in the Mediterranean and Middle East, vol 8. Cambridge University Press, New York, pp 4333–4355
9. Ambraseys N, Melville C, Adams R (1995) The seismicity of Egypt, Arabia and the Red Sea. In: Ambraseys NN, Melville CP, Adams RD (eds) The seismicity of Egypt, Arabia and the Red Sea, vol. 1. Cambridge University Press, Cambridge, UK, p 201. ISBN 0521391202
10. Ambraseys NN, Simpson KU, Bommer JJ (1996) Prediction of horizontal response spectra in Europe. Earthq Eng Struct Dyn 25(4):371–400
11. Ambraseys N, Douglas J, Sarma S, Smit P (2005a) Equations for the estimation of strong ground motions from shallow crustal earthquakes using data from Europe and the Middle East: horizontal peak ground acceleration and spectral acceleration. Bull Earthq Eng 3(1):1–53
12. Ambraseys NN, Melville CP, Adams RD (2005b) The seismicity of Egypt, Arabia and the Red Sea: a historical review. Cambridge University Press, New York
13. Ares AF (2010) Development of probabilistic seismic hazard analysis for international sites, challenges and guidelines. In: 2010 1st International proceedings of the nuclear & renewable energy conference (INREC). IEEE, Piscataway, pp 1–6
14. Atkinson GM, Boore DM (1995) Ground-motion relations for eastern North America. Bull Seismol Soc Am 85(1):17–30
15. Badawy A (1999) Historical seismicity of Egypt. Acta Geod Geophys Hung 34(1–2):119–135
16. Bommer JJ (2002) Deterministic vs. probabilistic seismic hazard assessment: an exaggerated and obstructive dichotomy. J Earthq Eng 6(S1):43–73
17. Bommer JJ, Scherbaum F (2008) The use and misuse of logic trees in probabilistic seismic hazard analysis. Earthq Spectra 24(4):997–1009
18. Boore DM, Joyner WB, Fumal TE (1997) Equations for estimating horizontal response spectra and peak acceleration from western North American earthquakes: a summary of recent work. Seismol Res Lett 68:128–153
19. Burkhard M, Grünthal G (2009) Seismic source zone characterization for the seismic hazard assessment project PEGASOS by the Expert Group 2 (EG1b). Swiss J Geosci 102(1):149–188
20. Campbell (2005) Overview of seismic PHSA approaches with emphasis on the management of uncertainties. In: 2nd workshop on earthquake engineering for nuclear facilities: uncertainties in seismic hazard

21. Campbell KW, Bozorgnia Y (2003) Updated near-source ground-motion (attenuation) relations for the horizontal and vertical components of peak ground acceleration and acceleration response spectra. Bull Seismol Soc Am 93(1):314–331
22. Campbell KW, Bozorgnia Y (2008) NGA ground motion model for the geometric mean horizontal component of PGA, PGV, PGD and 5 % damped linear elastic response spectra for periods ranging from 0.01 to 10 s. Earthq Spectra 24(1):139–171
23. Cotton F, Scherbaum F, Bommer JJ, Bungum H (2006) Criteria for selecting and adjusting ground-motion models for specific target regions: application to central Europe and rock sites. J Seismol 10(2):137–156
24. Deif A, Nofal H, Elenean KA (2009) Extended deterministic seismic hazard assessment for the Aswan High Dam, Egypt, with emphasis on associated uncertainty. J Geophys Eng 6(3):250
25. Deif A, El-Hussain I, Al-Jabri K, Toksoz N, El-Hady S, Al-Hashmi S, Al-Toubi K, Al-Shijbi Y, Al-Saifi M (2013) Deterministic seismic hazard assessment for Sultanate of Oman. Arab J Geosci 6(12):4947–4960
26. Dowrick D (2009) Seismic hazard assessment: earthquake resistant design and risk reduction, 2nd edn. Wiley, New York, pp 45–129
27. Fnais M, Abdelrahman K, E-Hady S, Abdel-monem E (2013) Seismicity and seismotectonics of the Jeddah area, Saudi Arabia. Earthq Resist Eng Struct IX 132:219
28. Gettings ME, Blank HR, Mooney WD, Healey JH (1986) Crustal structure of southwestern Saudi Arabia. J Geophys Res 91(B6):6491–6512
29. Guidoboni E, Comastri A (2005) Catalogue of Earthquakes and Tsunamis in the Mediterranean Area from the 11th to the 15th Century. Istituto nazionale di geofisica e vulcanologia
30. Guidoboni E, Comastri A, Traina G (1994) Catalogue of ancient earthquakes in the Mediterranean area up to the 10th Century, Vol 1. SGA
31. Gupta I (2002) The state of the art in seismic hazard analysis. ISET J Earthq Technol 39(4):311–346 (Paper, No. 428)
32. Gupta S, Kumar S, Wason H, Das R (2012) A statistical analysis of completeness of earthquake data around Dehradun city and its implications for seismicity evaluation. In: 15WCEE (Accepted)
33. Gutenberg B, Richter CF (1944) Frequency of earthquakes in California. Bull Seismol Soc Am 34(4):185–188
34. Hashemi M, Alesheikh AA, Zolfaghari MR (2013) A spatio-temporal model for probabilistic seismic hazard zonation of Tehran. Comput Geosci 58:8–18
35. Joyner WB, Boore DM (1981) Peak horizontal acceleration and velocity from strong-motion records including records from the 1979 Imperial Valley, California, earthquake. Bull Seismol Soc Am 71(6):2011–2038
36. Kijko A (2004) Estimation of the maximum earthquake magnitude M max. Pure Appl Geophys 161(8):1655–1681
37. Kijko A, Sellevoll M (1989) Estimation of earthquake hazard parameters from incomplete data files. Part I. Utilization of extreme and complete catalogs with different threshold magnitudes. Bull Seismol Soc Am 79(3):645–654
38. Kijko A, Sellevoll MA (1992) Estimation of earthquake hazard parameters from incomplete data files. Part II. Incorporation of magnitude heterogeneity. Bull Seismol Soc Am 82(1):120–134
39. Leonard M, Clark D, Collins C, McPherson A (2011) The 2012 Australian seismic hazard map-source zones and parameterisation. In: Proceedings of the 2011 Australian earthquake engineering society conference, Barossa Valley, SA (this volume)
40. Lin T, Baker J (2011) Probabilistic seismic hazard deaggregation of ground motion prediction models. In: Proceedings of the 5th International conference on earthquake geotechnical engineering, Santiago, Chile, p 12
41. Meletti C, Galadini F, Valensise G, Stucchi M, Basili R, Barba S, Vannucci G, Boschi E (2008) A seismic source zone model for the seismic hazard assessment of the Italian territory. Tectonophysics 450(1):85–108

42. Mignan A, Woessner J (2012) Estimating the magnitude of completeness for earthquake catalogs. Community online resource for statistical seismicity analysis. pp 1–45. http://www.corssa.org/export/sites/corssa/.galleries/articles-pdf/Mignan-Woessner-2012-CORSSA-Magnitude-of-completeness.pdf
43. Mohamed AE-EA, El-Hadidy M, Deif A, Elenean KA (2012) Seismic hazard studies in Egypt. NRIAG J Astron Geophys 1(2):119–140
44. Mokhtar T (2004) Variations of the crustal structure of Arabia. J King Abdulaziz Univ Earth Sci 15:1–27
45. Mokhtar T, Ammon C, Herrmann R, Ghalib H (2001) Surface wave velocities across Arabia. In: Monitoring the comprehensive nuclear-test-ban treaty: surface waves. Springer, pp 1425–1444
46. Pailoplee S, Sugiyama Y, Charusiri P (2010) Probabilistic seismic hazard analysis in Thailand and adjacent areas by using regional seismic source zones. Terr Atmos Ocean Sci 21(5):757–766
47. Peruš I, Fajfar P (2009) How reliable are the ground motion prediction equations? In: Proceedings of the 20th international conference on structural mechanics in reactor technology (SMiRT 20), Espoo, Paper, p 9
48. Poirier J, Taher M (1980) Historical seismicity in the near and Middle East, North Africa, and Spain from Arabic documents (VIIth–XVIIIth century). Bull Seismol Soc Am 70(6):2185–2201
49. Re M (2000) Natural catastrophes—the current position, special millennium issue. Munich Reinsurance Co, Munich
50. Rehman F (2016) Assessment of seismic hazard by microzonation in Bahra area, Wadi Fatima, Kingdom of Saudi Arabia. (unpublished doctoral dissertation), King Abdulaziz University, Jeddah, Saudi Arabia
51. Rodgers AJ, Walter WR, Mellors RJ, Al-Amri AM, Zhang Y-S (1999) Lithospheric structure of the Arabian shield and platform from complete regional waveform modelling and surface wave group velocities. Geophys J Int 138(3):871–878
52. Sabetta F, Pugliese A (1987) Attenuation of peak horizontal acceleration and velocity from Italian strong-motion records. Bull Seismol Soc Am 77(5):1491–1513
53. Sadigh K, Chang C-Y, Egan J, Makdisi F, Youngs R (1997) Attenuation relationships for shallow crustal earthquakes based on California strong motion data. Seismol Res Lett 68(1):180–189
54. Smith RP, Jackson SM, Hackett WR (1996) Paleoseismology and seismic hazards evaluations in extensional volcanic terrains. J Geophys Res 101(B3):6277–6292
55. Stepp JC, Wong I, Whitney J, Quittmeyer R, Abrahamson N, Toro G, Youngs R, Coppersmith K, Savy J, Sullivan T (2001) Probabilistic seismic hazard analyses for ground motions and fault displacement at Yucca Mountain, Nevada. Earthq Spectra 17(1):113–151
56. Stern RJ, Johnson PR (2008) Do variations in Arabian plate lithospheric structure control deformation in the Arabian-Eurasian convergence zone? In: Proceedings IOP conference series: earth and environmental science, vol 2. IOP Publishing, p 012005
57. Stern RJ, Johnson P (2010) Continental lithosphere of the Arabian Plate: a geologic, petrologic, and geophysical synthesis. Earth Sci Rev 101(1–2):29–67
58. Tavakoli H, Amiri MT, Abdollahzade G, Janalizade A (2013) Probabilistic and deterministic seismic hazard assessment: a case study in Babol, pp 265–270
59. Vipin K, Sitharam T (2013) Delineation of seismic source zones based on seismicity parameters and probabilistic evaluation of seismic hazard using logic tree approach. J Earth Syst Sci 122(3):661–676
60. Waseem M, Khan MA, Javed MW, Ali SM (2013) Deterministic seismic hazard analysis for Peshawar, Pakistan. J Himal Earth Sci 46(1):67–72
61. Wiemer S, Wyss M (2000) Minimum magnitude of completeness in earthquake catalogs: examples from Alaska, the western United States, and Japan. Bull Seismol Soc Am 90(4):859–869
62. Woessner J, Wiemer S (2005) Assessing the quality of earthquake catalogues: estimating the magnitude of completeness and its uncertainty. Bull Seismol Soc Am 95(2):684–698

63. Yazdani A, Kowsari M (2013) Earthquake ground-motion prediction equations for northern Iran. Nat Hazards 69(3):1877–1894

CHAPTER 13

Complete Coverage Path Planning for a Multi-UAV Response System in Post-Earthquake Assessment

Arman Nedjati [1,*], Gokhan Izbirak [1], Bela Vizvari [1] and Jamal Arkat [2]

[1]*Department of Industrial Engineering, Eastern Mediterranean University, Famagusta, North Cyprus, via Mersin 10, Famagusta 99450, Turkey*

[2]*Department of Industrial Engineering, University of Kurdistan, Pasdaran Boulevard, Sanandaj 15175-66177, Iran*

ABSTRACT

This paper presents a post-earthquake response system for a rapid damage assessment. In this system, multiple Unmanned Aerial Vehicles (UAVs) are deployed to collect the images from the earthquake site and create a response map for extracting useful information. It is an extension of well-known coverage path problem (CPP) that is based on the grid pattern map decomposition. In addition to some linear strengthening techniques, two mathematic formulations, 4-index and 5-index models, are proposed in the approach and coded in GAMS (Cplex solver). They are tested on a number of problems and the results show that the 5-index model outperforms the 4-index model. Moreover, the proposed system could be significantly improved by the solver-generated cuts, additional constraints, and the variable branching priority extensions.

Keywords: coverage path planning; rapid damage assessment; UAV monitoring; post-earthquake response

1. INTRODUCTION

Rapid damage assessment in post-earthquake situations plays an important role in the early response phase activities (i.e., evacuation of injured individuals, debris collection, and relief distribution). The ground-based post-earthquake inspection is extremely time-consuming, and unhelpful in severely damaged areas; therefore, at present, aerial systems are widely used for investigations. It is highly desirable for densely populated urban areas to have a pre-planned immediate and automated post-disaster mapping and monitoring system. In the past decade, scientists have made an attempt to improve high-resolution satellite imaging and laser scanning systems to evaluate the disaster damage and loss [1,2,3,4,5,6,7,8]. However, the satellite systems have many limitations for an efficient post-disaster imaging, such as weather conditions (cloudy or dust whirls), time constraints for acquiring images and uplinking the acquisition plan to the satellite, delay in satellite data delivery after collection, etc. [9,10]; therefore, much attention has been focused on utilizing small unmanned aerial vehicles (UAVs) for post-disaster mapping [11,12,13].

In recent times, UAVs have become a popular entertaining tool globally. UAVs of various sizes are available on the market that can carry film or photographic cameras for different applications. The classification of UAVs according to their classes, distinct applications, and characteristics is beyond the scope of this article, though interested readers can refer to earlier studies [13,14]. Moreover, Colomina and Molina [15] have reviewed significant applications of UAVs, such as imaging and remote sensing. This study focuses on the mapping applications of small UAVs in post-earthquake situations. In a UAV-mapping system, consecutive overlapping aerial images taken by the UAVs are processed to obtain a complete map or to extract useful information. In earthquake situations, the information, such as building and bridge destructions, road blockages, and population relocation, aids the managers in organizing a more effective post-earthquake response system. Within the first 30 min, the post-earthquake survival rate is 91%, decreasing to 36.7% by the second day [11]. Hence, in addition to accuracy, time is also an important factor in post-earthquake mapping. Based on this fact, it can be said that an optimal solution to the UAV coverage path-planning problem with a time-dependent objective function will have a great effect on the efficiency of the response system.

Coverage path problem (CPP), can be considered as a variant of Vehicle Routing Problem (VRP). In VRP, a fleet of vehicles start and end their tour at a single depot while visiting all nodes on the route. Unlike the classical VRP that considers a single main depot, in multi depot vehicle routing problem (MDVRP), there are several depots and the customers can be served from any of the depots. MDVRP has widely been studied in the literature and the interested readers may refer to the extensive survey of Montoya-Torres et al. [16]. Based on the literature the UAV routing problem is also formulated as a VRP model where additional constraints need to be added to reflect the characteristics of the problem. UAV routing problem has attracted much attention in the past decade for different applications [17,18]. CPP, tries to find an appropriate path for robots while covering the pre-defined nodes. The problem has a wide range of applications in automated harvesting, vacuum cleaning, mapping, demining, monitoring, etc. It is worth mentioning that CPP is different from the Covering Tour Problem (CTP). In CTP there are two types of nodes, those that must be visited by the tour and those that must be covered i.e., it lies within a pre-defined distance from a vertex of the tour [19]. Choset [20] and Galceran and Carreras [21] conducted comprehensive surveys on CPP methods, algorithms, and recent advances.

Immediately after an earthquake, the damage and location of afflicted people can be identified by imaging and then processing the obtained images. Since time factor is very important in post-earthquake response phase, UAVs are considered to be reliable for this task. Earthquake is not the only disaster that can be monitored by UAVs, but it is the most suitable one for imaging due to absence of information about the condition of each construction over the entire area. The imaging application of UAV based on CPP necessitates further considerations. One important issue is the direction of the path while passing through a node. Another important issue that arises due to earthquake situation is that, unlike the routing problem, the maximum mission time of vehicles must be minimized. UAVs start their route from a base and end their route at the same base for refueling or at the mission completion. This study extends the CPP for an application in the post-earthquake rapid damage assessment. The CPP features on a grid-based map are adopted on a multi-depot multi-tour vehicle routing problem for an emergency situation.

Zelinsky et al. [12] were the first to investigate the CPP on a robot, starting from an origin and ending at a goal point to minimize the length, energy consumption, and travel time. Later Carvalho et al. [22] proposed an algorithm for a mobile robot in an industrial environment such that the obstacles were not pre-specified on the map. The subject of path planning has attracted a lot of attention due to the recent development of UAV systems. Li et al. [23] studied an exact cellular decomposition method for UAV path planning in a polygon region. For the purpose of precision agriculture mapping, Barrientos et al. [24] performed an experiment using an integrated tool. Initially, they subdivided the polygon area and then conducted path planning for each subarea using a multi-UAV system. Torres et al. [25] presented a path planning algorithm for a single UAV with the aim of reducing battery usage and minimizing the number of turns, coping with both convex and non-convex regions. In a study by Wang et al. [26], an algorithm was used that minimized the consumed energy by the UAV for covering 3D terrain. In addition, a distinct study [27] introduced the multi-robot boundary coverage problem with the application of inspection of blade surfaces inside a turbine. Furthermore, Galceran and Carreras [21] conducted a comprehensive survey on CPP.

Most CPP algorithms are based on boustrophedon path strategy. In this strategy, the back and forth motion in a sweep direction of a polygon tries to cover an area with a minimum number of turns. Another approach divides the decomposing area into sub-areas and then finds the visiting sequence of the sub-areas in a sweep direction base [28], describing how to find the optimal sweep direction for robots in a polygon. Huang and York claimed that the minimization of the number of turns leads to the most efficient solution (Figure 1). Li et al. [23] showed that the path with a lower number of turning motions is a more efficient coverage path for UAVs with regards to energy, distance, and travel time.

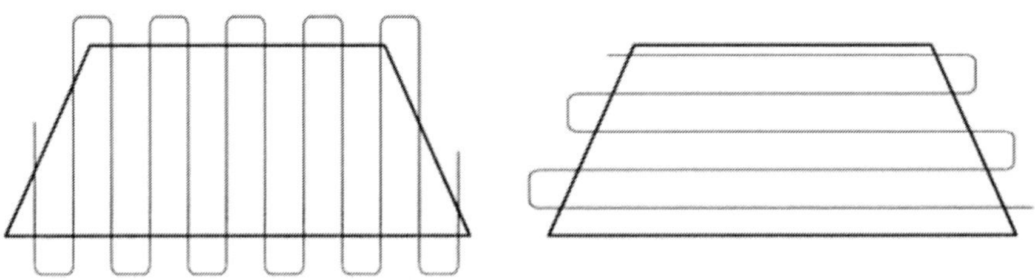

(a) (b)

Figure 1. The right polygon is covered by lesser waste (lesser number of turns) than the left one. (**a**) The vertical direction with 11 turns; (**b**) The horizontal direction with 4 turns.

Furthermore, the map decomposition and sweep direction method were applied for a multi-UAV system [29]. Due to lack of time in a post-earthquake situation, using a multi-UAV system is potentially more beneficial and practical. Avellar et al. [30] studied a multi-UAV CPP in which the nodes were pre-defined corresponding to the boustrophedon motion. Similar to earlier studies, they decomposed the region into several sub-regions and used a VRP model to find the optimal path. However, the travel time between the sub-areas and the base to a region, as well as the possibility of turning back from the middle of a region to the base for refueling, were neglected.

Although numerous studies have been conducted on the sweep direction method, path covering based on grid-based decomposition has not yet received much attention. In a study by Wang and Li [31], the desired region was divided into grids and the amount of information collected by UAV and the path-length were maximized. According to their algorithm, passing through the center of a grid is equivalent to covering the grid; however, to have a complete image of the grid, the covering is considered as entering a grid from one side and exiting from the opposite side. Therefore, the turning motions in the grids are not considered as a part of the covered area (Figure 2). The grid size can vary depending on the altitude of the UAVs and focal length of the camera; however, by increasing the number of grids the complexity of the problem will increase.

In this study, it is assumed that the number of available UAVs, potential location of UAV bases, and the possible number of open bases are based on the restrictions prescribed by the decision maker. In addition, the total mission time limit should be specified earlier according to the endurance of UAVs. We have tried to address the following questions in the proposed post-earthquake mapping problem: (1) which potential UAV base must be open; (2) how many UAVs among all the available ones should be assigned to each open base; (3) what must be the passing sequence of squares in each route; and (4) what is the minimum time required for covering all residential squares with the existing limitations. The proposed problem has a Minimax objective function that minimizes the maximum traveling time of each UAV in order to find the fastest possible mapping schedule. Each UAV belongs to a UAV base from where its mission starts and ends, and battery replacement or refueling is performed at the same base. Each UAV's mission time cannot exceed a maximum flight time (MT) due to its endurance. To the best of our knowledge, this problem has not been studied thus far, and the formulations presented here are the first mixed integer linear programming (MILP) mathematical models of the grid-based CPP problem. We call this problem multi-UAV multi-tour location coverage path planning (MUTLCPP). The application of this model is a post-earthquake mapping system due to the Minimax objective function, and a grid-based decomposition of the residential areas.

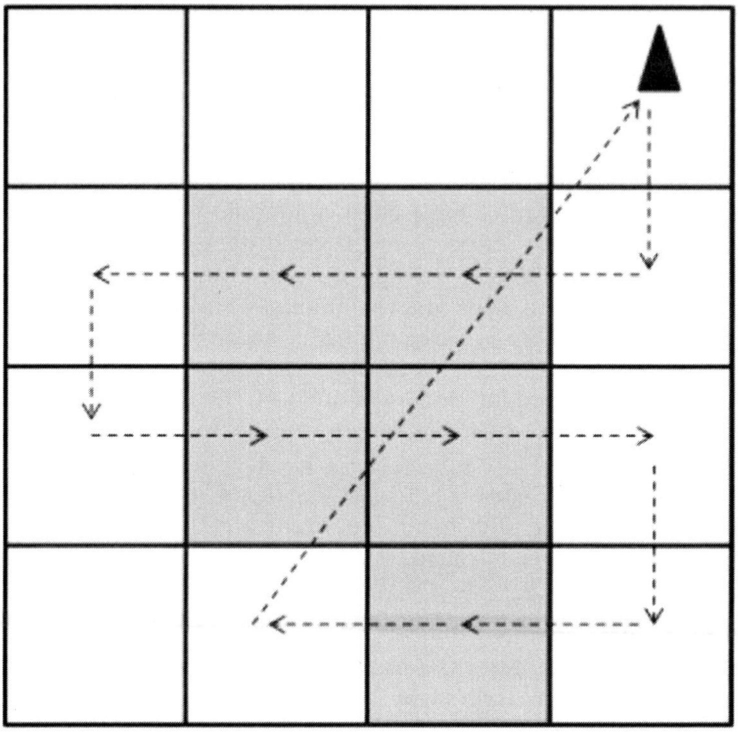

Figure 2. The gray cells indicate the cells that have been covered by the unmanned aerial vehicle (UAV), according to entrance and exit from the opposite sides. The picture depicts a feasible solution of grid decomposition-based coverage path problem (CPP).

The remainder of this study is organized as follows: Section 2 represents the problem definition and two 4-index and 5-index mathematical models. Section 3 contains the extensions and techniques used to improve the performance of the branch and cut algorithm. Section 4 includes the test problems and computational results, and Section 5 contains conclusions and future prospects of CPP.

2. PROBLEM DEFINITION AND MATHEMATICAL FORMULATIONS

Associated with the MUTLCPP problem, the sets, parameters, and decision variables are defined as follows (in Table 1):

Table 1. The definition of sets, parameters, and variables.

Sets:	
S_C	The set of residential squares that must be covered
S_D	The set of potential AUV base (depot) squares
S_E	The set of nonresidential squares
U	The set of available UAVs
L	The set of routes
R	The set of moving directions from one square to the other
Parameters:	
t_{ij}	The travel time from square i to j
d_{ij}	The distance between square i and j
TD	Total number of UAV bases (depots) that asked to be open
MT	The possible flight time of each UAV
Variables:	
w_{iju}^{lr}	Binary decision variable equals to 1, if UAV u exists from r side of square i to square j in route l, and 0 otherwise
w_{ij}^{ul}	Binary decision variable equals to 1, if UAV u travels from square i to square j in route l, and 0 otherwise
v_{ij}^{ul}	Positive integer variable determines how many nodes are left to cover in route l by UAV u, after visiting i and just before visiting j
y_i	Binary variable equals to 1, if potential depot i is open and 0 otherwise
x_{ij}^{lr}	Binary variable equals to 1, if a UAV travels from r side of square i to square j in route l
p_{iu}	Binary variable equals to 1, if UAV u belongs to depot i and 0 otherwise
g_{iu}^{lr}	Binary variable equals to 1, if a UAV enters from r side of i and exits from the opposite side in route l
g_i^{lr}	Binary variable equals to 1, if a UAV enters from r side of i and exits from the opposite side in route l
T	The maximum mission time of UAVs

The map of earthquake struck region was decomposed into equal sized squares, according to the UAV's camera footprint. The camera footprint (CF) depends on the camera sensor size (CSS), lens focal length (FL), and flight altitude (FA). Figure 3shows the camera footprint, which can be calculated as follows.

$$CF = FA\frac{CSS}{FL} \quad (1)$$

The camera plays a very important role in CPPs especially for a post-earthquake situations. Despite the resolution demanded by the decision maker, the focal length of the lens and the sensor size of the camera have a great impact on the mapping schedule length. A smaller focal length with a larger sensor size ends up in a larger camera footprint (Equation (1)) and consequently a lower number of squares in the map and schedule length. Also, the horizontal flight disposition is preferred for covering residential areas by a sequence of overlapping images, although the new UAV systems allow changes in the viewing angle of the installed camera.

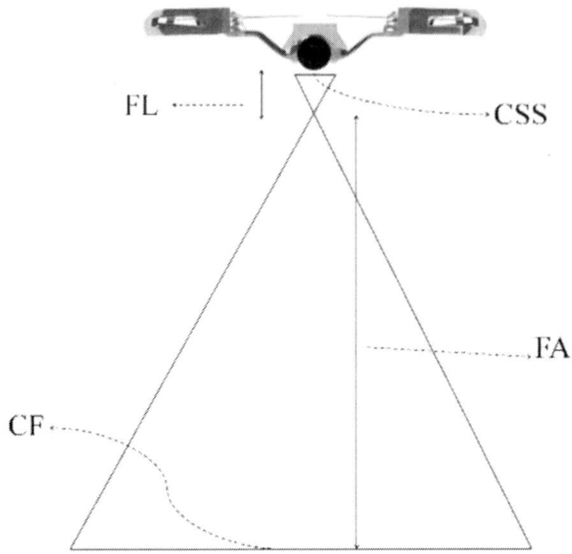

Figure 3. Camera footprint.

The ground sampling distance (GSD) is a scale of resolution in aerial photogrammetry. It is the distance between two successive pixel centers on the ground. The smaller value of the GSD leads to the higher resolution of the image and more differentiable details. At a fixed focal length, decreasing the altitude results in a smaller GSD, and at a fixed flight altitude, by increasing the focal length, the GSD value will decrease. In an earthquake situation, the GSDs of several meters may suffice for damage assessment of non-urban regions. However, it is quiet insufficient when analyzing highly populated earthquake-struck cities and detailed images of at least 1 m GSD are essential [32].

While the geographical mapping might not require a quick scheduling process, in post-earthquake mapping, time factor has a great importance. Since the use of UAVs for mapping is in its early days, there are potential improvements and considerable innovations related to unseen problems and technical challenges.

To ease the image processing, an overlap from each side is essential; thus, if the overlap is α% from each side, then the square edge must be equal to (1−2α)% of the camera footprint length. The overlap is required to discover the common details among the images, however; the minimum and maximum overlaps depend on the post-image-processing software. Considering the fact that image-processing discussion is beyond the scope of this study, and there is no universally accepted overlap standard [33], we only mention that the overall overlap for a specified-path mapping must be at least 20% of the image width [34].

The MUTLCPP is defined on an undirected graph G(S,A), where S=SO∪ SD is a set of squares and A={(i,j)|i,j∈S} is the set of arcs. SD is a set of potential depot squares and SO=SC∪SE, where SC is the set of residential squares that must be covered and SE is the set of empty squares. Non-negative travel time, tij, and travel distance, dij, are associated with each arc with a triangular inequality (i.e., tik+tkj≥tij). Moreover, the set of homogenous UAVs is denoted by U={1,2,3,..,u} and the set of routes by L={1,2,3,...,l}. There is a set R={Left, Right, Up, Down, Direct} that specifies the leaving direction form one square to another. All the squares belonging to SC must be covered, at least once, such that the UAV enters and exists from opposite sides (i.e., Right-Left, Left-Right, Up-Down, and Down-Up). Among the |SD| number of potential depots, TD number of them can be opened, according to the restrictions prescribed by the decision maker. The flight time limitation for each UAV is specified by MT. All UAVs start and end their tours at the same base. Because time plays a vital role in the post-disaster response phase against cost, the proposed model must minimize the maximum mission time of all UAVs.

When M is a sufficiently large positive number, the 4-index and 5-index models of MUTLCPP are presented as follows:

2.1. 4-Index Formulation

Model:

$$Min\ T \tag{2}$$

The objective Equation (1) is to minimize the maximum mission time of UAVs. This objective function leads to the minimum mapping schedule length for all UAVs. The MUCLTPP's constraints are as follows:

$$\text{S.t.} \sum_{i \in S_D} y_i = TD \tag{3}$$

$$\sum_{i \in S_D} p_{iu} = 1\ \forall u \in U \tag{4}$$

Equation (3) ensures that the total number of open bases is TD, and Equation (4) imposes that each UAV has to be allocated to one depot only.

$$\sum_{i \in S} \sum_{u \in U} w_{ij}^{ul} \leq 1 \; \forall j \in S, i \neq j, l \in L \tag{5}$$

Equation (5) ensures that each square is passed at most once in each route.

$$\sum_{j \in S} w_{ij}^{ul} - \sum_{j \in S} w_{ji}^{ul} = 0 \; \forall i \in S, i \neq j, l \in L, u \in U \tag{6}$$

Equation (6) guarantees that if a UAV enters a square then it must leave that square. In other words, the total number of entrances to a square is equal to the total number of exits from that square.

$$\sum_{j \in S_O} \sum_{l \in L} \sum_{u \in U} w_{ij}^{ul} - My_i \leq 0 \; \forall i \in S_D \tag{7}$$

Equation (7) assures that the UAVs exit from an open base only. This constraint forces the wulij variable to take the value of zero if the base is close (yi=0).

$$v_{ij}^{ul} - Mw_{ij}^{ul} \leq 0 \; \forall i \& j \in S, i \neq j, \; u \in U, \; l \in L \tag{8}$$

Equation (8) assigns zero to vulij variables according to the movement variable of wulij.

$$\sum_{i \in S} v_{ij}^{ul} - \sum_{i \in S} v_{ji}^{ul} - \sum_{i \in S} w_{ij}^{ul} = 0 \; \forall j \in S_O, i \neq j, u \in U, l \in L \tag{9}$$

Equation (9) indicates that if a UAV pass through square i∈SO, then the remaining number of the nodes in route l will decrease by one.

$$\sum_{i \in S} \sum_{j \in s, j \neq i} \sum_{l \in L} t_{ij} w_{ij}^{ul} \leq MT \; \forall l \in L \tag{10}$$

Equation (10) ensures that the duration of each route does not exceed the UAV's flight time limitation MT.

$$\sum_{i\in S}\sum_{j\in s,j\neq i}\sum_{l\in L} t_{ij}w_{ij}^{ul} + \sum_{i\in S_D}\sum_{j\in S_O}\sum_{l\in L} STw_{ij}^{ul} \leq T\ \forall u \in U \tag{11}$$

The value of T which is the maximum mission time of UAVs is determined by Equation (11).

$$\sum_{j\in S_O}\sum_{l\in L} w_{ij}^{ul} - Mp_{iu} \leq 0\ \forall i \in S_D, u \in U \tag{12}$$

$$\sum_{i\in S_D}\sum_{j\in S_O}\sum_{u\in U} w_{ij}^{ul} \leq 1\ \forall l \in L \tag{13}$$

Equation (12) indicates that UAVs can only leave their own depot and Equation (13) eliminates the illegal routes.

$$\sum_{l\in L}\sum_{r\in R,r\neq Direct}\sum_{j\in S} x_{ij}^{lr} = 0\ \forall i \in S,\ \ d(i,j) > dm \tag{14}$$

Equation (14) prevents the formation of arcs with the length more than dm for the horizontal and vertical directions. dmis the distance between the center of two consecutive squares; the length of horizontal or vertical arcs cannot exceed it.

$$\sum_{j\in S}(x_{ij}^{lr} + x_{ji}^{lr}) - 2g_i^{lr} \geq 0\ \forall i \in S_C, i \neq j, l \in L, r \in R | r \neq Direct \tag{15}$$

Equation (15) specifies that the value of glri can be one if the entrance and exit are in the same horizontal or vertical directions.

$$\sum_{l\in L}\sum_{r\in R|r\neq Direct} g_i^{lr} \geq 1\ \forall i \in S_C \tag{16}$$

$$\sum_{r\in R} x_{ij}^{lr} - \sum_{u\in U} w_{ij}^{ul} = 0\ \forall i \neq i, j \in S, l \in L \tag{17}$$

Equation (16) ensures that all residential squares must be covered by a UAV in either horizontal or vertical directions and Equation (17) defines the value of variable xlrij.

$$x_{ij}^{l\ Up} = 0\ \forall i, j \in S \Big| i - 1 \leq j, l \in L \tag{18}$$

$$x_{ij}^{l\ Down} = 0\ \forall i, j \in S \Big| i + 1 \geq j, l \in L \tag{19}$$

$$x_{ij}^{l\ Left} = 0\ \forall i, j \in S \Big| i \neq j + 1, l \in L \tag{20}$$

$$x_{ij}^{l\ Right} = 0\ \forall i, j \in S \Big| i \neq j - 1, l \in L \tag{21}$$

Equations (18)–(21) imposes the value of zero for the wrong horizontal and vertical directions. For example Equation (18) ensures that the travel in the upward (up) direction from i to j is zero unless i–1 is less than or equal j. The travel from one square to its right square must be impossible for the *up*, *down*, and *left* directions and it is shown in Equation (21).

$$w_{ij}^{ul},\ x_{ij}^{lr}\ \in \{0,1\}\ \forall i \& j \in S, u \in U, l \in L, r \in R \tag{22}$$

$$y_i,\ p_{iu},\ g_j^{lr} \in \{0,1\}\ \forall i \in S_D, j \in S_C, l \in L, r \in R \tag{23}$$

Finally, Equations (22) and (23) are used as the integrality constraints.

2.2. 5-Index Formulation

The 5-index model development is presented in the same structure of the 4-index formulation as follows:

Min T (1) S.t (2), (3), and

$$\sum_{i \in S} \sum_{r \in R} \sum_{u \in U} w_{iju}^{lr} \leq 1\ \forall j \in S, i \neq j, l \in L \tag{24}$$

$$\sum_{j \in S} \sum_{r \in R} w_{iju}^{lr} - \sum_{j \in S} \sum_{r \in R} w_{jiu}^{lr} = 0\ \forall i \in S, i \neq j, l \in L, u \in U \tag{25}$$

$$\sum_{j\in S_O}\sum_{l\in L}\sum_{u\in U}\sum_{r\in R} w_{iju}^{lr} - My_i \le 0 \quad \forall i \in S_D \tag{26}$$

$$v_{ij}^{ul} - \sum_{r\in R} Mw_{iju}^{lr} \le 0 \ \forall i\&j \in S, i \ne j,\ u \in U,\ l \in L \tag{27}$$

$$\sum_{i\in S} v_{ij}^{ul} - \sum_{i\in S} v_{ji}^{ul} - \sum_{i\in S}\sum_{r\in R} w_{iju}^{lr} = 0 \ \forall j \in S_O, i \ne j, u \in U, l \in L \tag{28}$$

$$\sum_{i\in S}\sum_{j\in s, j\ne i}\sum_{l\in L}\sum_{r\in R} t_{ij} w_{iju}^{lr} \le MT \ \forall l \in L \tag{29}$$

$$\sum_{i\in S}\sum_{j\in s, j\ne i}\sum_{l\in L}\sum_{r\in R} t_{ij} w_{iju}^{lr} + \sum_{i\in S_D}\sum_{j\in S_O}\sum_{l\in L}\sum_{r\in R} STw_{iju}^{lr} \le T \ \forall u \in U \tag{30}$$

$$\sum_{j\in S_O}\sum_{l\in L}\sum_{r\in R} w_{iju}^{lr} - Mp_{iu} \le 0 \quad \forall i \in S_D, u \in U \tag{31}$$

$$\sum_{i\in S_D}\sum_{j\in S_O}\sum_{u\in U}\sum_{r\in R} w_{iju}^{lr} \le 1 \quad \forall l \in L \tag{32}$$

$$\sum_{l\in L}\sum_{u\in U}\sum_{j\in S}\sum_{r\in R, r\ne Direct} w_{iju}^{lr} = 0 \quad \forall i \in S,\ d(i,j) > dm \tag{33}$$

$$\sum_{j\in S} \left(w_{iju}^{lr} + w_{jiu}^{lr}\right) - 2g_{iu}^{lr} \ge 0 \ \forall i \in S_C, i \ne j, l \in L, u \in U,\ r \in R | r \ne Direct \tag{34}$$

$$\sum_{l\in L}\sum_{r\in R|r\ne Direct}\sum_{u\in U} g_{iu}^{lr} \ge 1 \ \forall i \in S_C \tag{35}$$

$$w_{iju}^{l\ Up} = 0 \ \forall i, j \in S \big| i - 1 \le j, l \in L, u \in U \tag{36}$$

$$w_{iju}^{l\ Down} = 0 \ \forall i, j \in S \big| i + 1 \ge j, l \in L, u \in U \tag{37}$$

$$w_{iju}^{l\ Left} = 0\ \forall i, j \in S \Big| i \neq j+1, l \in L, u \in U \tag{38}$$

$$w_{iju}^{l\ Right} = 0\ \forall i, j \in S \Big| i \neq j-1, l \in L, u \in U \tag{39}$$

$$w_{iju}^{lr} \in \{0,1\}\ \forall i\&j \in S, u \in U, l \in L, r \in R \tag{40}$$

$$y_i,\ p_{iu},\ g_{ju}^{lr} \in \{0,1\}\ \forall i \in S_D, j \in S_C, l \in L, r \in R, u \in U \tag{41}$$

Equations (24)–(41) are the same equations as Equations (5)–(16) and Equations (18)–(23) in the 4-index formulation. In this problem, the size of |L| is problem dependent, and an increase in the number of routes results in an exponential increase in the execution time of the problem. The minimum size of |L| equals to the size of |U|; the maximum size can be calculated according to any feasible solution that the problem solver decides. However, it is preferable to use solutions that connect the squares (horizontally or vertically) with minimum turns.

3. METHODOLOGY

The 4-index and 5-index models were coded in GAMS and the Cplex solver was selected. To improve the algorithm and make it more effective, some extensions and techniques were found to be appropriate for our problem. Branch-and-Bound is the main algorithm used for solving MIP problems in Cplex. Using well-known techniques based on the model's features can effectively reduce the execution time of the problem and improve the performance of MIP for effective collaborative decision-making in disaster management. The following subsections shortly describe the techniques that were effective for the models.

3.1. Solver Generated Cuts and Heuristics

In the early sixties, Gomory [35] introduced the procedure of fractional cuts. The procedure was considered impractical until a study [36] proved that Gomory's cuts can be effectively distributed over different branches of the search tree in a branch and cut framework for a mixed 0–1 problem. In the present time, most optimizers use Gomory's cuts for MILP problems. In the Cplex optimizer, the fraccuts option decides whether or not Gomory's fractional cuts should be generated for the problem; and the parameters 0, 1 and 2 control the cuts to be generated automatically, moderately, and aggressively, respectively. In the case of lack of progress in the best bound, one effective solution uses more aggressive levels of cut generation to further tighten the formulation [37]. As initial observations confirm the lack of progress in the best-found-bound, parameter 2 (aggressive cut generation) was found to be more effective for the proposed models. According to the structure of the mathematical formulations, cover cuts are also useful. Therefore, Gomory's fractional cuts and cover cuts were aggressively generated for the models.

GAMS (General Algebraic Modeling System) is a high level modeling language designed for programming the linear, nonlinear, and mixed integer optimization models. GAMS can support different forms of linear and nonlinear programming problems. The system has a language compiler and an integrated high-performance commercial solver, e.g., Cplex. Cplex is a famous commercial optimizer that can solve linear, mixed integer, quadratic, and quadratically constrained programming problems. GAMS provides access to the Cplex solver with a variety of options to interact with it and improve the solution procedure. Cplex uses several alternative algorithms such as primal simplex, dual simplex, barrier, or sifting algorithms. For mixed integer programming problems, Cplex uses the branch-and-cut algorithm.

Cplex takes advantage of different heuristic techniques to avoid tree pollution, increasing the diversity of the branch-and-bound search, achieving integer feasible solution in shorter time, and fastening the final optimality. Based on the characteristics of the proposed problem, two heuristic features from the different heuristic procedures offered by Cplex were utilized, i.e., local branching heuristic (LBHeur), and node heuristic (NHeur). The LBHeur was introduced by Fischetti and Lodi [38] and has been widely used by commercial optimizers. We can infer the following facts based on the ILOG Cplex explanations:

- LBHeur attempts to enhance new incumbents found throughout an MIP search. The LBHeur algorithm will be appealed only if a new incumbent is identified. In cases where several incumbents are identified at a single node, the last one will be considered.
- NHeur implements the techniques to find a feasible solution from the present node in branch and bound algorithm. The parameter HeurFreq restricts this offer. A positive number specifies the frequency of invoking NHeur (in the number of nodes).

3.2. Additional Constraints

User cuts or additional constraints are the polynomial constraints that strengthen the mathematical formulation and reduce the solution space. These cuts are not supposed to affect the feasible solutions space. The following constraints eliminate unfavorable fractional solutions and expedite the execution process:

$$\sum_{i\in S_D}\sum_{j\in S_O} v_{ij}^{ul} - \sum_{i\in S}\sum_{j\in S_O}\sum_{r\in R} w_{iju}^{lr} = 0 \; \forall u \in U, l \in L \tag{42}$$

$$\sum_{i\in S_O}\sum_{j\in S_D}\sum_{u\in U}\sum_{l\in L} v_{ij}^{ul} = 0 \tag{43}$$

Equation (42) imposes that the total number of squares that are to be visited by UAV u in route l right after leaving the depot is equal to the total number of squares allocated to UAV u in route l. Equation (43) ensures that the value is equal to 0 just before ending the tour at the depot. Equations (9), (42) and (43) are adapted from a previous study [39].

$$p_{iu} - y_i \leq 0 \; \forall i \in S_D, u \in U \tag{44}$$

Equation (44) indicates that if a UAV belongs to a depot, then the depot must be open.

$$\sum_{i\in S_D}\sum_{l\in L}\sum_{u\in U}\sum_{r\in R}\sum_{j\in S_O} w_{iju}^{lr} \leq |L| \tag{45}$$

Equation (45) ensures that the total number of arcs leaving the depots is less than or equal to the number of the available routes.

$$\sum_{i\in S_N}\sum_{l\in L}\sum_{u\in U}\sum_{r\in R|r\neq direct}\sum_{j\in S_N} w_{iju}^{lr} = 0 \tag{46}$$

According to the logic of the problem and the grid-based decomposed map, the UAVs must travel directly through the nodes of the SE set, and it is applied in Equation (46). As mentioned earlier, only the residential squares must be covered by horizontal or vertical movements. In other words, in favor of the objective function value, and in order to find the minimal path, the UAVs should travel in the *direct* direction while passing through the nonresidential squares.

$$\sum_{r\in R}\sum_{u\in U}\sum_{i\in S_C} g_{iu}^{lr} - \sum_{i\in S}\sum_{j\in S_C}\sum_{u\in U}\sum_{r\in R} w_{iju}^{lr} \le 0 \; \forall l \in L \qquad (47)$$

Equation (47) ensures that the number of times a UAV enters a square in each route is more than or equal to the number of times the square is covered in that route.

3.3. Variable Branching Priority

The variable branching priority based on the mathematical formulation's specifications is one of the most efficient techniques that speeds up the solution procedure and tightens the model. Without using this feature, the solver automatically decides the appropriate variable to be branch on. It is imperative to select a suitable priority among the existing variables depending on the variables' dependence, relevance, and significance in the proposed mathematical model. In this study there are 4-index and 5-index mathematical formulations of a single problem. In this study, 4-index and 5-index mathematical formulations were used for a single problem. Based on the model's characteristics and variables' significance, the following priorities were suggested:

I. For the 4-index model, the variable branching priority order must be glri, xlrij, wulij, vulij, piu and yi.

II. The branching priority of the 5-index model should be glri, vulij, wlriju, piu, and yi.

4. COMPUTATIONAL EXPERIMENTS AND RESULTS

In this section, some instances are designed to test the validity of the models and features proposed in Section 3. Based on the problem's characteristics, it is assumed that the number of available UAVs, number of possible open depot squares, and maximum limit for the total traveling time of each UAV are known. Each test instance was named in the X-Y-Z-L format, where X refers to the total number of squares in the map, Y refers to the total number of residential squares, Z represents the number of potential UAV bases, and L is the number of available UAVs. The test instances are shown in Figure 4. For instance, in problem 36-11-2-3 map, where 36 is the total number of squares, 11 squares are residential areas, two squares are potential depots, and three represents the available UAVs.

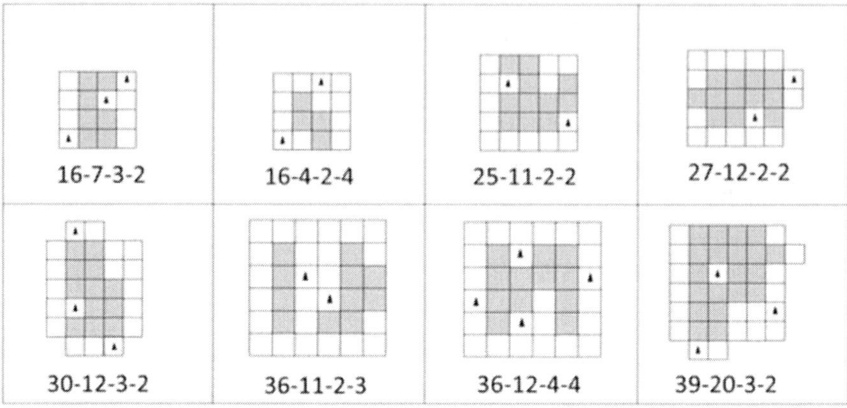

Figure 4. Test instance maps (the gray squares are assumed to be residential areas that should be covered by UAVs and squares with black triangles are potential UAV bases).

It is assumed that each small square side is one in terms of length unit and the speed of the UAVs is also one in terms of speed unit. The mathematical models were coded in GAMS 24.7.3 and solved using the Cplex 12.6.3.0 solver. The models were run on a laptop with the configuration of Intel® Core™ i7–2620M CPU 2.70 GHz, RAM 8.00 GB, and windows 10 Enterprise operating system.

To evaluate the efficiency of the techniques mentioned in Section 3, each technique was considered alone in an individual run at the first step. Subsequently, the combination of efficient ones was used. All the problems were solved in a 5-h time limit, irrespective whether a feasible or optimal solution was found or not. The reason for the 5-h time limit is that no significant improvement would occur by increasing this limit. The calculation time increases exponentially by increase in the number of squares or the number of routes. Table 2 and Table 3 show the results of the 4-index and 5-index models, respectively, in a combination of techniques mentioned in Section 3.

Table 2. Effects of techniques described in Section 3.1, Section 3.2 and Section 3.3 on the 4-index mathematical model.

	4-Index Model														
	Original Model			3.1, 3.2			3.1, 3.3			3.2, 3.3			3.1, 3.2, 3.3		
Instance	CPU	Results	Gap (%)	CPU	Results	Gap (%)	CPU	Results	Gap (%)	CPU	Results	Gap (%)	CPU	Results	Gap (%)
16-7-3-2	254	21.3	0	136	21.3	0	281	21.3	0	216	21.3	0	284	21.3	0
16-4-2-4	18,000	13.15	15.2	3850	13.15	0	4300	13.15	0	5400	13.15	0	3650	13.15	0
25-11-2-2	1400	24	0	310	24	0	343	24	0	162	24	0	84	24	0
27-12-2-2	18,000	35.98	35.1	1325	28	0	2104	28	0	1206	28	0	1618	28	0
30-12-3-2	12,587	26.32	0	2877	26.32	0	2754	26.32	0	3107	26.32	0	2639	26.32	0
36-11-2-3	18,000	NA	-	18,000	NA	-	18,000	NA	-	18,000	NA	-	18,000	NA	-
36-12-4-4	18,000	NA	-	18,000	NA	-	18,000	NA	-	18,000	NA	-	18,000	24.4	59.4
39-20-3-2	18,000	NA	-	18,000	NA	-	18,000	NA	-	18,000	NA	-	15,669	36	0

Table 3. Effects of techniques described in Section 3.1, Section 3.2 and Section 3.3 on the 5-index mathematical model.

	5-Index Model														
	Original Model			3.1, 3.2			3.1, 3.3			3.2, 3.3			3.1, 3.2, 3.3		
Instance	CPU	Results	Gap (%)	CPU	Results	Gap (%)	CPU	Results	Gap (%)	CPU	Results	Gap (%)	CPU	Results	Gap (%)
25-5-3-3	888	21.3	0	467	21.3	0	782	21.3	0	159	21.3	0	129	21.3	0
16-4-2-4	11,320	13.15	0	1359	13.15	0	5527	13.15	0	650	13.15	0	479	13.15	0
25-11-2-2	850	24	0	286	24	0	186	24	0	226	24	0	48	24	0
27-12-2-2	18,000	30.82	30.9	966	28	0	18,000	30.24	30.1	570	28	0	559	28	0
30-12-3-2	6875	26.32	0	2472	26.32	0	3830	26.32	0	2958	26.32	0	1421	26.32	0
36-11-2-3	18,000	NA	-	18,000	NA	-	18,000	NA	-	18,000	21.3	13.7	18,000	20.79	6.3
36-12-4-4	18,000	NA	-	18,000	NA	-	18,000	NA	-	18,000	NA	-	18,000	22.82	49.2
39-20-3-2	18,000	NA	-	18,000	42.82	27.1	18,000	NA	-	18.000	36	11.7	10,554	36	0

According to the results presented in Table 2 and Table 3, it can be concluded that the 5-index model shows better solutions in terms of CPU time and the absolute gap. Furthermore, as previously anticipated, the combination of the proposed LP strengthening techniques offered better solutions in a shorter amount of time. The CPU time of the 5-index problem 16-4-2-4 was 11,320 s in the original model, which decreased to 479 s by adding the techniques. In addition, for problem 36-12-4-4, a feasible solution was not obtained in the original model; however, with a combination of techniques, a solution with 6.3% relative gap was acquired. Furthermore, while no feasible solution was found for the original model of problem 39-20-3-2 in the 5-h time limit, the optimal solution was found under the influence of the techniques in 10,554 s. Figure 5 depicts the optimal solutions for problems 39-20-3-2 and 36-11-2-3 test instances as examples.

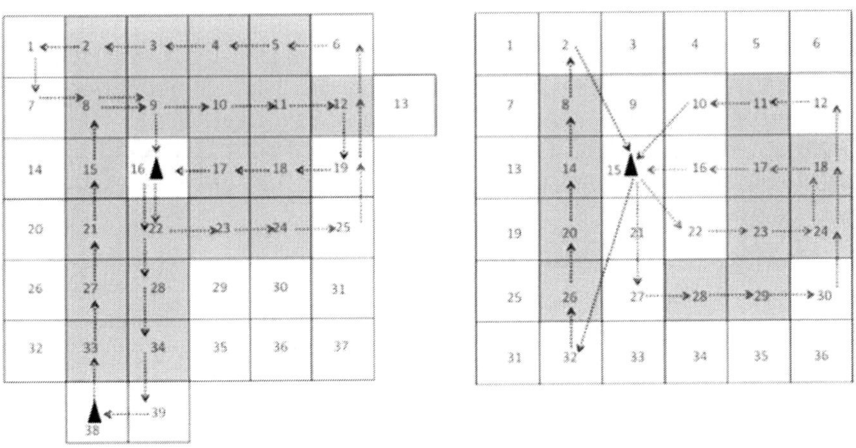

(a) (b)

Figure 5. The optimal solutions of problems 39-20-3-2 and 36-11-2-3. (**a**) There are two available UAVs and each belongs to one base; (**b**) There are three available UAVs and base number 15 is selected to be open.

5. CONCLUSIONS AND FUTURE PROSPECTS

A multi-UAV and multi-tour CPP problem based on the grid map decomposition was investigated. MUTLCPP supports the application of post-earthquake CPP where high accuracy is required. Moreover, due to the speedy decrease in the survival rate in post-earthquake situations, the Minimax objective function was considered for the optimization problem. Two mathematical models of MUTLCPP were presented and coded in GAMS, and the 5-index model outperformed the 4-index formulation in terms of solution quality and execution time. The models were improved by additional constraints and other techniques described in Section 3. On the basis of the results of the test instances, it can be affirmed that the combination of the proposed techniques sped up the execution procedure and gained solutions with a lower gap in a shorter amount of time. In the small size instance of 16-7-3-2, the combination of techniques decreased the calculation time by 85% for the 5-index model but increased by 7% for the 4-index formulation. The results depicted in Table 2 and Table 3 show that the linear strengthening techniques are more effective for the 5-index model. Based on the results of Table 2 and Table 3, the proposed techniques decreased the CPU time by 85%, 95%, 94%, 96%, and 79% for the first five instances of the 5-index model and decreased the time by 79%, 94%, more than 91%, and 79% for the corresponding five instances of the 4-index model. For the further three instances, while the solution is not available in the original models, by the intervention of the linear strengthen techniques optimal solutions or very good feasible solutions were found. Generally, by comparing the results (Table 2 and Table 3), it is clear that the 5-index model performs much better that the 4-index formulation. Another conclusion based on the obtained results is that the solver generated cuts (Section 3.1) have less effect, and additional constraints (Section 3.2) have the maximum effect on decreasing the CPU time.

There are several interesting directions for further investigations. Some new valid inequalities can be added to the model to strengthen the MILP formulations. In addition, exact methods such as the lagrangian relaxation can be developed to find the optimal or near-optimal solutions to large real size problems. The most interesting subject for further investigations is designing a heuristic or meta-heuristic algorithm to solve large-scale problems of MUTLCPP. Each randomly generated individual solution in a heuristic algorithm can be represented in such a way that each cell takes an integer <6 indicating the UAV direction (set R). Finally, one can extend the problem by reflecting the map rotation possibility to test the different covering directions.

ACKNOWLEDGMENTS

The authors would like to thank the anonymous reviewers for their valuable comments.

AUTHOR CONTRIBUTIONS

Arman Nedjati conceived the problem, designed the mathematical models, and wrote the paper. Gokhan Izbirak, Bela Vizvari and Jamal Arkat revised the models and the manuscript.

CONFLICTS OF INTEREST

The authors declare no conflict of interest.

REFERENCES

1. Pesaresi, M.; Gerhardinger, A.; Haag, F. Rapid damage assessment of built-up structures using VHR satellite data in tsunami-affected areas. Int. J. Remote Sens. 2007, 28, 3013–3036.
2. Chang, K.T.; Wang, E.H. Developing procedures for post-earthquake structural evaluation by laser scanning techniques. Insight-Non-Destr. Test. Cond. Monit. 2012, 54, 562–567.
3. Szantoi, Z.; Malone, S.; Escobedo, F.; Misas, O.; Smith, S.; Dewitt, B. A tool for rapid post-hurricane urban tree debris estimates using high resolution aerial imagery. Int. J. Appl. Earth Obs. Geoinf. 2012, 18, 548–556.
4. Olsen, M.J.; Chen, Z.; Hutchinson, T.; Kuester, F. Optical techniques for multiscale damage assessment. Geomat. Nat. Hazards Risk 2013, 4, 49–70.
5. Kashani, A.G.; Crawford, P.S.; Biswas, S.K.; Graettinger, A.J.; Grau, D. Automated Tornado Damage Assessment and Wind Speed Estimation Based on Terrestrial Laser Scanning. J. Comput. Civ. Eng. 2015, 29, 4014051.
6. Jiang, S.; Friedland, C.J. Automatic urban debris zone extraction from post-hurricane very high-resolution satellite and aerial imagery. Geomat. Nat. Hazards Risk 2016, 7, 933–952.
7. Janalipour, M.; Mohammadzadeh, A. Building Damage Detection Using Object-Based Image Analysis and ANFIS From High-Resolution Image (Case Study: BAM Earthquake, Iran). IEEE J. Sel. Top. Appl. Earth Obs. Remote Sens.2016, 9, 1937–1945.
8. Fernandez Galarreta, J.; Kerle, N.; Gerke, M. UAV-based urban structural damage assessment using object-based image analysis and semantic reasoning. Nat. Hazards Earth Syst. Sci. 2015, 15, 1087–1101.
9. Al-Wassai, F.A.; Kalyankar, N.V. Major Limitations of Satellite images. Available online: https://arxiv.org/ftp/arxiv/papers/1307/1307.2434.pdf (accessed on 29 November 2016).

10. Grandoni, D. Advantages and Limitations of Using Satellite Images for Flood Mapping. Available online: http://www.copernicus.eu/sites/default /files /documents/User_uptake/Emergency_Events/Athens/03_Session_01_AdvantageAn dLimitationsOfUsingSatelliteImagery.pdf (accessed on 29 November 2016).
11. Qi, J.; Song, D.; Shang, H.; Wang, N.; Hua, C.; Wu, C.; Qi, X.; Han, J. Search and Rescue Rotary-Wing UAV and Its Application to the Lushan Ms 7.0 Earthquake. J. Field Robot. 2016, 33, 290–321.
12. Zelinsky, A.; Jarvis, R.A.; Byrne, J.C.; Yuta, S. Planning Paths of Complete Coverage of an Unstructured Environment by a Mobile Robot. In Proceedings of the International Conference on Advanced Robotics, Tokyo, Japan, 1–2 November 1993; pp. 533–538.
13. Nedjati, A.; Vizvari, B.; Izbirak, G. Post-earthquake response by small UAV helicopters. Nat. Hazards 2015, 80, 1669–1688.
14. Kendoul, F. Survey of Advances in Guidance, Navigation, and Control of Unmanned Rotorcraft Systems. J. Field Robot. 2012, 29, 315–378.
15. Colomina, I.; Molina, P. Unmanned aerial systems for photogrammetry and remote sensing: A review. ISPRS J. Photogramm. Remote Sens. 2014, 92, 79–97.
16. Montoya-Torres, J.R.; López Franco, J.; Nieto Isaza, S.; Felizzola Jiménez, H.; Herazo-Padilla, N. A literature review on the vehicle routing problem with multiple depots. Comput. Ind. Eng. 2015, 79, 115–129.
17. Oberlin, P.; Rathinam, S.; Darbha, S. Today's Traveling Salesman Problem. IEEE Robot. Autom. Mag. 2010, 17, 70–77.
18. Ho, H.-M.; Ouaknine, J. The Cyclic-Routing UAV Problem is PSPACE-Complete. In Foundations of Software Science and Computation Structures; Springer: Berlin/Heidelberg, Germany, 2015; pp. 328–342.
19. Gendreau, M.; Laporte, G.; Semet, F. The Covering Tour Problem. Oper. Res. 1997, 45, 568–576.
20. Choset, H. Coverage for robotics—A survey of recent results. Ann. Math. Artif. Intell. 2001, 31, 113–126.
21. Galceran, E.; Carreras, M. A survey on coverage path planning for robotics. Robot. Auton. Syst. 2013, 61, 1258–1276.
22. De Carvalho, R.N.; Vidal, H.A.; Vieira, P.; Ribeiro, M.I. Complete coverage path planning and guidance for cleaning robots. In Proceedings of the 1997 IEEE International Symposium on Industrial Electronics, Guimaraes, Portugal, 7–11 July 1997; pp. 677–682.
23. Li, Y.; Chen, H.; Joo Er, M.; Wang, X. Coverage path planning for UAVs based on enhanced exact cellular decomposition method. Mechatronics 2011, 21, 876–885.
24. Barrientos, A.; Colorado, J.; Cerro del, J.; Martinez, A.; Rossi, C.; Sanz, D.; Valente, J. Aerial remote sensing in agriculture: A practical approach to area coverage and path planning for fleets of mini aerial robots. J. Field Robot.2011, 28, 667–689.
25. Torres, M.; Pelta, D.A.; Verdegay, J.L.; Torres, J.C. Coverage path planning with unmanned aerial vehicles for 3D terrain reconstruction. Expert Syst. Appl. 2016, 55, 441–451.
26. Wang, X.; Sun, T.; Li, D. Energy-optimal coverage path planning on topographic map for environment survey with unmanned aerial vehicles. Electron. Lett. 2016, 52, 699–701.
27. Easton, K.; Burdick, J. A Coverage Algorithm for Multi-robot Boundary Inspection. In Proceedings of the 2005 IEEE International Conference on Robotics and Automation; Barcelona, Spain, 18–22 April 2005; pp. 727–734.
28. Huang, W.H.; York, N. Optimal Line-sweep-based Decompositions for Coverage Algorithms. In Proceedings of the 2001 IEEE International Conference on Robotics & Automation, Seoul, Korea, 21–26 May 2001; pp. 27–32.
29. Maza, I.; Ollero, A. Multiple UAV cooperative searching operation using polygon area decomposition and efficient coverage algorithms. In Distributed Autonomous Robotic Systems 6; Springer: Tokyo, Japan, 2007; pp. 221–230.

30. Avellar, G.; Pereira, G.; Pimenta, L.; Iscold, P. Multi-UAV Routing for Area Coverage and Remote Sensing with Minimum Time. Sensors 2015, 15, 27783–27803.
31. Wang, X.; Li, D. Coverage path planning for UAVs in unknown directional regions. Int. J. Wirel. Mob. Comput. 2015, 8, 285–293.
32. Lemmens, M. Geo-information Technology—What It Is, How It Was and Where It Is Heading to. In Geo-Information; Springer: Dordrecht, The Netherlands, 2011; pp. 1–22.
33. Greenwood, F. How To Make Maps With Drones. In Drones and Aerial Observation: New Technologies for Property Rights, Human Rights, and Global Development a Primer; New America: Washington, DC, USA, 2015; pp. 35–48.
34. Falkner, E.; Morgan, D. Aerial Mapping: Methods and Applications; Lewis Publishers: Boca Raton, FL, USA, 2002.
35. Gomory, R.E. An Algorithm for the Mixed Integer Problem; RAND Corporation: Santa Monica, CA, USA, 1960.
36. Balas, E.; Ceria, S.; Cornuéjols, G.; Natraj, N. Gomory cuts revisited. Oper. Res. Lett. 1996, 19, 1–9.
37. Klotz, E.; Newman, A.M. Practical guidelines for solving difficult mixed integer linear programs. Surv. Oper. Res. Manag. Sci. 2013, 18, 18–32.
38. Fischetti, M.; Lodi, A. Local branching. Math. Program. 2003, 98, 23–47.
39. Karaoglan, I.; Altiparmak, F. A memetic algorithm for the capacitated location-routing problem with mixed backhauls. Comput. Oper. Res. 2015, 55, 200–216.

CHAPTER 14

Short-Term Earthquake Forecast with the Seismic Sequence Hierarchization Method

Giulio Riga[1], Paolo Balocchi[2]

[1]*Geologist, Independent Researcher, Lamezia Terme, Italy*

[2]*Geologist, Independent Researcher, Modena, Italy*

ABSTRACT

All strong earthquakes are preceded by branching structures having different durations whose development scheme is partly largely predictable because it follows a well organized and recognizable pattern. By using a seismic sequence hierarchization method, this study graphically explains the preparation process of an earthquake, called "branching structure". In addition, criteria apt to distinguish the structures that will produce shocks of average magnitude from strong earthquakes' will be defined. Based on the temporal oscillations of the magnitude values, we explain the procedure for identifying the developmental stages that characterize the energy accumulation stage of the branching structure, in order to early detect the energy release stage's trigger point and obtain information on how it will develop over time. The study identifies also some pre-signals (trigger points) of various magnitudes in the energy release stage, which allows us to early predict the foreshocks and mainshock time position. The method we developed constitutes a truly innovative approach for the earthquake forecasting analysis, which dramatically differs from those developed so far, as it considers the structure of the seismic sequence not only as a magnitude values' oscillation, but also as a sequence of developmental stages that may begin much earlier.

Keywords: Branching Structure, Earthquake, Microsequence, Hierarchization, Foreshock

1. INTRODUCTION

In some cases the main shock is triggered by one or more lower magnitude early events, called foreshocks, which in turn can trigger additional seismic events. Even a strong earthquake is followed in time and space by groupings of earthquakes with a lower magnitude, called aftershocks, which generally are triggered by the same main event.

These aspects related to the aftershocks behavior have been incorporated in the triggering models, where the probability of occurrence of an earthquake in a given time and place is associated with the past history of the earthquakes occurred in the surrounding areas. The most famous model is the short term ETAS (Epidemic Type Aftershock Sequence) [1] [2] , where each event, regardless of its magnitude, increases the probability of triggering seismic events in the surrounding areas. In this system, a sequence of aftershocks may be seen as a "trigger cascade" [3] that extends geographical scope of the main initial shock. Conversely, according to our opinion, this approach highlights how the seismic sequence is a seqeunce of energy release and accumulation phases that develop before and after the strongest shocks [4] .

A long term challenge consists in determining which earthquakes are connected, directly or indirectly, and in identifying the earthquakes sequences based on causal connection [5] . In the past, various declustering algorithms were proposed to subdivide earthquakes into main shock and aftershocks [6] - [11] . Some of these methods are arbitrary rules characterized by various parameters which jeopardise their objectivity. To solve these issues, a series of methods that seek to evaluate the correlations between the earthquakes has been developed [6] - [11] .

We believe that the study and identification of branched structures, without resorting to a physical modeling, can provide valuable information on the evolution of the seismic sequence and on the process of strong earthquake preparation as well as on triggering relationships between earthquakes.

An unquestionably crucial aspect in the analysis of the magnitude values' historical series is the identification of the graph patterns that may help to explain some of the dynamics issues related to the preparation of an earthquake, especially those anticipating an energy release cycle and allow us to identify the critical stages.

For example, the branching structure is a pattern that not only allows us to evaluate graphically the extent of the preparation process of an earthquake but also contributes to outline the past and future evolution of an area's seismicity.

In the initial developmental stages, the branching structures develop slowly, while in the final phases before the energy release stage, the decrease of the magnitude values is very rapid.

These are identified by considering that after an energy shock, the magnitude values gradually decrease up to the trigger point by following a hierarchization process.

Normally, each branching structure is connected with one or more medium-strong earthquakes concerning a certain area. In some cases, after the branching structure's trigger point formation, various orders foreshocks with increasing magnitude may form, which play a crucial role in the preparation phase of an earthquake.

In this case, the energy release stage is characterized by one or more foreshocks as well as by one or more mainshocks [4] [12] .

By using the graphic and quantitative analysis, we can learn the developmental stages of the branching structure, assess whether the phenomenon develops according to the expected standard scenario and know early the "hazard degree" (Short-Term Earthquake Probabilities) of the region examined, which heavily depends on space-time relationships between each event [13] . Therefore, the hazard degree is a function of the seismic sequence's developmental stage. If the area is located in the first developmental stages, the hazard degree of the area will be relatively low compared to the last developmental stages where it will feature a greater degree.

The analysis of the seismic sequence structure is performed using: 1) a datasets spatial coverage involving either parts of a region or several countries (from a few tens of km to thousands of km), 2) a magnitude values range of 2.0 - 10 m, 3) a depths range of 1 - 50 km.

2. BRANCHING STRUCTURE DEVELOPMENTAL PATTERN

The strongest earthquakes usually trigger aftershocks aimed at re-adjusting the new stress condition resulting from the strongest shock. The aftershocks show a peculiar pattern that develops over time following a hierarchization process identified as branching structure. Such structures are not born equal, but since the first developmental stage we can understand if they will generate a medium or strong earthquake. Indeed, the energy that a branching structure can release in the energy release stage, depends on the number and duration of the developmental stages and the average magnitude of the first developmental stage. The greater the duration, the number of the developmental stages and the average magnitude of the first developmental stage, the greater the released energy.

The entire branching structure preparation process can be divided into three phases: 1) energy accumulation; 2) triggering; 3) energy release (Figure 1) [4] .

1) Energy accumulation stage. It consists of five developmental stages related to temporal amplitude, magnitude and number of seismic events decreasing over time, which may even last for several years and in some cases constitutes 90% of the branching structure.

The most energetic shocks take place during the first developmental stage.

2) Trigger stage. It consists of small magnitude shocks organized in an easily detectable micro-sequence called "Double Bottom-Three Small Earthquakes" (DB-3SE). This stage is easily recognizable when the last shock coincides with the point in which the elastic deformation probably reaches its peak local values (trigger point).

3) Energy release stage. During this stage, the energy is released immediately after the trigger point (Flash earthquakes type) or in one or more cycles of increasing amplitude and magnitude (Progressive earthquakes type), characterised by a seismic activity acceleration known as foreshocks [14] [15] . Foreshocks depend on different, still unclear physical mechanisms [16] - [18] and start to develop even hours/years before a strong earthquake

In summary, in the pattern of the branching structure the following apply:

1) The most energetic earthquakes occur, at irregular intervals, during the first part of the energy accumulation stage and in the final part of the energy release phase;

2) The amplitude of the branching structure's developmental stages tends to decrease over time in the energy accumulation stage and to increase in the energy release stage;

3) The energy can be abruptly (Flash earthquakes type) or gradually (Progressive earthquakes type) released, alternating standby with peak activity periods;

4) The energy release stage is activated by small magnitude earthquakes organized in simple DB-3SE micro- sequences wherein the last shock is the trigger point;

5) In the energy accumulation and release stages, multiple seismogenetic structures, at regional or interregional level, may be involved [19] - [21] ;

6) The aftershocks and foreshocks contribute to dissipate tectonic energy. Conversely, the cumulative effect of aftershocks increases it while foreshocks' lower it [22] .

Figure 1 shows the branching structure's pattern: we see that the first developmental stage (the widest) starts to develop after a strong earthquake, while the subsequent ones are related to the magnitude values decrease over time. The branching structure's trigger point determines the transition from the accumulation to the energy release stage that develops in a shorter time compared to the first.

3. METHODS

3.1. Linear Graph

The graph is one of the most important components in making both long- and short-term predictions process.

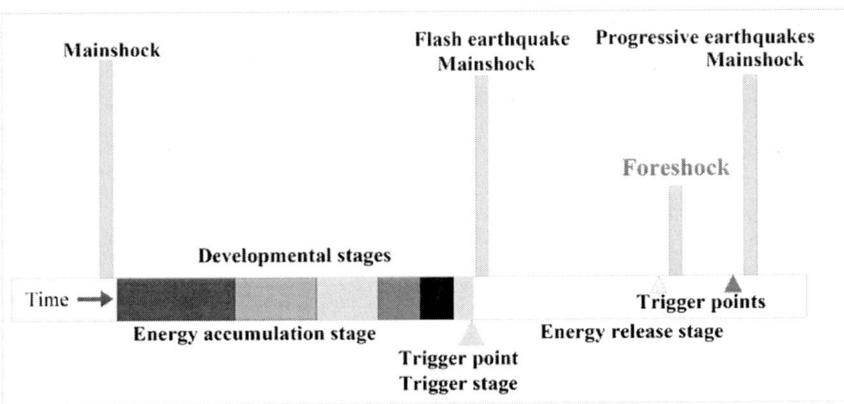

Figure 1. Branching structure's schematic representation. The variously colored rectangles indicate the developmental stages of the energy accumulation stage.

By representing the magnitude values as a linear graph, we can see how the magnitude values behave in the time window considered, identify the energy accumulation and release stage, detect the premonitory patterns [2] , set projections related to future magnitude values and strong earthquakes occurrence times.

The time frame used to process a short term graph can be daily/weekly, while for the medium/long-term is monthly/yearly. Obviously, the wider the reference window, the greater the details shown in the graph.

The line segment that connects two points on the graph shows a local stage, which can be visually interpreted compared to other lines' by means of a coded color.

Figure 2 shows the linear graph of the magnitude values whose reading is quite simple. As we note, the seismic sequence develops by alternating energy release with adjustment stages. Black-colored circles indicate the magnitude points M in which the reverse of the cyclical adjustment (red-colored lines) and energy (green- colored lines) stages occur. Both the energy release and adjustment stages may consist of one or more seismic events.

Although this type of graph is perfect for a quick and immediate glimpse of the seismic sequence developmental structure, we need to examine it in depth to obtain more detailed information.

For an early identification of the energy accumulation and release stages, we need to join the relative maximum points with a straight line both in case of downward and upward magnitude values (Figure 3).

These straight lines are called dynamic trendlines (magnitude values vary with the time) and clearly indicate the direction in which the values develop, both in the forthcoming and in the long term period.

The upward breaking of the breakout trendline activates the energy release stage, which is characterized by one or more mainshoks (Flash earthquakes type) or by a first foreshock (Progressive earthquakes type) followed a first adjustment shock, which involves the magnitude values' return to somewhere close or below the energy accumulation trendline. Then the values begin to rise again until a second foreshock occurs, which allows us to understand the direction of the energy release trandline.

Typically, although with the Progressive earthquakes type we can record up to four foreshocks, most frequently the pattern consists of two foreshocks and one mainshock.

Besides indicating the ongoing trend, the trendlines temporarily prevent the magnitude values from rising by allowing, in some cases, the early detection of certain levels of magnitude that will be achieved in the future.

Actually, by analyzing the graph shown in Figure 3, we can see how the knowledge of the magnitude levels achieved in points 1 and 2 allows us to calculate the magnitude value in points 3 and 4 of the energy accumulation stage or the magnitude value in point 6 of the energy release stage after the the points 4 and 5 have formed. The point 4, which is common to the two phases, is the transition point from the energy accumulation and the energy release stage.

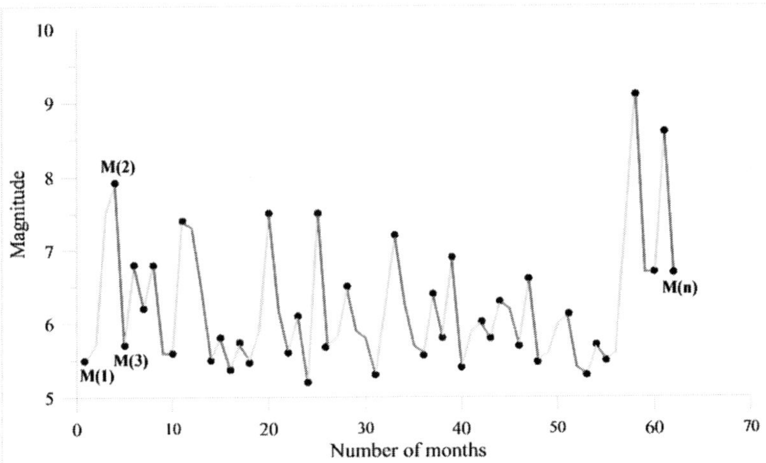

Figure 2. Earthquake linear graph, Japan, 2011.

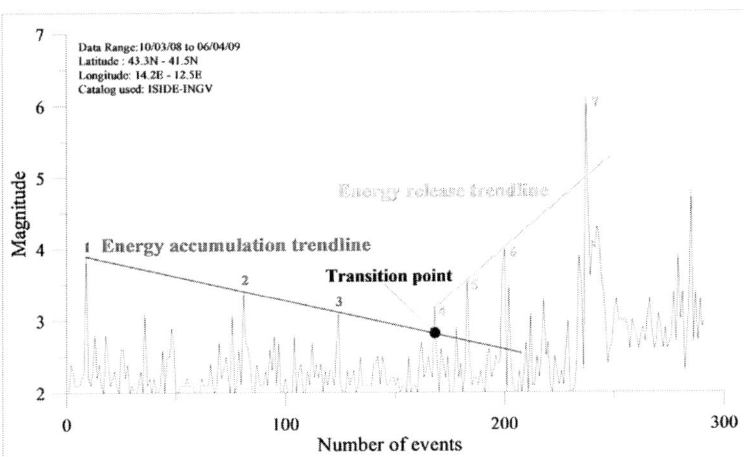

Figure 3. Earthquake linear graph, L'Aquila (Italy), 2009. The red-colored line indicates the energy accumulation trendline, while the green-colored one indicates the energy release trendline. The black-colored circles represent the transition point between the accumulation and the energy release stage.

3.2. DB-3SE Microsequence-Generated Trigger Point

Some studies performed on seismic sequences show how earthquakes of any magnitude can trigger other earthquakes [23] [24] that have different triggering properties [25] . Generally, low magnitude earthquakes are greater in number compared to those of higher magnitude: therefore, they collectively can trigger also big seismic events [24] [26] due to the stress in the surrounding area [27] . Through the study of seismic sequence, and with appropriate calculation algorithms it is possible to distinguish the seismic events triggered by the background activity [8] .

By analyzing the magnitude values trend occurring before the start of an energy release stage, we can identify a small shocks microsequence that may be used for the prediction of immediate and short term medium/high magnitude earthquakes. In fact, the energy release stage's trigger point coincides with the third shock in the DB-3SE microsequence, which consists of three, small magnitude shocks and is a very reliable stage-reverse signal.

Once completed the last phase of the branching structure's energy accumulation stage, in order to locate the first trigger point and the following ones, in the case where the energy release stage develops according to a Progressive earthquakes-type pattern, it is necessary that the magnitude values of the shocks that make up the temporary DB-3SE microsequence (Figure 4(a)) meet the following conditions:

$$M(n-1) \le M(n) > M(n+1) \tag{1}$$

$$M(n-1) < M(n) \ge M(n+1) \tag{2}$$

and

$$M(n) < M(n-2) \tag{3}$$

where M(n) is the second shock's magnitude value, while M(n – 1) and M(n + 1) are the magnitude values of the first and third shock, respectively.

For the completion of the DB-3SE microsequence, we need a subsequent shock whose magnitude must be:

$$M(n+2) \ge M(n)$$ (Figure 4(b)). (4)

Figure 4 shows both temporary and final DB-3SE microsequences, while the red-colored triangle indicates the the energy release stage's trigger point.

The trigger point coincides with the M(n + 1) magnitude shock in the DB-3SE microsequence, while the subsequent final shock often corresponds to a more energetic event (foreshock or mainshock). Figure 5 shows the four possible patterns of the final DB-3SE microsequence, depending on the magnitude values of the three shocks which it consists of (red-colored circles).

3.3. Seismic Sequence Hierarchization

A seismic sequence may be considered as a succession of cyclical shocks with variable amplitude over time where events occur following a particular hierarchization process during which each earthquake can generate several earthquakes [28] .

The herarchization process's basic concept is that a source event (strong earthquake), can generate seismic events of decreasing magnitude during the energy accumulation stage. In essence, these seismic events related to the source event and classified as secondary [29] , may also, in turn, generate other minor seismic events, thus triggering a process that continues up to the energy release phase's trigger point and may end after a strong event.

The number of seismic events included in a seismic cycle (adjustment and energy release stages) depends on the seismicity bottom value (i.e., the minimum value in the analyzed magnitude values' time series). The greater the bottom value, the smaller the number of events the cycle will consist of.

The use of a herarchization process in the seismic sequence analysis, allowed us to develop a graph pattern that effectively contributes to outline the past and future development of the seismic sequence. In fact, all the adjustment and energy release stages included in a seismic sequence constitute the "seismic grid", whose development is probably related to various mechanisms involving, among the others, the lithospheric rocks' deformation process and the interaction between the different seismogenetic structures on regional or interregional scale.

The seismic grid can be represented by some geometric entities, such as the "seismic nodes", i.e. the contact points between the two phases, the "reverse phase nodes" that are the transition points between the adjustment and the energy release stage, and, finally, the energy release (r) and accumulation (a) "seismic branches" that make up a seismic cycle (Figure 6).

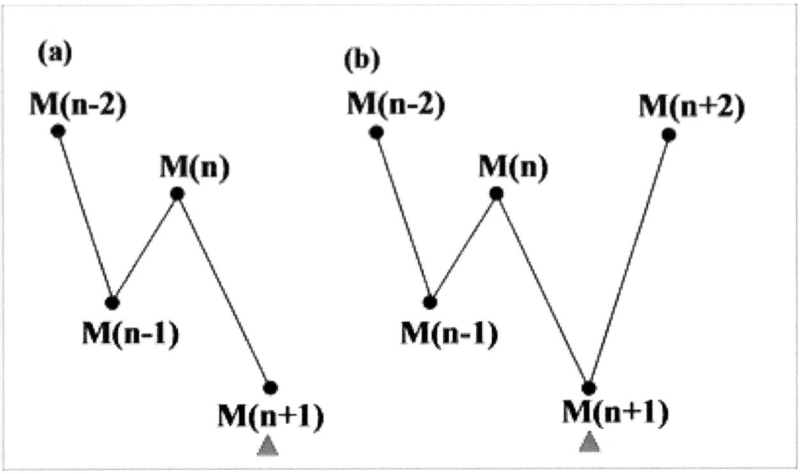

Figure 4. Schematic representation of DB-3SE microsequence: (a) temporary microsequence, (b) final microsequence.

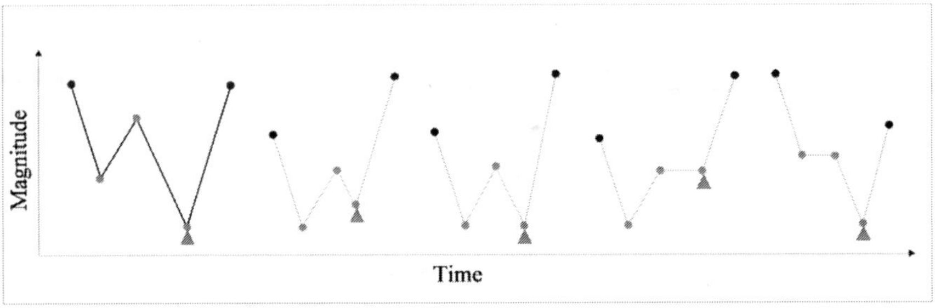

Figure 5. Schematic representation of microsequence DB-3SE. The red-colored circles indicate the "DB-3SE" microsequence while the red-colored triangle indicates the trigger point. The black-colored circles represent the microsequence's initial and final shocks.

As to the branches arrangement, the seismic grid can be arranged by applying a hierarchization method that consists in assigning an order number to the various branches that make up the seismic grid according to the following rules:

- One or more branches of n order energy accumulation followed by one or more n order energy release branches result in a n + 1 order branch (the hierarchical order increase occurs);
- The magnitude value of the "reverse phase nodes" increases as the cycle order increases;
- The higher order branch (5A) is the higher order seismic cycle.

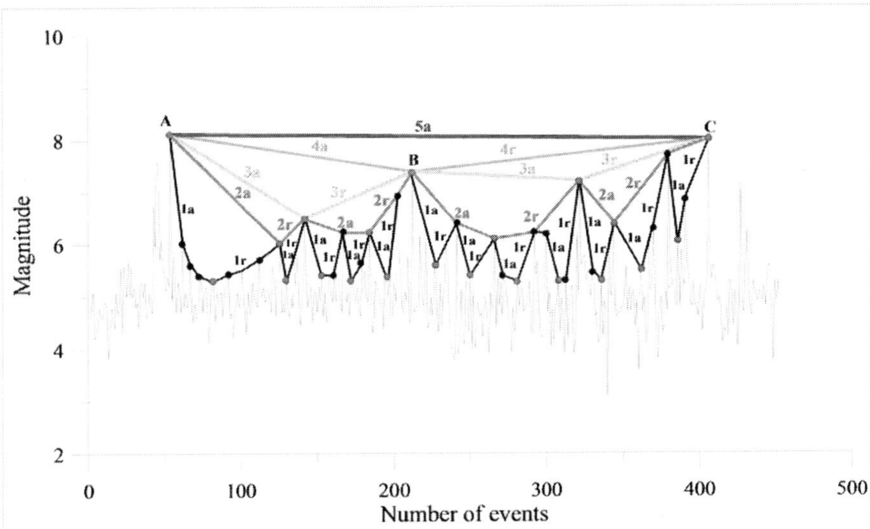

Figure 6. Schematic representation of the seismic sequence hierarchy. Each circle represents a seismic node, while the red-colored ones are the reverse phase nodes. The lines of different thickness and color are the earthquakes' branches. The higher order AC branch represents the higher order seismic cycle consisting of the energy accumulation (4a) and release (4r) stages. We see that the magnitude value is greater in the reverse nodes A, B and C related to higher order seismic branches.

In general, in the various cycles we can observe the proportionality principle, i.e. the higher is the order of the energy accumulation seismic branch, the greater the expected magnitude in the energy release stage.

Moreover, the smaller the inclination of the higher order energy accumulation branch, the greater the time or the number of shocks needed for the energy accumulation stage to be completed.

The end of a (n) order seismic cycle usually takes place after the amplitude of the (n – 1) order energy release branch has reached a value equal to the amplitude of the (n – 1) order energy accumulation branch plus 50% of it.

We can obtain more accurate data on the developmental level of a seismic grid by removing the seismic branches related to the energy release stage (Figure 7). The main branching structure obtained (S,A,B,C,D) retains the remote seismic events and the pre-earthquake seismic context's effects which we can obtain information from in order to determine its future development.

Note how the various orders seismic branches begin to form after a strong earthquake (source node S-source node). and are related to the magnitude values decrease over time. We also see how from the early stages of the seismic grid formation, it is possible to identify the branching structures that, depending on higher order seismic branch, will produce shocks of average magnitude or strong earthquakes in the energy release stage (the greater the order of the first seismic branch that originates from a source point, the greater the expected magnitude). Actually, the branching structures are not all the same but may take different appearances: their duration and the energy they release in the energy release stage, depend on the magnitude values of higher order branch and the time arrangement of low order subsequent seismic branches.

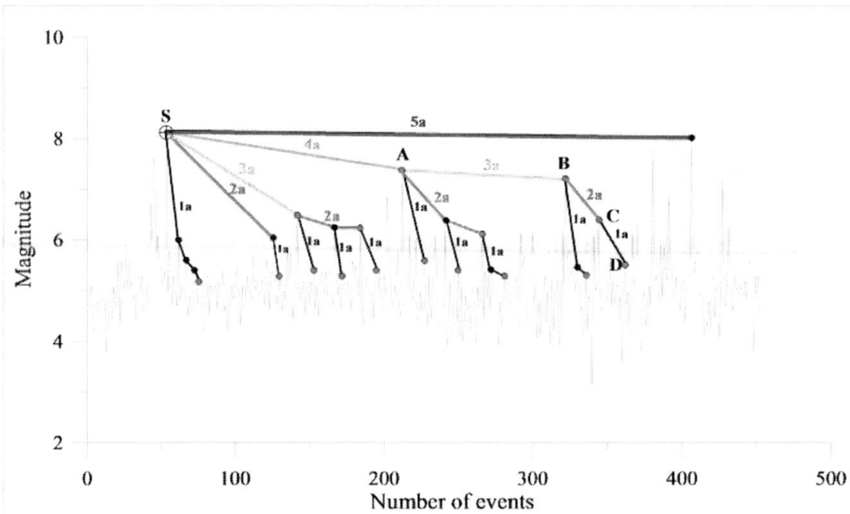

Figure 7. Schematic representation of the branching structure obtained by removing the energy release stage's seismic branches. The red-colored circles represent the reverse node of all the branching structure identified, while the letters S, A, B, C and D indicate those in the main branching structure (the point S is the source-point).

A branching structure is perfectly hierarchical when each (n) order branch is followed by another that features an immediately lower order (n − 1). If this is not the case, a seismic anomaly occurs. In addition, when the seismic branches' order decreases, the time or the number of shocks that we need to wait for the trigger stage to be formed, which precedes the energy release stage, decreases as well.

This means that only after the formation of the (1a) first order branch, the branching structure enters a critical stage where the DB-3SE microsequences generate the trigger point that can be profitably used for short term forecast.

3.4. Graphical Procedure for Branching Structures Identification

The long-term, monthly hierarchization of the seismic sequence (from 1973 to 2004) in Indonesia, which consists of any single maximum magnitude value monthly recorded by assuming a bottom value of 2.0 M, can be implemented by identifying the reverse nodes and following the multi-step procedure as follows:

1) Locate the first order reverse nodes P1(n) that meet the following condition:

$$M(n-1) < M1(n) \geq M(n+1) \tag{5}$$

where, M1(n) is the value of the first order reverse node magnitude and M(n − 1) and M(n + 1) are the magnitude values of seismic events that precede and follow the first reverse seismic node (Figure 8), respectively;

2) Locate the second order reverse nodes P2(n) that meet the following condition:

$$M1(n-1) < M2(n) \geq M1(n+1) \tag{6}$$

where, M2(n) is the magnitude value of the second order reverse node and M1(n − 1) and M1(n + 1) are the magnitude values of the first order reverse nodes that precede and follow the second order reverse node;

3) Draw the black-colored line joining the second order seismic nodes;

4) Repeat the same procedure to locate the second, third and fourth order reverse nodes (magenta-, green- and pastel blue-colored lines) (Figure 9);

5) Locate the line joining the second order reverse nodes (black-colored lines), P2m(n) second order minimum reverse nodes whose magnitude must comply with the following condition:

$$M2(n-1) \geq M2m(n) < M2(n+1) \tag{7}$$

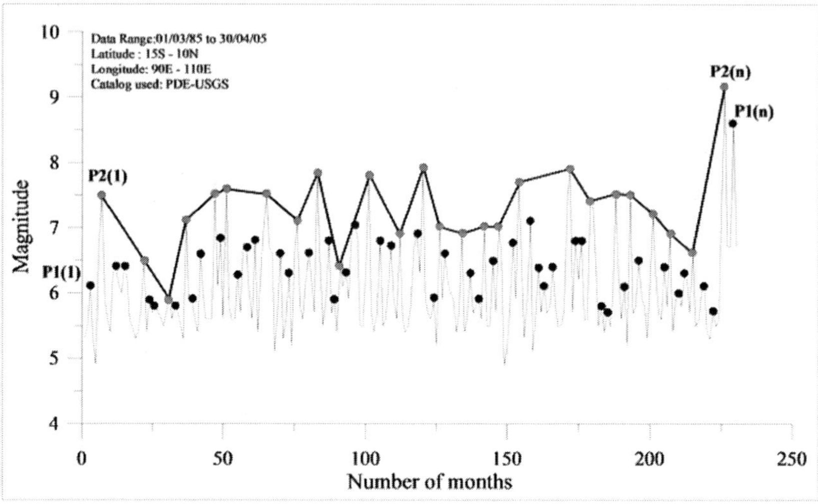

Figure 8. Earthquake's seismic sequence, Indonesia, 2004. Representation of the first order reverse nodes (black- and red- colored circles) and the second order reverse nodes (red-colored circles).

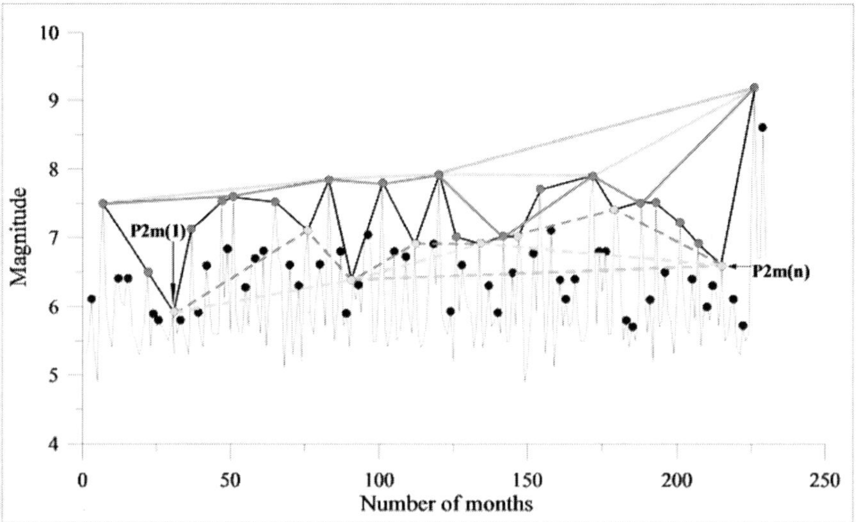

Figure 9. Fluctuation of various orders reverse nodes. The solid-colored line indicates various orders seismic branches that join the maximum reverse nodes, while the dashed colored lines indicate various orders seismic branches that join minimum reverse nodes.

where M2m(n) is the magnitude value of the second order, minimum reverse node and M2(n – 1) and M2(n + 1) are respectively the magnitude values of the second order reverse nodes that precede and follow the second order minimum reverse node (Figure 8);

6) Join the second order minimum reverse nodes with a line (magenta-colored dashed line);

7) Repeat the same procedure to draw higher order green, pastel blue and blue-colored dashed lines.

In order to identify various order branching structures we need to remove from the graph the energy release seismic branches as well as the descending seismic branches that join the minima (dashed lines).

Moreover, we need to identify the triggering stages (TS) of the branching structures, by drawing the line joining the lower extremity of the first order seismic branch with the trigger point generated by the DB-3SE microsequence (light green—colored line).

The colors assigned to the seismic branches and triggering stage let us determine the development status achieved by the various order branching structures.

3.5. Results

Figure 10 displays several various orders branching structures (BR1, BR2 and BR3) that have developed over time. The greatest (BR3) consists of an energy accumulation stage characterized by four developmental stages and a flash earthquake-type energy release stage featuring a mainshock with a magnitude of 9.1 Mw on 26-12-2004.

During the branching structures' early development and completion stages, several various orders trigger points were generated. In the graph, the yellow-colored triangles, placed at the extremities of the minimum ascending branches (magenta-colored dashed lines), represent the first order trigger point, while the red-colored triangles indicate the second order ones generated by the DB-3SE mirosequence.

In the lower branching structure (A,B and C), we note the abnormal formation of a fourth order branch that generated a trigger point in the node C (yellow triangle) just before the mainshock.

The same procedure can be used to locate the greater magnitude aftershocks in the energy accumulation phase.

Figure 11 shows the branching structures that have developed after the earthquake occurred Indonesia on 26 December 2004 (source node).

In particular, we observe a fourth order branching structure, which on 28 March 2005 the first significant aftershock developed from with a magnitude of 8.6 Mw, that originated the first, fourth order seismic branch.

3.6. Modified Aroon Oscillator

The Aroon oscillator [30] was modified so as to be able to obtain adequate information on the type of branching structure (primary or secondary) being formed and on the developmental stages' hierarchization degree (whether a branching structure is located in an accumulation or release stage).

The oscillator features a red line that indicates the energy accumulation stage and a green line indicating the energy release stage (Figure 12). The scale used by the oscillator ranges from 0% to 100%.

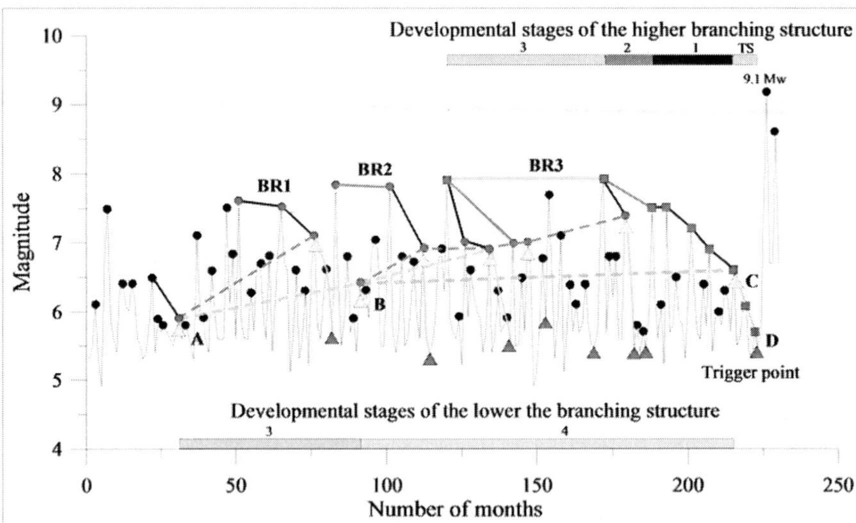

Figure 10. Earthquake's seismic sequence, Indonesia, 2004. The red colored squares show the higher order nodes of the branching structure, while the red-colored circles indicate the lower order ones, and the yellow- and red-colored triangles indicate the first and the second order trigger points, respectively. The upper part of the graph shows the developmental stages of the higher branching structure, while in the lower part the developmental stages of the lower branching structure less are reported.

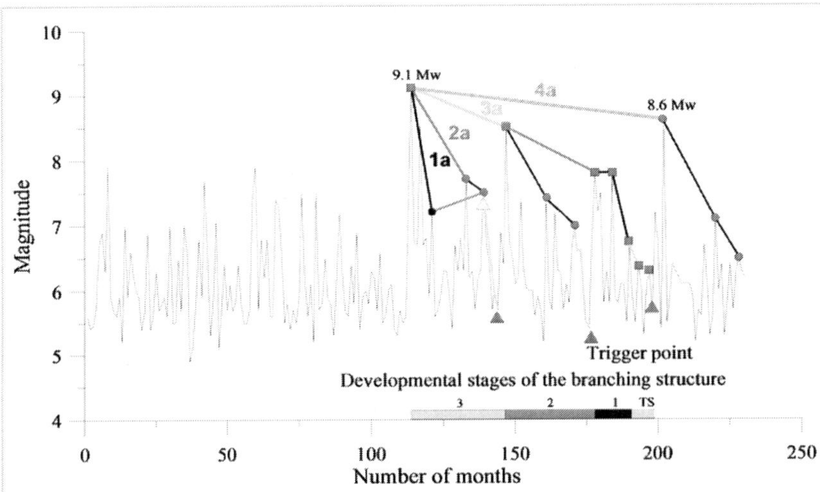

Figure 11. Earthquake's seismic sequence, Indonesia, 2004. Pattern of the branching structure related to the energy accumulation stage. The red-colored square indicates the nodes of the higher order branching structure, while the yellow and red triangles indicate the first and second order trigger points, respectively. The lower part of the graph shows the developmental stages of the branching structure associated with the first significant aftershock.

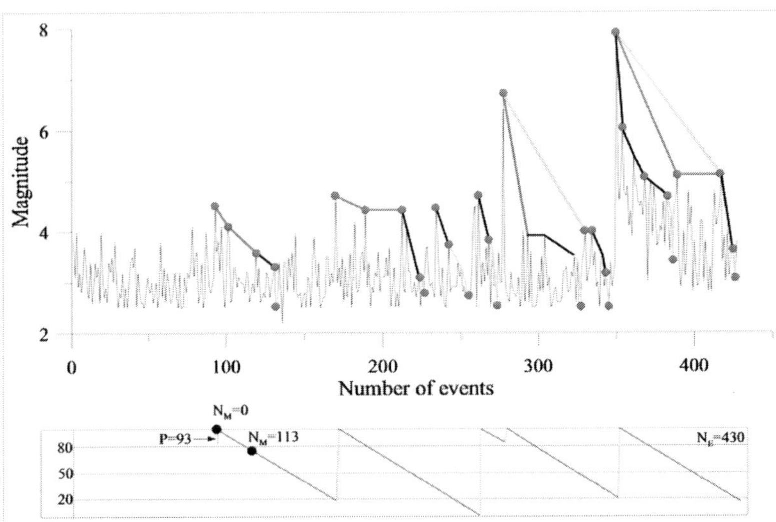

Figure 12. Schematic representation of modified Aroon oscillator. The red-colored lines indicate the energy accumulation stages, while the green lines indicate the energy release stages. The black-colored circles indicate the calculation points.

When the values are increasing, the oscillator indicates an energy release stage in progress, while when the values are decreasing, it indicates that an accumulation energy stage is being formed. The farther the oscillator from the value of 50%, the stronger the current phase. Usually, the oscillator is decreasing if seismic branching with the same order is subsequent to each other.

There are other basic situations for interpreting the indicator:

1) When the oscillator decreases from 100% to 80%, it indicates the beginning of an energy accumulation stage.

2) When the oscillator decreases 20% to 0%, it indicates that energy accumulation stage is being completed.

3) When the oscillator increases from 80% to 100%, it indicates that the energy release stage is being strengthened;

The formula to calculate the modified Aroon oscillator is the following:

$$AR = \left[\left(P - \left(N - N_M\right)\right)/P\right] \cdot 100 \tag{8}$$

where P (reference period) is the number of seismic events, days or months from the beginning of the seismic sequence to the first reference absolute or relative maximum value (mainshock or average magnitude shock), N is the oscillator's calculation point in relation to the beginning of the seismic sequence and N_M is the number of seismic events, days or months from the absolute or relative maximum value that has formed during the reference period with respect to the beginning of the sequence.

Usually the period P must be such that the already formed source points coincide with the points at which the oscillator reaches 100% (default p = 42).

The algorithm provides for a calculation cycle from N = P up to N_E (total number of events in the seismic sequence) in which we need to establish the position of the reference period maximum with respect to the origin of the seismic sequence through a second internal cycle of calculation from $N_M = N - (P - 1)$ up to N, and subsequently calculate the modified Aroon oscillator.

For example, if the period consists of 93 events and the maximum (N_M) and the calculation point (N) of the oscillator are equal to zero, we will have $AR = \left[\left(93 - (0 - 0)\right)/93\right] \cdot 100 = 100\%$. If the maximum ($N_M$) occurred in 93 past events and the oscillator's calculation point is N = 113, then the oscillator's value is $AR = \left[\left(93 - (113 - 93)\right)/93\right] \cdot 100 = 78\%$.

3.7. Maximum Magnitude Assessment Procedure

Figure 13 shows the branching structures' patterns, which have been identified in the Japan's monthly seismic sequence from 1973 to 31 August 2015.

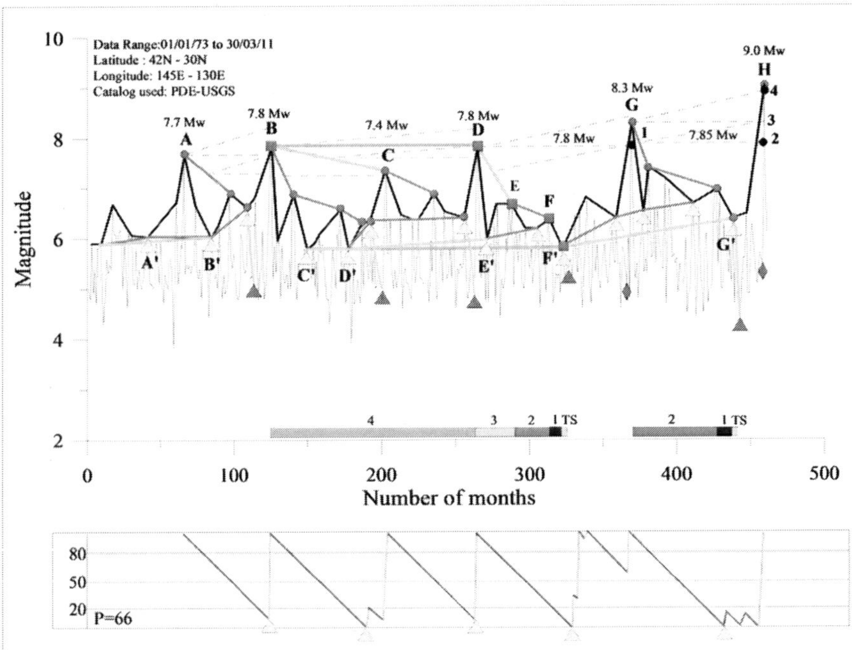

Figure 13. Japan's monthly sequence. The red-colored squares indicate the higher order nodes of the branching structure of higher order, the red circles the minor order ones, while the red and triangles and diamonds indicate the first, second and third order trigger points, respectively. In the lower part of the graph the modified Aroon oscillator is displayed.

The red squares indicate the main branching structure's nodes, while the red circles indicate the minor branching structures' ones.

The letters A, B, D, G and H denote the higher seismic nodes that correspond to the strongest earthquakes recorded in the time interval considered, while the letters A', B', C', D', E', F' and G' are the nodes relating to minor, second order ascending seismic branches included in the lower branching structure that correspond to first order trigger points (yellow-colored triangles).

The red-colored triangles and diamonds, generated by the DB-3SE microsequence upon the higher branching structure's completion and during the energy release stage, indicate the second and third order trigger points.

The development dynamics of the branching structure suggests to compute the magnitude value of the strongest earthquake in correspondence with the highest order branch.

From the mid-point of the higher order branch (point 1) included in the branching structure that preceded the earthquake occurred on 11 March 2011, we need to draw an horizontal, dashed line extending to the right up to the calculation point of the expected magnitude value (point 2). The intersection of this half line with the vertical line passing through the set calculation point, provides the minimum magnitude value (red-colored dashed line).

To calculate the maximum value we need to draw a horizontal dashed line from the point G that extends to the right up to the calculation point (point 3). Then we join the point 3 with the mid-point of the 2 order branch (point 1); from the point G we finally draw a parallel line to the segment 1 - 3 (black-colored dashed line).

The intersection of this parallel line with the vertical line passing through the calculation point, gives the maximum magnitude value (point 4).

We should however bear in mind that not always these minimum and maximum values are accurately reflected in the energy release stage. In the case where the range between the minimum and the maximum value is very wide, we can consider the mean value between the minimum and maximum values computed as the maximum value.

Based on Figure 14, which shows in details the energy release stage of the big earthquake occurred on 11 March 2011, with a magnitude of 9.0 Mw, we can formulate some interesting considerations about the characteristics of the higher and lower branching structures that preceded the earthquake.

First, we can observe that the first order trigger point coincided with the extremity of the third order lower seismic branch (green line) and that the second and third order trigger points (red box-included red triangle and diamond) that preceded the mainshock were generated by the DB-3SE microsequence.

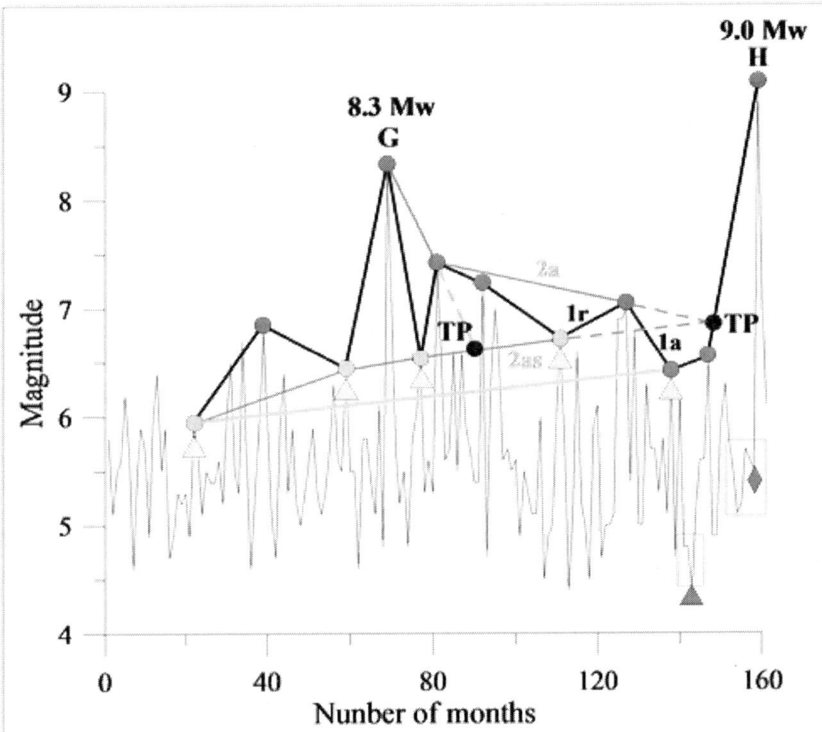

Figure 14. Energy release stage (Japan's monthly sequence). The red circles indicate the higher order nodes of the branching structure, the green circles represent the lower order ones, while the yellow and red triangles and the red diamonds indicate the first, second and third order trigger points. The black-colored circles indicate the trigger points generated by the intersection between the higher second order branches of energy accumulation (2a) lower and ascending magenta-colored ones (2as).

In addition, by means of a simple graphic procedure we identified a further trigger point (TP) generated by second order higher branches of the energy accumulation stage (2a) and lower ascending, magenta-colored branches (2as). In this case, the next DB-3SE microsequence (red diamond) represents a second order trigger point.

If the TP falls before the lower extremity of the first order energy accumulation branch (1a), we consider the latter as the trigger point.

Figure 15 shows the branching structures that have developed in the time interval between 1973 and 31-08-2015 in the Southern California's seismic sequence. In particular, we note the well hierarchized branching structure that has generated the earthquake occurred in Landers on 28-06-1992, with a magnitude of 7.3 Mw, which consisted of two developmental stages with the origin in the node B (source).

The TP was followed by first, second and third order trigger points. The maximum magnitude value calculated with the graphical method starting from the second order branch's mid-point was close to the one recorded.

The Landers earthquake has subsequently generated a third order branching structure (C,E,F,G,E') from which, after the second order trigger point, a hierarchical abnomaly has developed in the node H (seismic branches FH-HF') consistinting of a third order branch directly followed by a first order one. The energy release stage that followed was of Progressive earthquakes-type, where the strongest aftershocks, including the 7.2 Mw ones occurred on 04-04-2010 (Baja California, Mexico), were preceded by first and second order trigger points (yellow- and red-colored triangles). The TP was generated shortly before the strongest shock and after the second order trigger point. Subsequently, the earthquake on 04-04-2010 has activated a new branching structure characterized by a first, fourth order final seismic related to the energy accumulation (CD branch).

The minimum temporary magnitude value associated with this seismic bomb developmental phase is 7.15 M.

4. EXAMPLES OF BRANCHING STRUCTURES

The examples shown in Figure 16, processed from the data in Table 1, show the main branching structures that have developed before some strong earthquakes happened in the past in various parts of the world.

The branching structures, identified with the long term, seismic sequence hierarchization, show that big earthquakes can trigger additional earthquakes with decreasing magnitude during the energy accumulation stage. After the trigger point formation (yellow triangle), we notice some energy release stages of flash earthquakes (Figures 16 (a)-(e)) or progressive earthquakes type (Figures 16(f)-(l)). Comprehensive retrospective tests show that the proposed methodology is reasonably stable upon changes in the size of the seismic area and therefore it is able to locate all the branching structures that develop during the energy accumulation and release stages.

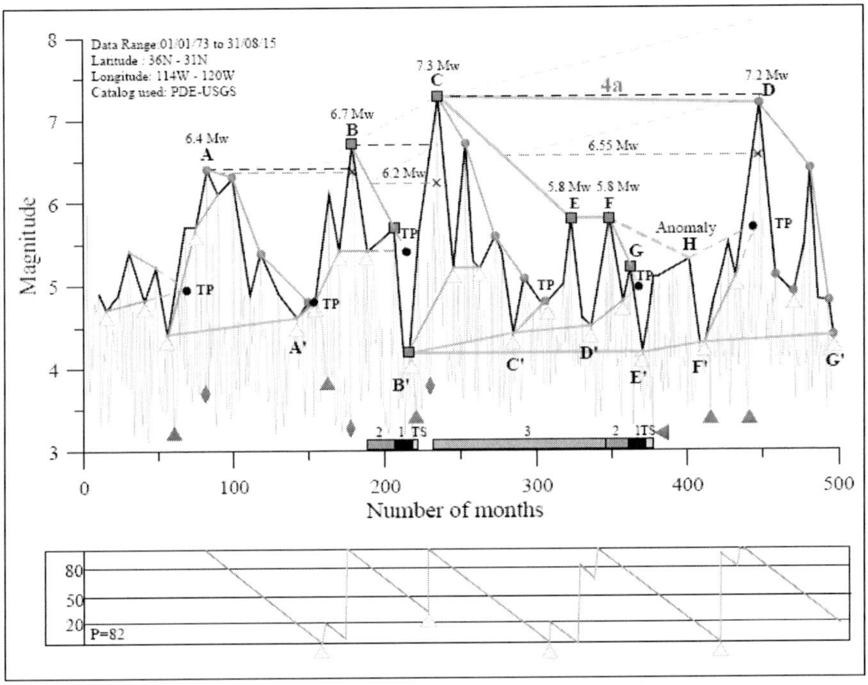

Figure 15. Southern California (US)'s monthly sequence. Red colored squares represent the higher order branching structure nodes, red circles the minor order ones, while the yellow and red triangles and the red diamonds indicate the first, second and third order trigger points, respectively. In the lower part of the graph developmental stages of the main branching structure are reported.

Table 1. Earthquakes, date range, coordinates of the region analyzed and catalogue used.

Earthquakes	Date Range	Latitude	Longitude	Catalog Used
L'Aquila—Italy	27/04/2007 - 06/04/2009	43.2 N - 41.5 N	14.2 E - 12.5 E	ISIDE-INGV
Loma Prieta—Ca, USA	24/09/1982 - 18/10/1989	38.0 N - 36.0 N	120.0 W - 123.0 W	PDE-USGS
Turkey	09/04/1988 - 24/10/2011	44.0 N - 36.0 N	44.0 E - 36.0 E	PDE-USGS
Japan	25/09/2003 - 11/03/2011	42.0 N - 30.0 N	130.0 E - 145.0 E	PDE-USGS
Chile	03/09/2006 - 27/02/2010	30.0 S - 40.0 S	63.0 W - 80.0 W	PDE-USGS
Samoa	09/01/1995 - 29/09/2009	12.0 S - 20.0 S	168.0 W - 176.0 W	PDE-USGS
Sumatra	01/01/1981 - 30/04/2005	90.0 S - 130.0 S	10.0 E - 15.0 E	PDE-USGS
Nepal	18/07/2002 - 12/05/2015	32.0 N - 26.0 N	88.0 E - 81.0 E	PDE-USGS
Tonga	22/03/2003 - 03/05/2006	16.0 S - 24.0 S	179.0 W - 178.0 W	PDE-USGS
Russia	28/06/1990 - 05/12/1997	59.0 N - 51.0 N	170.0 E - 154.0 E	PDE-USGS

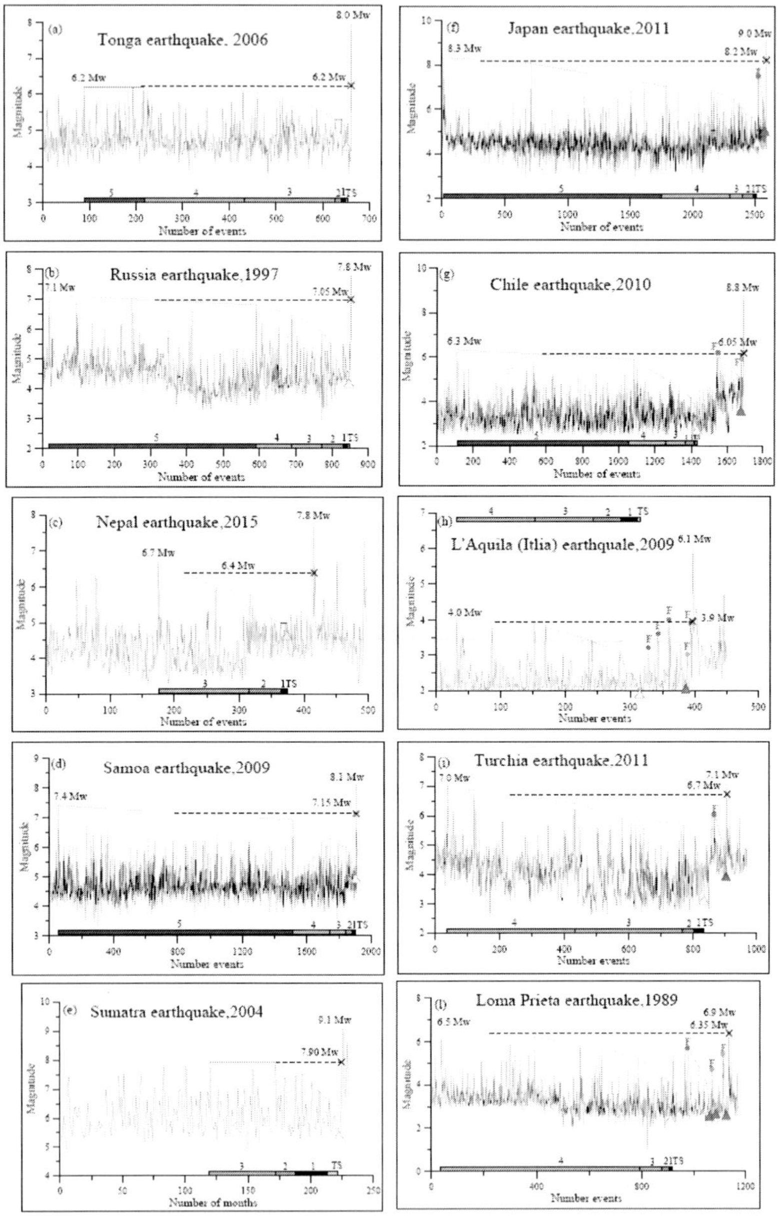

Figure 16. Examples of branching structure. The colored lines indicate the developmental stages, the letter "F" the foreshocks, the red circles the first order foreshocks and the green circles the second order foreshocks. The yellow and red triangles indicate the rigger points of main and secondary branching structures.

5. CONCLUSIONS

This study has the aim to propose a model for the graphical analysis of seismic sequences, starting from the assumption that the sequences do not develop in an unpredictable manner, but follow easily identifiable trends.

As a result, we can say that some developments recur cyclically, making it possible, if identified on time, the formulation of hypotheses on the future development of a seismic sequence.

The proposed model, based on the earthquakes time series' hierarchization, has shown that the seismic sequences develop according to similar developmental stages and that the mainshock magnitude value associated with a seismological branching sequence depends on the first developmental stage (higher order seismic branch).

This approach develops a dramatically innovative earthquake forecasting model starting from the seismicity analysis conducted on an area of different width, which provides a reasonable estimate of the expected magnitude.

CITE THIS PAPER

Giulio Riga, Paolo Balocchi (2016) Short-Term Earthquake Forecast with the Seismic Sequence Hierarchization Method. *Open Journal of Earthquake Research*,**05**,79-96.

REFERENCES

1. Ogata, Y. (1999) Seismicity Analysis through Point-Process Modeling: A Review. Pure and Applied Geophysics, 155, 471-507.
2. Helmstetter, A. and Sornette, D. (2002) Diffusion of Epicenters of Earthquake Aftershocks, Omori's Law, and Generalized Continuous-Time Random Walk Models. Physical Review E, 66, Article ID: 061104.
3. Gu, C., Schumann, A.Y., Baiesi, M. and Davidsen, J. (2013) Triggering Cascades and Statistical Properties of Aftershocks. Journal of Geophysical Research: Solid Earth, 118, 4278-4295.
4. Riga G. and Balocchi, P. (2016) Seismic Sequence Structure and Earthquakes Triggering Patterns. Open Journal of Earthquake Research, 5, No. 1. http://www.scirp.org/Journal/PaperInformation.aspx?PaperID=63538
5. Parsons, T. and Velasco, A.?A. (2009) On Near-Source Earthquake Triggering. Journal of Geophysical Research, 114, Article ID: B10307.
6. Zhuang, J., Ogata, Y. and Vere-Jones, D. (2002) Stochastic Declustering of Space-Time Earthquake Occurrences. Journal of the American Statistical Association, 97, 369-380.

7. Zhuang, J., Ogata, Y. and Vere-Jones, D. (2004) Analyzing Earthquake Clustering Features by Using Stochastic Reconstruction. Journal of Geophysical Research, 109, Article ID: B05301.
8. Baiesi, M. and Paczuski, M. (2004) Scale-Free Networks of Earthquakes and Aftershocks. Physical Review E, 69, Article ID: 066106.
9. Baiesi, M. and Paczuski, M. (2005) Complex Networks of Earthquakes and Aftershocks. Nonlinear Processes in Geophysics, 12, 1-11.
10. Zaliapin, I., Gabrielov, A., Keilis-Borok, V. and Wong, H. (2008) Clustering Analysis of Seismicity and Aftershock Identification. Physical Review Letters, 101, Article ID: 018501.
11. Marsan, D. and Lengliné, O. (2008) Extending Earthquakes' Reach through Cascading. Science, 319, 1076-1079.
12. Mogi, K. (1963) Some Discussions on Aftershocks, Foreshocks and Earthquake Swarms—The Fracture of a Semi- Infinite Body Caused by an Inner Stress Origin and Its Relation to the Earthquake Phenomena (Third Paper). Bulletin of the Earthquake Research Institute, 41, 615-658.
13. Van Stiphout, T., Wiemer, S. and Marzocchi, W. (2010) Are Short-Term Evacuations Warranted? Case of the 2009 L'Aquila Earthquake. Geophysical Research Letters, 37, L06306.
14. Jones, L.M. and Molnar, P. (1976) Frequency of Foreshocks. Nature, 262, 677-679.
15. Abercrombie, R.E. and Mori, J. (1996) Occurrence Patterns of Foreshocks to Large Earthquakes in the Western United States. Nature, 381, 303-307.
16. Keilis-Borok, V.I. and Malinovskaya, L.N. (1964) One Regularity in the Occurrence of Strong Earthquakes. Journal of Geophysical Research, 69, 3019-3024.
17. Jaumé, S.C. and Sykes, L.R. (1999) Evolving Towards a Critical Point: A Review of Accelerating Seismic Moment/ Energy Release Prior to Large and Great Earthquakes. In: Wyss, M., Shimazaki, K. and Ito, A., Eds., Seismicity Patterns, Their Statistical Significance and Physical Meaning, Birkhauser, Basel, 279-305. http://link.springer.com/chapter/10.1007%2F978-3-0348-8677-2_5
18. Sammis, S.G. and Sornette, D. (2002) Positive Feedback, Memory and the Predictability of Earthquakes. Proceedings of the National Academy of Sciences of the United States of America, 99, 2501-2508. http://arxiv.org/pdf/cond-mat/0107143v1.pdf
19. Ishibashi, K. (1988) Two Categories of Earthquake Precursors, Physical and Tectonic, and Their Roles in Intermediate- Term Earthquake Prediction. Pure and Applied Geophysics, 126, 687-700.
20. Shebalin, P. (2006) Increased Correlation Range of Seismicity before Large Events Manifested by Earthquake Chains. Tectonophysics, 424, 335-349.
21. Mantovani, E., Babbucci, D., Tamburelli, C. and Viti, M. (2009) A Review on the Driving Mechanism of the Tyrrhenian-Apennines System: Implications for the Present Seismotectonic Setting in the Central-Northern Apennines. Tectonophysics, 476, 22-40.
22. Ziv, A. and Schmittbuhl, J. (2003) The Seismic Cycle and the Difference between Foreshocks and Aftershocks in a Mechanical Fault Model. Geophysical Research Letters, 30, 2237.https://eost.unistra.fr/fileadmin/upload/EOST/Jean_Schmittbuhl/Publications/Articles/Zi-Sc03.pdf
23. Helmstetter, A. and Sornette, D. (2003) Foreshocks Explained by Cascades of Triggered Seismicity. Journal of Geo- bphysical Research: Solid Earth, 108, 2457.
24. Helmstetter, A., Kagan, Y.Y. and Jackson, D.D. (2005) Importance of Small Earthquakes for Stress Transfers and Earthquake Triggering. Journal of Geophysical Research: Solid Earth, 110, Article ID: B05S08.
25. Shearer, P.M. (2012) Space-Time Clustering of Seismicity in California and the Distance Dependence of Earthquake Triggering. Journal of Geophysical Research: Solid Earth, 117, B10306.

26. Helmstetter, A. (2003) Is Earthquake Triggering Driven by Small Earthquakes? Physical Review Letters, 91, Article ID: 058501.
27. Felzer, K.R. and Brodsky, E.E. (2006) Decay of Aftershock Density with Distance Indicates Triggering by Dynamic Stress. Nature, 441, 735-738.
28. Marzocchi, W. and Lombardi, A.M. (2008) A Double Branching Model for Earthquake Occurrence. Journal of Geo- physical Research: Solid Earth, 113, Article ID: B08317.
29. Werner, M.J. (2008) On the Fluctuations of Seismicity and Uncertainties in Earthquake Catalogs: Implications and Methods for Hypothesis Testing. University of California, Los Angeles. https://www.ethz.ch/content/dam/ethz/special-interest/mtec/chair-of-entrepreneurial-risks-dam/documents/dissertation/Max_Werner_thesis_final_Dec07.pdf
30. Chande, T. (1995) A Time Price Oscillator. Technical Analysis of Stocks & Commodities, 13, 369-374.ftp://80.240.216.180/Transmission/%D0%A4%D0%B0%D0%B9%D0%BB%D1%8B/S&C%20on%20DVD%2011.26/VOLUMES/V13/C09/THETIME.pdf

CHAPTER 15

Earthquake Damage Visualization (EDV) Technique for the Rapid Detection of Earthquake-Induced Damages Using SAR Data

Ram C. Sharma [1,2,*], Ryutaro Tateishi [1], Keitarou Hara [2], Hoan Thanh Nguyen [3], Saeid Gharechelou [1,4] and Luong Viet Nguyen [5]

[1]*Center for Environmental Remote Sensing (CEReS), Chiba University, 1-33 Yayoi-cho, Inage-ku, Chiba 263-8522, Japan*

[2]*Department of Informatics, Tokyo University of Information Sciences, 4-1 Onaridai, Wakaba-ku, Chiba 265-8501, Japan*

[3]*Institute of Geography, Vietnam Academy of Science and Technology, 18 Hoang Quoc Viet Str., Cau Giay Dist., Hanoi 10000, Vietnam*

[4]*Department of Civil Engineering, Shahrood University of Technology, Shahrood 3619995161, Iran*

[5]*Space Technology Institute, Vietnam Academy of Science and Technology, 18 Hoang Quoc Viet Str., Cau Giay Dist., Hanoi 10000, Vietnam*

ABSTRACT

The damage of buildings and manmade structures, where most of human activities occur, is the major cause of casualties of from earthquakes. In this paper, an improved technique, Earthquake Damage Visualization (EDV) is presented for the rapid detection of earthquake damage using the Synthetic Aperture Radar (SAR) data. The EDV is based on the pre-seismic and co-seismic coherence change method. The normalized difference between the pre-seismic and co-seismic coherences, and vice versa, are used to calculate the forward (from pre-seismic to co-seismic) and backward (from co-seismic to pre-seismic) change parameters, respectively. The backward change parameter is added to visualize the retrospective changes caused by factors other than the earthquake. The third change-free parameter uses the average values of the pre-seismic and co-seismic coherence maps. These three change parameters were ultimately merged into the EDV as an RGB (Red, Green, and Blue) composite imagery. The EDV could visualize the earthquake damage efficiently using Horizontal transmit and Horizontal receive (HH), and Horizontal transmit and Vertical receive (HV) polarizations data from the Advanced Land Observing Satellite-2 (ALOS-2). Its performance was evaluated in the Kathmandu Valley, which was hit severely by the 2015 Nepal Earthquake. The cross-validation results showed that the EDV is more sensitive to the damaged buildings than the existing method. The EDV could be used for building damage detection in other earthquakes as well.

Keywords: earthquake damage; coherence; visualization; EDV; 2015 Nepal Earthquake; ALOS-2; SAR; cross-validation; buildings

1. INTRODUCTION

Earthquakes are one of the most catastrophic natural disasters. More than 23 million deaths have been caused by earthquakes during the period of 1902–2011 alone globally (http://earthquake.usgs.gov, last accessed 15 September 2016). Earthquake-resistant buildings and infrastructures, and emergency services during disasters are vital for protecting against earthquake damage. However, mass deaths and injuries caused by earthquakes over the past decades undermine the present situation. While urbanization is inevitable as a cost of the rising population, damage reduction strategies, particularly in vulnerable fault zones, are mandatory. Satellite remote sensing is an important technology for the monitoring of disasters.

Buildings' damage information is essential for rescue, humanitarian and reconstruction operations in the disaster area [1]. The space-borne optical and Synthetic Aperture Radar (SAR) data have been used for building damage assessment in a number of places—for example, the 1995 earthquake in Kobe, Japan [2], the 1999 earthquake in Izmit,

Turkey [3], the 2001 earthquake in Gugrat, India [4,5], the 2003 earthquake in Bam, Iran [6,7,8,9], the 2005 earthquake in Azad Kasmir, Pakistan [10], the 2006 earthquake in Java, Indonesia [11], the 2008 earthquake in Sichuan, China [12,13,14,15], the 2009 earthquake in L'Aquila, Italy [16], and the 2010 earthquake in Haiti [17,18,19]. Building damage has been ranked in the field using damage scales [20,21,22]; however, remotely sensed images are mostly capable of detecting heavy damage levels only, i.e., totally collapsed buildings [23,24].

The near real-time disaster mapping using satellite data is important for minimizing the effects of natural disasters [25]. Very-high-resolution optical imagery can visualize earthquake damage, and it has been used by many researchers [9,12,26,27]. Major techniques used for retrieving the earthquake-induced building damage information are object based classification [28,29,30], template matching and pattern recognition [31], and supervised classification [32,33,34]. Airborne imagery is another source of building damage information [35,36,37,38,39,40]. In some research, a combination of SAR and optical data has provided better results [14,41]. However, availability of the optical images when urgently needed can not be guaranteed due to clouds and atmospheric conditions. SAR data with negligible atmospheric effects are considered more appropriate for the detection and mapping of the earthquake damage. SAR backscattering intensity data have shown some successes in detecting damage [42,43,44]. More recently, Yun et al. [45] have produced the NASA Damage Proxy Map (NDPM) based on pre-seismic and co-seismic coherence change methods using SAR data.

This paper presents an improved technique called the Earthquake Damage Visualization (EDV) for rapid detection of earthquake induced damage. The EDV technique builds on the existing NDPM method by using additional polarization and more change parameters. The performance of the EDV technique is demonstrated in the case of the 2015 Nepal Earthquake using the ALOS-2 SAR data and compared it to the existing NDPM method.

2. METHODOLOGY

2.1. Study Area

This research was carried out in the urban areas of the Kathmandu Valley, which was severely hit by a 7.8 Mw earthquake on 25 April 2015. The Kathmandu Valley is a bowl-shaped valley standing at 1425 m above sea level, and surrounded by mountainous ranges. The location map of the study area is displayed in Figure 1. The major shake areas based on Mercalli intensity data of the earthquake (http://earthquake.usgs.gov, last accessed 15 September 2016) are also plotted in Figure 1, which shows that the Kathmandu Valley was shaken severely by the earthquake.

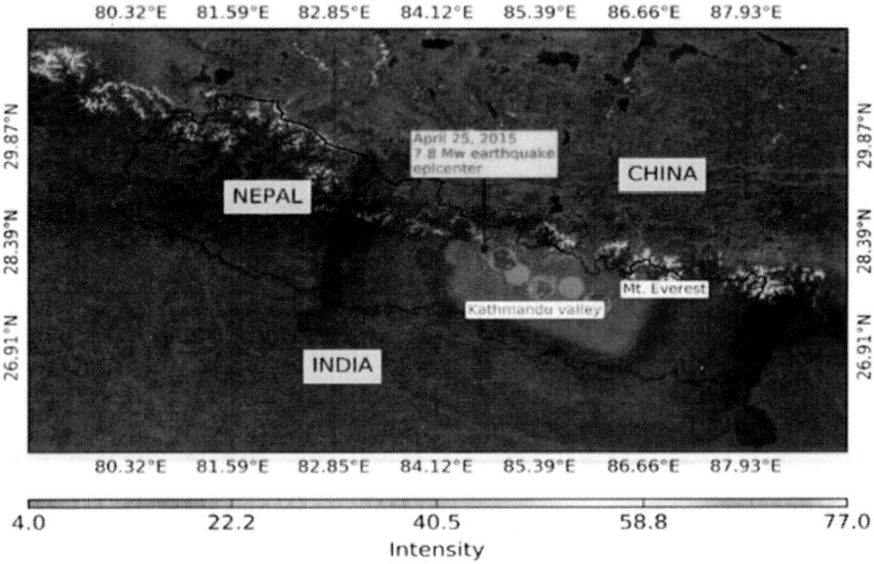

Figure 1. Location map of the study area, the Kathmandu Valley (red polygon) displayed with the shake areas.

2.2. Proposal of the EDV

The EDV is based on estimates of the coherence changes between pre-seismic and co-seismic timings. A pair of before earthquake (pre-seismic) images are used to calculate the pre-seismic coherence; whereas a pair of before and after earthquake (co-seismic) images are used to calculate the co-seismic coherence. A coherence map shows similarity (cross-correlation) between two radar images pixel by pixel. High coherence values means that the ground objects remained the same between the two timings of the radar image acquisition, whereas low coherence means that the objects changed during that time. The pre-seismic and co-seismic coherence maps are then used to calculate forward (from pre-seismic to co-seismic) change, backward (from co-seismic to pre-seismic) change, and no change (between pre-seismic and co-seismic) parameters. The normalized difference between the pre-seismic and co-seismic coherences, and vice versa, are calculated to derive the forward and backward change parameters, respectively. The backward change parameter is added to see the retrospective changes caused by factors other than the earthquake. The average values of the pre-seismic and co-seismic coherences are calculated as the third change-free parameter. These three change parameters are ultimately merged as the RGB (Red Green and Blue) color composite imagery, called EDV.

Advanced Land Observing Satellite-2 (ALOS-2) based L-band Single Look Complex (SLC) SAR images available from the Japan Aerospace Exploration Agency (JAXA) were used in the research. A pair of pre-seismic (4 October 2014 and 21 February 2015) images and another pair of co-seismic (21 February 2015 and 2 May 2015) images were used to calculate pre-seismic and co-seismic coherences, respectively. The pre-seismic and co-seismic coherences were calculated separately for the Horizontal transmit and Horizontal receive (HH), and Horizontal transmit and Vertical receive (HV) polarizations and averaged together for creating the EDV. The procedure of calculating the EDV for a single polarization is described in Figure 2.

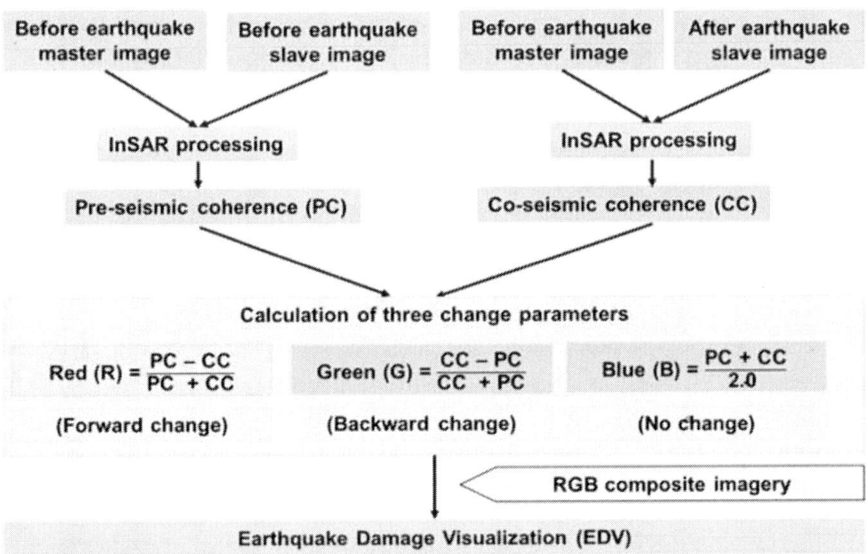

Figure 2. Flow chart showing creation of the Earthquake Damage Visualization (EDV).

2.3. Validation Approach

The performance of the EDV was evaluated by using the k-fold cross-validation method. The k-fold method divides all the samples into k groups of samples, called folds. The k-1 folds are used for learning, whereas the remaining one fold is used for the validation. The total number of folds used in this validation was 10, so it is called the 10- fold cross-validation method. Four commonly used supervised classifiers (k-Nearest Neighbors, Gaussian Naive Bayes, Random Forests, and Support Vector) were tested for the cross-validation. Readers are referred to the OpenCV (http://opencv.org)—an optimized C/C++ programming library for computer vision, machine learning and robotics—for in-depth description of the supervised classifiers used.

For the cross-validation, ground truth data were prepared considering the European Macroseismic Scale 1998 (EMS-1998, [46]). The EMS-1998 categorizes strength of an earthquake that affects a specific location, and it has 12 divisions. However, from the viewpoint of casualties and injuries caused by earthquake induced damage, and considering the capability of currently available satellite based SAR data, we generalized two classes in the research: safe and lethal. Divisions 1 (Not felt) to 7 (Damaging) were grouped as a safe class; whereas divisions 8 (Heavily damaging) to 12 (completely devastating) were grouped as the lethal class. Ground truth polygons belonging to lethal and safe blocks of buildings, historical monuments, towers, and temples were prepared through visual inspection. Each class consists of 90 polygons. The distribution of the ground truth polygons is shown in Figure 3 over the line of sight displacement map resulting from the differential interferometric processing carried out in the research. The displacement map shows that the ground surface uplifted up to 1.4 m in the northern part of the Kathmandu Valley due to the earthquake (Figure 3).

The performance of the EDV was compared to the NASA Damage Proxy Map (NDPM) using the ground truth polygons. The NDPM map is available online for downloading from http://aria-share.jpl.nasa.gov /events/20150425-Nepal_EQ/DPM/ (last accessed 15 September 2016). Cross-validation of the EDV and NDPM methods were carried out based on 90 ground truth polygons available for each class (lethal or safe). The ground truth polygons belonging to two classes (lethal or safe) were laid over the raster images (EDV or NDPM), and the pixels delineated by a polygon were extracted and their median values were calculated. The number of pixels falling inside a polygon varies according to the size of building block. The median pixel data calculated for each layer of the EDV and NDPM images were used for the cross-validation. The EDV has three layers, whereas the NDPM has a single layer. Google Earth based time-lapse images were used to visualize performance of the EDV and NDPM images.

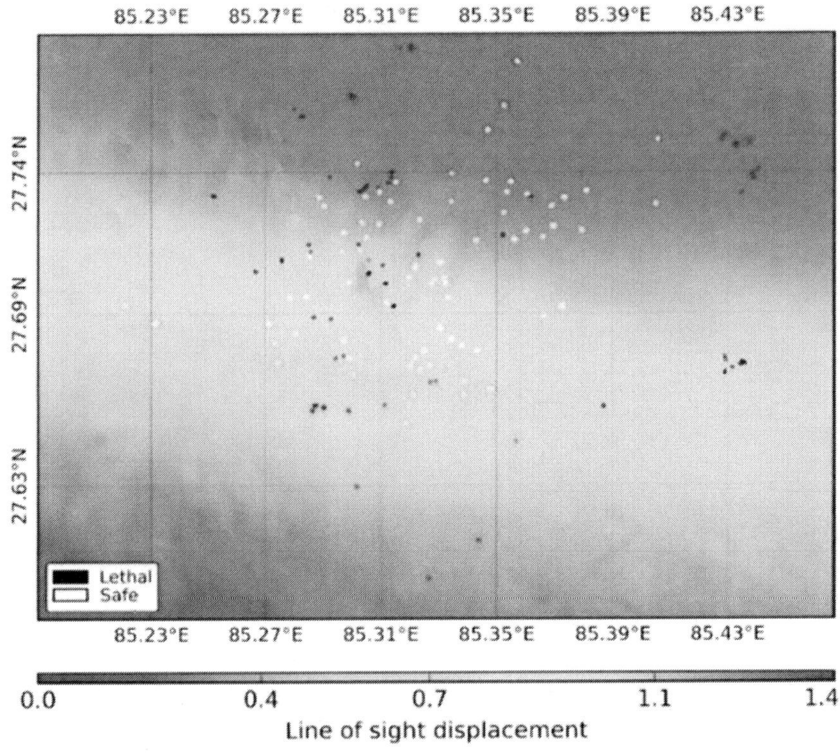

Figure 3. Distribution of the ground truth polygons in the Kathmandu Valley displayed over the line of sight displacement image resulted from differential interferometric processing in the research.

3. RESULTS AND DISCUSSION

3.1. Performance of the EDV

The EDV created from the ALOS-2 SAR data over the Kathmandu Valley is displayed in Figure 4. The EDV has three components: red, green, and blue. The coherence changes in the forward direction (from pre-seismic to co-seismic) are measured by the red component, whereas the green component measures the backward (from co-seismic to pre-seismic) changes, and the blue means relatively no changes (between pre-seismic and co-seismic). As the buildings can maintain high coherence over the time, the decrease in co-seismic coherence from the pre-seismic coherence is used as a proxy of the earthquake induced damage. In this research, a new component called the backward change is added to visualize the retrospective changes caused by factors other than the earthquake. Higher average values of the pre-seismic and co-seismic coherences accounted for by the third change-free parameter implies lower changes in both directions (backward or forward).

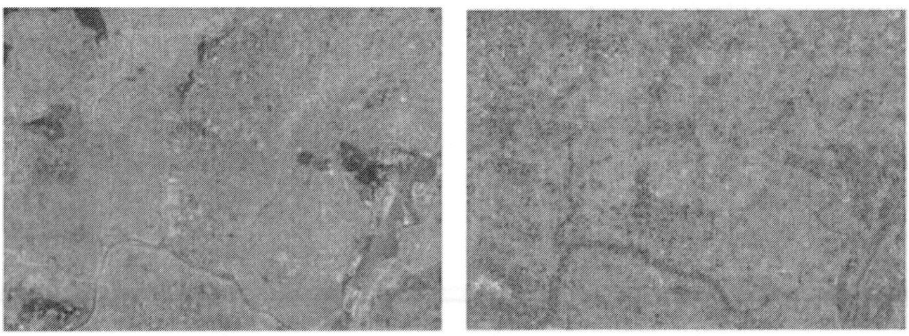

(a) (b)

Figure 4. Earthquake Damage Visualization (EDV) imagery showing forward (red), backward (green), and change-free (blue) components: (**a**) Google map imagery of the Kathmandu Valley dated 3 May 2015; (**b**) the corresponding EDV imagery.

The performance of the EDV is demonstrated in a number of highly damaged (lethal) locations (Figure 5 and Figure 6). The redness of the EDV imagery indicates severity of the building damage. As most of the lethal polygons have a higher amount of redness (than greenness or blueness), the EDV has visualized the building damage efficiently. Nevertheless, in some locations (Figure 5a and Figure 6b), EDV is dim to visualize the damage.

It should be noted that the red pixels of the EDV imagery (Figure 4) are not always matched with the highly damaged areas. Similar to the earthquake induced building damage, growth of trees can also drop the coherence, and it can be confused with building damage. Therefore, detection of building damage seems more difficult if buildings are intermixed with trees.

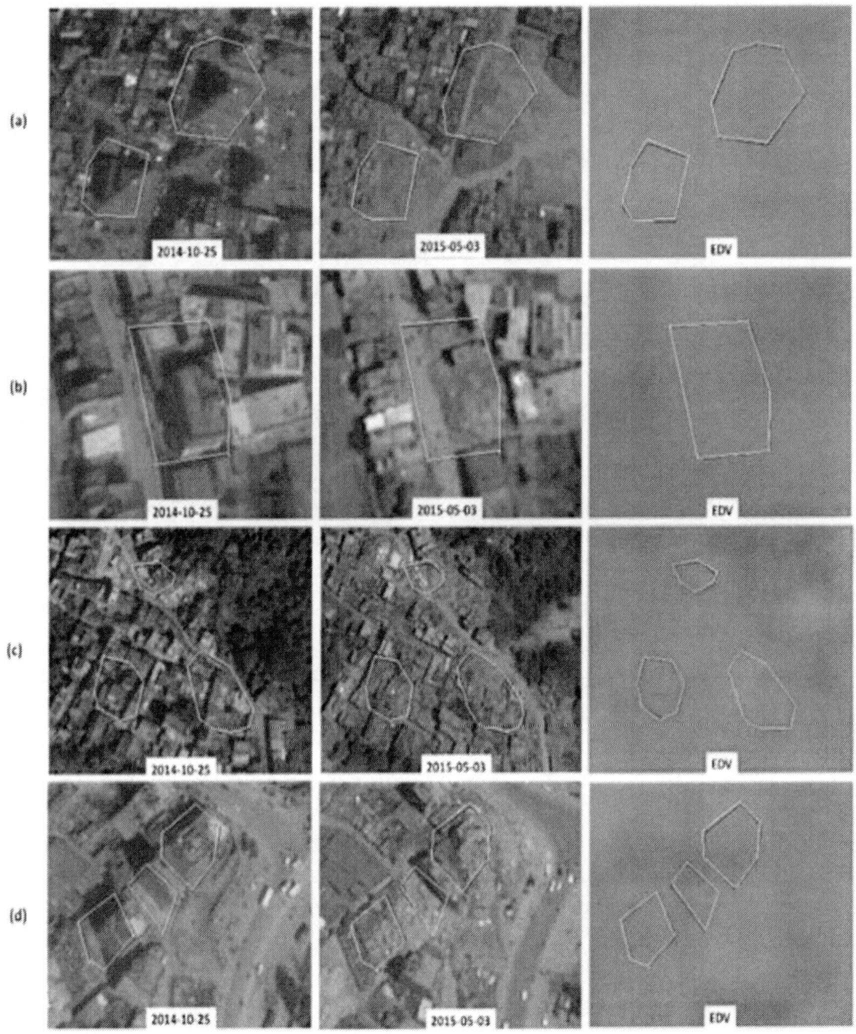

Figure 5. Performance of the EDV in different locations (**a–d**) that were highly damaged by the earthquake. The blocks of highly damaged (lethal) buildings are delineated by yellow polygons in each image. The left and middle columns show pre-seismic and post-seismic Google Earth images, whereas the right column shows the EDV image. The date of the Google Earth image is labeled in each image. The amount of redness in the EDV indicates severity of the building damage.

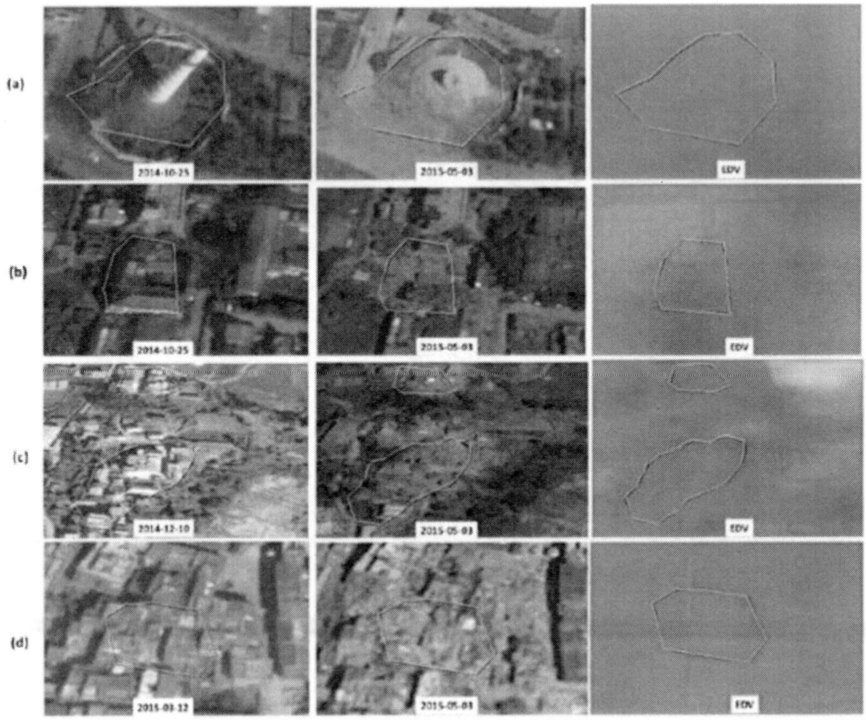

Figure 6. Performance of the EDV in different locations (**a–d**) that were highly damaged by the earthquake. The blocks of highly damaged (lethal) buildings are delineated by yellow polygons in each image. The left and middle columns show pre-seismic and post-seismic Google Earth images, whereas the right column shows the EDV image. The date of the Google Earth image is labeled in each image. The amount of redness in the EDV indicates severity of the building damage.

3.2. Comparison to the NDPM

The performance of the NDPM in a number of locations is demonstrated in Figure 7. The amount of redness reflects the severity of building damage. However, many highly damaged locations are lacking redness completely or partially. While comparing the Figure 6b with the label 6b of the Figure 7, NDPM is missing the lethal blocks completely. Contrary to the EDV, the NDPM is very dim for all highly damaged locations except for the location 5d. As the same scale of the redness is used, it is difficult to distinguish the damaged buildings using the NDPM. However, the EDV has demonstrated high contrast in most of the locations (Figure

5 and Figure 6), and it is more sensitive to the damaged buildings than the NDPM.

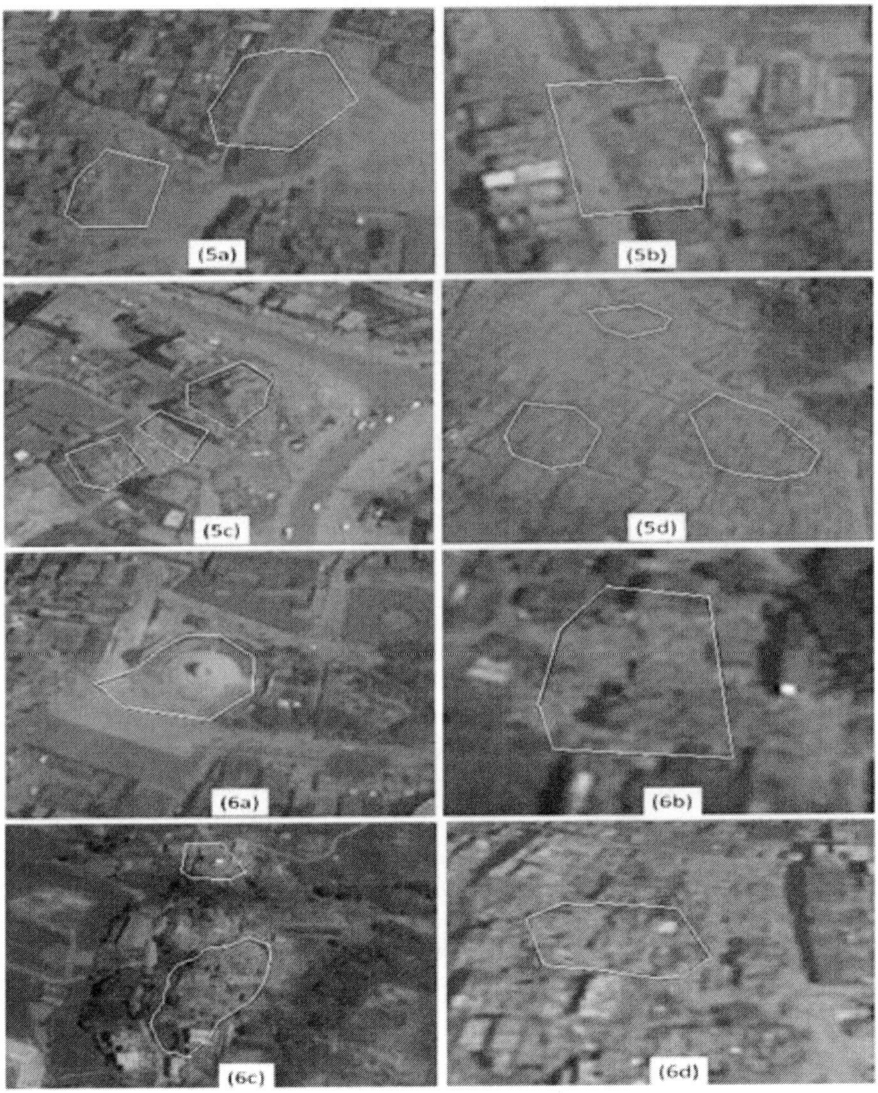

Figure 7. The NASA Damage Proxy Map (NDPM) in a number of highly damaged locations overlaid on the Google Earth imagery dated 3 May 2015 (post-seismic). The labels 5a to 5d and 6a to 6d in this figure denote the corresponding locations described in Figure 5 and Figure 6, respectively. The amount of redness in the NDPM imagery indicates severity of building damage.

3.3. Cross-Validation Results

Following the visual performance in previous sections, the performance of the EDV was compared to the NDPM quantitatively by using all ground truth polygons (90 for each class) prepared. The overall accuracy (kappa coefficient) of EDV and NDPM calculated from 10-fold cross-validation method using four supervised classifiers (k-Nearest Neighbors, Gaussian Naïve Bayes, Random Forests, and Support Vector Machine) are shown in Table 1.

Table 1. Performance of Earthquake Damage Visualization (EDV) and NASA Damage Proxy Map (NDPM) based on statistical metrics, overall accuracy (kappa coefficient) calculated from 10-fold cross-validation method using four supervised classifiers.

Methods	k-Nearest Neighbors	Gaussian Naïve Bayes	Random Forests	Support Vector Machine
EDV	0.82 (0.64)	0.86 (0.71)	0.76 (0.52)	0.69 (0.38)
NDPM	0.60 (0.20)	0.71 (0.41)	0.61 (0.21)	0.59 (0.17)

Among the supervised classifiers tested, the EDV performed better than the NDPM in terms of overall accuracy and kappa coefficient. Out of four classifiers, Gaussian Naïve Bayes showed the best performance for both the EDV (overall accuracy = 0.86, Kappa coefficient = 0.71) and NDPM (overall accuracy = 0.71, Kappa coefficient = 0.41) followed by the k-Nearest Neighbors and Random Forests.

There is a significant difference between the performance of EDV and NDPM; the highest kappa coefficient achieved with the Gaussian Naïve Bayes is 0.41 in the case of NDPM, whereas the EDV achieved 0.71 kappa coefficient using the same classifier. The NDPM used only the HH polarization (single band information) based pre-seismic and co-seismic coherence change method. On the contrary, the EDV deployed both the HH and HV polarizations; normalized differences of the pre-seismic and co-seismic coherences, and vice versa, as well as the third change-free parameter. The accuracy metrics, overall accuracy (kappa coefficient) based on different supervised classifiers: k-Nearest Neighbors, Gaussian Naïve Bayes, Random Forests, and Support Vector Machine were 0.69 (0.39), 0.77 (0.53), 0.61 (0.22), and 0.71 (0.41), respectively, using a single HH polarization based forward change parameter as in the case of NDPM. The overall accuracy (Kappa coefficient) obtained from a single HH polarization based forward change parameter is significantly lower than the overall accuracy (Kappa coefficient) obtained from the EDV. The

EDV provides improved detection of the building damage by adding additional polarization and more change parameters over the NDPM.

4. CONCLUSIONS

An improved technique, the EDV was proposed in the research for visualizing the earthquake-induced building damage. The EDV consists of the forward (from pre-seismic to co-seismic), backward (from co-seismic to pre-seismic), and change-free (between pre-seismic and co-seismic) parameters. The performance of the EDV was assessed in the Kathmandu Valley, which was severely shaken by the 2015 Nepal Earthquake. The cross-validation results showed that the EDV is more sensitive to the building damage than the existing NDPM method. The improvement was achieved by using an additional polarization (HV) and more change parameters. False detection of the buildings damage in forested areas due to growth of trees over the time was observed. Forested areas could be masked out in advance or the false detection could be minimized if the short time span images are available.

The earthquake induced hazards such as ground shaking/displacement, soil liquefaction, flooding/tsunami, and fires can damage buildings and man-made structures. The major cause of casualties of the earthquake is the damage of buildings and man-made structures where most of the human activities occur. Therefore, detection of building/structure damage is important for identifying the location of trapped victims during the immediate post-disaster period, and assisting the deployment of emergency services in the trapped houses. Though EDV provides better and rapid damage information based on coherence change measures using multiple SAR polarizations, tracking the individual damaged buildings is very challenging. Availability of sub-meter resolution SAR data is expected in the future to overcome the challenge. However, the availability of near-real time SAR data cannot be assured. For instance, a post-earthquake ALOS-2 SAR image was captured on 2 May, which is seven days after the strike of the 2015 Nepal Earthquake. This limits the usefulness of the satellite data in terms of detecting damage and saving life, as the best time has already passed. Moreover, planning the orbit of satellite to capture earthquake damage is difficult due to the uncertain nature of the earthquakes. Near real-time observations of the earthquake-prone fault zones using geostationary satellites or satellites with rapid revisits are suggested. Airborne lidar imaging can also be a promising technique in this kind of situation. Nevertheless, development of local telecommunication technology is inevitable for saving lives during the disaster. On the other hand, better building damage information obtained from the EDV is expected to provide new insights into disaster-preparedness planning such as updating building codes in the vulnerable fault zones, and allocating open spaces and evacuation zones.

ACKNOWLEDGMENTS

The authors are thankful to JAXA for the ALOS-2 SAR data used in the research in the framework of the ALOS Research program, PI No. 1172. This research is partially supported by the Vietnam National Foundation for Science and Technology Development (NAFOSTED) under Grant No. 105.08-2015.31.

AUTHOR CONTRIBUTIONS

Ram C. Sharma designed the research, processed the SAR data, and wrote the manuscript. Hoan Thanh Nguyen, Saeid Gharechelou and Luong Viet Nguyen assisted in the preparation of ground truth data and the processing of the SAR data. Ryutaro Tateishi and Keitarou Hara revised the manuscript. All authors contributed to and approved the final manuscript before submission.

CONFLICTS OF INTEREST

The authors declare that there is no conflict of interest.

REFERENCES

1. Dong, L.; Shan, J. A comprehensive review of earthquake-induced building damage detection with remote sensing techniques. ISPRS J. Photogramm. Remote Sens. 2013, 84, 85–99.
2. Matsuoka, M.; Yamazaki, F. Use of Satellite SAR Intensity Imagery for Detecting Building Areas Damaged Due to Earthquakes. Earthq. Spectra 2004, 20, 975–994.
3. Turker, M.; San, B.T. SPOT HRV data analysis for detecting earthquake-induced changes in Izmit, Turkey. Int. J. Remote Sens. 2003, 24, 2439–2450.
4. Yusuf, Y.; Matsuoka, M.; Yamazaki, F. Damage assessment after 2001 Gujarat earthquake using Landsat-7 satellite images. J. Indian Soc. Remote Sens. 2001, 29, 17–22.

5. Saito, K.; Spence, R.J.S.; Going, C.; Markus, M. Using High-Resolution Satellite Images for Post-Earthquake Building Damage Assessment: A Study Following the 26 January 2001 Gujarat Earthquake. Earthq. Spectra 2004, 20, 145–169.
6. Arciniegas, G.A.; Bijker, W.; Kerle, N.; Tolpekin, V.A. Coherence- and Amplitude-Based Analysis of Seismogenic Damage in Bam, Iran, Using Envisat ASAR Data. IEEE Trans. Geosci. Remote Sens. 2007, 45, 1571–1581.
7. Gamba, P.; Dell'Acqua, F.; Trianni, G. Rapid Damage Detection in the Bam Area Using Multitemporal SAR and Exploiting Ancillary Data. IEEE Trans. Geosci. Remote Sens. 2007, 45, 1582–1589.
8. Hoffmann, J. Mapping damage during the Bam (Iran) earthquake using interferometric coherence. Int. J. Remote Sens.2007, 28, 1199–1216.
9. Chini, M.; Pierdicca, N.; Emery, W.J. Exploiting SAR and VHR Optical Images to Quantify Damage Caused by the 2003 Bam Earthquake. IEEE Trans. Geosci. Remote Sens. 2009, 47, 145–152.
10. Chini, M.; Cinti, F.R.; Stramondo, S. Co-seismic surface effects from very high resolution panchromatic images: the case of the 2005 Kashmir (Pakistan) earthquake. Nat. Hazards Earth Syst. Sci. 2011, 11, 931–943.
11. Miura, H.; Midorikawa, S.; Kerle, N. Detection of Building Damage Areas of the 2006 Central Java, Indonesia, Earthquake through Digital Analysis of Optical Satellite Images. Earthq. Spectra 2013, 29, 453–473.
12. Ehrlich, D.; Guo, H.D.; Molch, K.; Ma, J.W.; Pesaresi, M. Identifying damage caused by the 2008 Wenchuan earthquake from VHR remote sensing data. Int. J. Digit. Earth 2009, 2, 309–326.
13. Balz, T.; Liao, M. Building-damage detection using post-seismic high-resolution SAR satellite data. Int. J. Remote Sens.2010, 31, 3369–3391.
14. Brunner, D.; Lemoine, G.; Bruzzone, L. Earthquake Damage Assessment of Buildings Using VHR Optical and SAR Imagery. IEEE Trans. Geosci. Remote Sens. 2010, 48, 2403–2420.
15. Pan, G.; Tang, D. Damage information derived from multi-sensor data of the Wenchuan Earthquake of May 2008. Int. J. Remote Sens. 2010, 31, 3509–3519.
16. Dell'Acqua, F.; Bignami, C.; Chini, M.; Lisini, G.; Polli, D.A.; Stramondo, S. Earthquake Damages Rapid Mapping by Satellite Remote Sensing Data: L'Aquila April 6th, 2009 Event. IEEE J. Sel. Top. Appl. Earth Obs. Remote Sens. 2011, 4, 935–943.
17. Corbane, C.; Saito, K.; Dell'Oro, L.; Bjorgo, E.; Gill, S.P.D.; Emmanuel Piard, B.; Huyck, C.K.; Kemper, T.; Lemoine, G.; Spence, R.J.S.; et al. A Comprehensive Analysis of Building Damage in the 12 January 2010 Mw7 Haiti Earthquake Using High-Resolution Satelliteand Aerial Imagery. Photogramm. Eng. Remote Sens. 2011, 77, 997–1009.
18. Tian, J.; Nielsen, A.A.; Reinartz, P. Building damage assessment after the earthquake in Haiti using two post-event satellite stereo imagery and DSMs. Int. J. Image Data Fusion 2015, 6, 155–169.
19. Miura, H.; Midorikawa, S.; Matsuoka, M. Building Damage Assessment Using High-Resolution Satellite SAR Images of the 2010 Haiti Earthquake. Earthq. Spectra 2016, 32, 591–610.
20. Hill, M.; Rossetto, T. Comparison of building damage scales and damage descriptions for use in earthquake loss modelling in Europe. Bull. Earthq. Eng. 2008, 6, 335–365.
21. Dolce, M.; Goretti, A. Building damage assessment after the 2009 Abruzzi earthquake. Bull. Earthq. Eng. 2015, 13, 2241–2264.
22. Schwarz, J.; Abrahamczyk, L.; Leipold, M.; Wenk, T. Vulnerability assessment and damage description for RC frame structures following the EMS-98 principles. Bull. Earthq. Eng. 2015, 13, 1141–1159.
23. Yamazaki, F.; Yano, Y.; Matsuoka, M. Visual Damage Interpretation of Buildings in Bam City Using QuickBird Images Following the 2003 Bam, Iran, Earthquake. Earthq. Spectra 2005, 21, 329–336.

24. Kerle, N. Satellite-based damage mapping following the 2006 Indonesia earthquake—How accurate was it? Int. J. Appl. Earth Obs. Geoinform. 2010, 12, 466–476.
25. Ge, L.; Ng, A.H.-M.; Li, X.; Liu, Y.; Du, Z.; Liu, Q. Near real-time satellite mapping of the 2015 Gorkha earthquake, Nepal. Ann. GIS 2015, 21, 175–190.
26. Meslem, A.; Yamazaki, F.; Maruyama, Y. Accurate evaluation of building damage in the 2003 Boumerdes, Algeria earthquake from Quickbird satellite images. J. Earthq. Tsunami 2011, 5, 1–18.
27. Tong, X.; Hong, Z.; Liu, S.; Zhang, X.; Xie, H.; Li, Z.; Yang, S.; Wang, W.; Bao, F. Building-damage detection using pre- and post-seismic high-resolution satellite stereo imagery: A case study of the May 2008 Wenchuan earthquake. ISPRS J. Photogramm. Remote Sens. 2012, 68, 13–27.
28. Conners, R.W.; Trivedi, M.M.; Harlow, C.A. Segmentation of a high-resolution urban scene using texture operators. Comput. Vis. Graphics Image Process. 1984, 25, 273–310.
29. Gusella, L.; Adams, B.J.; Bitelli, G.; Huyck, C.K.; Mognol, A. Object-Oriented Image Understanding and Post-Earthquake Damage Assessment for the 2003 Bam, Iran, Earthquake. Earthq. Spectra 2005, 21, 225–238.
30. Tiede, D.; Lang, S.; Füreder, P.; Hölbling, D.; Hoffmann, C.; Zeil, P. Automated Damage Indication for Rapid Geospatial Reporting. Photogramm. Eng. Remote Sens. 2011, 77, 933–942.
31. Parape, C.; Premachandra, C.; Tamura, M. Optimization of structure elements for morphological hit-or-miss transform for building extraction from VHR airborne imagery in natural hazard areas. Int. J. Mach. Learn. Cyber. 2015, 6, 641–650.
32. Li, P.; Xu, H.; Guo, J. Urban building damage detection from very high resolution imagery using OCSVM and spatial features. Int. J. Remote Sens. 2010, 31, 3393–3409.
33. Vetrivel, A.; Gerke, M.; Kerle, N.; Vosselman, G. Identification of damage in buildings based on gaps in 3D point clouds from very high resolution oblique airborne images. ISPRS J. Photogramm. Remote Sens. 2015, 105, 61–78.
34. Taskin, G.; Ersoy, O.K.; Kamasak, M.E. Earthquake-induced damage classification from postearthquake satellite image using spectral and spatial features with support vector selection and adaptation. J. Appl. Remote Sens. 2015, 9, 096017.
35. Turker, M.; Cetinkaya, B. Automatic detection of earthquake-damaged buildings using DEMs created from pre- and post-earthquake stereo aerial photographs. Int. J. Remote Sens. 2005, 26, 823–832.
36. Gerke, M.; Kerle, N. Automatic Structural Seismic Damage Assessment with Airborne Oblique Pictometry© Imagery. Photogramm. Eng. Remote Sens. 2011, 77, 885–898.
37. Maruyama, Y.; Tashiro, A.; Yamazaki, F. Detection of collapsed buildings due to earthquakes using a digital surface model constructed from aerial images. J. Earthq. Tsunami 2014, 8, 1450003.
38. Cao, C.; Liu, D.; Singh, R.P.; Zheng, S.; Tian, R.; Tian, H. Integrated detection and analysis of earthquake disaster information using airborne data. Geomat. Nat. Hazards Risk 2016, 7, 1099–1128.
39. Hong, Z.; Tong, X.; Cao, W.; Jiang, S.; Chen, P.; Liu, S. Rapid three-dimensional detection approach for building damage due to earthquakes by the use of parallel processing of unmanned aerial vehicle imagery. J. Appl. Remote Sens. 2015, 9, 097292.
40. Fernandez Galarreta, J.; Kerle, N.; Gerke, M. UAV-based urban structural damage assessment using object-based image analysis and semantic reasoning. Nat. Hazards Earth Syst. Sci. 2015, 15, 1087–1101.
41. Stramondo, S.; Bignami, C.; Chini, M.; Pierdicca, N.; Tertulliani, A. Satellite radar and optical remote sensing for earthquake damage detection: Results from different case studies. Int. J. Remote Sens. 2006, 27, 4433–4447.

42. Bazi, Y.; Bruzzone, L.; Melgani, F. An unsupervised approach based on the generalized Gaussian model to automatic change detection in multitemporal SAR images. IEEE Trans. Geosci. Remote Sens. 2005, 43, 874–887.
43. Dong, Y.; Li, Q.; Dou, A.; Wang, X. Extracting damage caused by the 2008 Ms 8.0 Wenchuan earthquake from SAR remote sensing data. J. Asian Earth Sci. 2011, 40, 907–914.
44. Trianni, G.; Gamba, P. Damage detection from SAR imagery: Application to the 2003 Algeria and 2007 Peru earthquakes. Int. J. Navig. Obs. 2008, 2008, 762378.
45. Yun, S.-H.; Hudnut, K.; Owen, S.; Webb, F.; Simons, M.; Sacco, P.; Gurrola, E.; Manipon, G.; Liang, C.; Fielding, E.; et al. Rapid Damage Mapping for the 2015 Mw 7.8 Gorkha Earthquake Using Synthetic Aperture Radar Data from COSMO–SkyMed and ALOS-2 Satellites. Seismol. Res. Lett. 2015, 86, 1549–1556.
46. Grünthal, G. European Macroseismic Scale 1998: EMS-98; Imprimerie Joseph Beffort: Helfent-Bertrange, Luxembourg, 1998.

CHAPTER 16

Seismic Performance of Historical and Monumental Structures

Halil Sezen[1] and Adem Dogangun[2]

[1] *Department of Civil, Environmental, and Geodetic Engineering, The Ohio State University, Columbus, Ohio, USA*

[2] *Department of Civil Engineering, Uludag University, Bursa, Turkey*

1. INTRODUCTION

Historical, religious and monumental structures and their susceptibility to damage in recent earthqaukes in Turkey are presented and discussed in this chapter. Turkey has a very large number of historical structures and is located in one of the most seismically active regions of the world. Some of these historical and monumental masonry and reinforced concrete structures suffered substantial damage or collapsed during two major earthquakes in 1999. The Kocaeli (M_w7.4) and Düzce (M_w7.2) earthquakes occurred on August 17 and November 12, 1999, and ruptured approximately 110 km and 40 km of the 1550-km-long North Anatolian fault, respectively. This chapter describes briefly the construction materials and techniques for historical religious and monumental structures and state-of-practice in Turkey, and presents dynamic analyses of a masonry minaret example. The seismic performance of the mosques and minarets (tall slender towers) during the 1999 earthquakes is presented.

2. SEISMIC DESIGN AND CONSTRUCTION PRACTICE IN TURKEY

Prior to the 1999 earthquakes, two codes governed the design and construction of reinforced concrete and masonry buildings in Turkey: Turkish Earthquake Code (1998) and the Turkish Building Code, TS-500 (1985). The earthquake code included procedures for calculating earthquake loads on buildings. The ductility requirements and details described in the earthquake code were rarely observed in religious or monumental structures inspected by the authors after the 1999 earthquakes. Details of the seismic design and building construction practice prior to the 1999 earthquakes are provided in Sezen et al. (2003).

The descriptions of ground motion characteristics, structural damage, and performance of structures during these earthquakes are provided in Sezen et al. (2003). Response spectra for selected acceleration histories for 5% damping are presented in Figure 1. The ground motions included in Figure 1 were recorded at: SKR station in Adapazari (Peak Ground Acceleration, PGA 0.41g, stiff soil); YPT station in Yarimca (PGA 0.23g, soft soil); and DZC station in Düzce (PGAs: EW-Aug.17 0.36g, NS-Aug.17 0.31g, EW-Nov.12 0.54g, and NS-Nov.12 0.35g, soft soil). Figure 1 also provides a comparison between the linear elastic acceleration response spectra calculated for rock and that calculated for soft soil sites using the provisions of the Uniform Building Code (UBC, 1997) and those of the Turkish seismic code (1998) with 5 percent damping for the highest seismicity in the United States and in Turkey, respectively. From Figure 1, it can be concluded that, for a structure with given periods of vibration, the difference between the base shear calculated using the 1998 Turkish seismic code and the base shear demand obtained from the recorded acceleration response spectra do not differ significantly. If the static lateral load distribution over the height is suggested by the code is assumed to be credible, structures designed and detailed according to the Turkish code should not have collapsed or suffered severe damage during the 1999 earthquakes.

Specifically, according to Figure 1, a structure with a fundamental period of 0.5 seconds would be subjected to seismic forces larger than those specified in the code. The period of older masonry structures with thick and shorter walls tends to be relatively small, probably 0.5 seconds or less. On the other hand, the minarets or very slender towers tend to be very flexible with relatively large fundamental periods. Figure 1 shows that the recorded spectral accelerations are significantly large at large periods. The authors' field observations after the 1999 earthquakes, especially the November 12 Duzce earthquake (DZC), showed more damage in slender structures like minarets. In addition, among other factors leading to inadequate

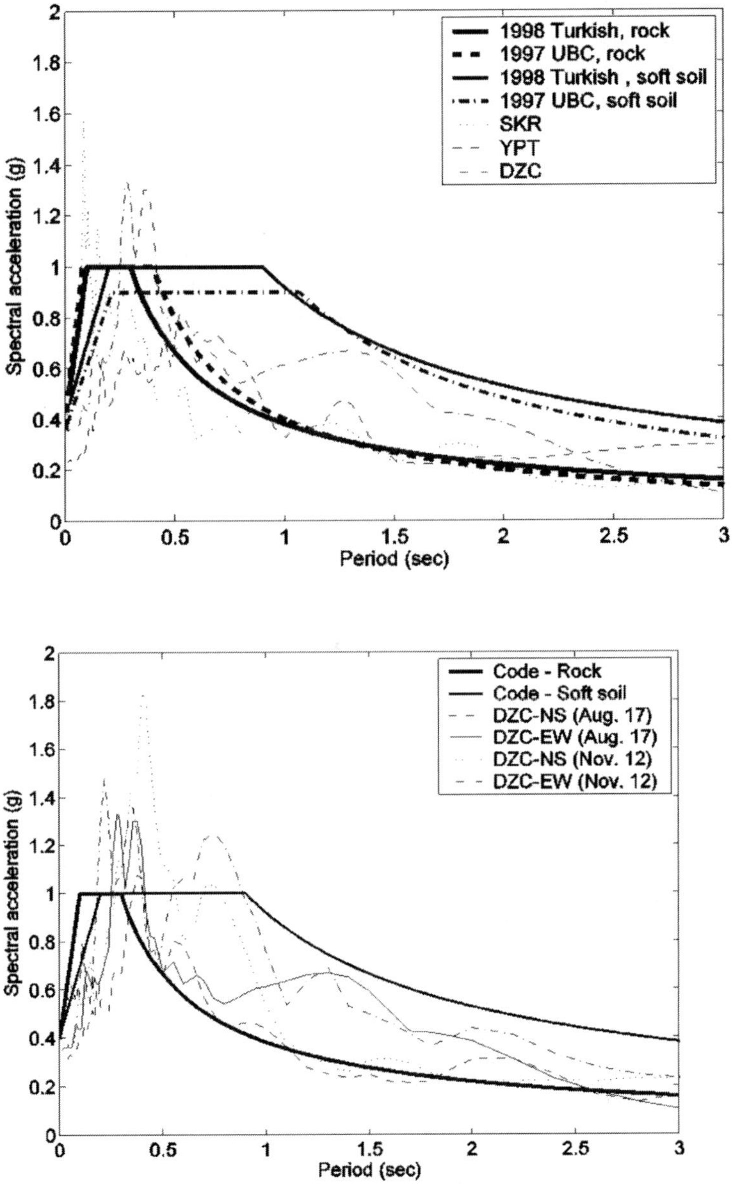

FIGURE 1. Absolute Acceleration Response Spectra

performance, the extent of damage observed in most reinforced concrete structures including minarets after the 1999 earthquakes is probably related to poor engineering and lack of conformance to the relatively new Turkish seismic code.

2.1. Construction Materials And Deterioration Of Historical Structures

In the region affected by the earthquakes, in general, construction materials include stone or brick masonry and wood for older mosques and minarets, steel and reinforced concrete in more recently constructed structures. Typically locally available stone blocks were used. A special mortar called Horasan mortar was used in historical masonry structures. The Horasan mortar is made by mixing lime with baked clay powder that could be obtained by grinding clay tiles or bricks. The Horasan mortar was commonly used in the Middle East, where use of earthenware, pottery and brick construction were widespread (Çamlıbel 1988).

Hollow clay tiles or lightweight concrete blocks are used as infill wall material in recently built reinforced concrete (RC) moment frame mosques. Solid clay bricks were used in infill walls of RC mosques built before early 1970s. Almost all historical religious structures in Turkey were constructed using cut stone, masonry blocks or combination of these two materials. Timber construction is rare and reinforced concrete is relatively recent. The structural and geometrical properties of each masonry structure depend on many factors including the structural knowledge and applications at the time of construction, experience of the architect or engineer, seismicity of the region, and availability of construction materials in that area. Recent earthquakes in Turkey have shown that most masonry monumental structures as well as buildings (Ural et al. 2012) in high seismic regions are vulnerable to structural damage and collapse.

Historical structures deteriorate over time mainly due to environmental effects, and therefore may experience failure or collapse under gravity loads or seismic loads lower than those predicted by the design codes. As discussed in Dogangun and Sezen (2012) and Sezen and Dogangun (2009), in many cases, the following factors trigger or exacerbate the damage: (1) Surface or rain water runoff. If the structure and drainage system are not maintained properly, grass or fungus may grow and weaken the structural materials. Water accumulated on or penetrated into structural members may cause cracks due to freezing and thawing. (2) Soil settlement and relative movement of foundation. (3) Insufficient material strength. Layers of clay or other impure materials inside stone blocks may eventually lead to wearing, spalling or cracking. In stone masonry structures, the properties of the mortar significantly influence the strength of the entire structural component such as a load bearing wall. Deterioration of mortar binding the stone blocks, especially poor quality mortar including mud or low quality lime, can reduce the strength and stiffness of the wall considerably. (4) Other problems. Historical structures can be subjected to various environmental and loading conditions depending on their use and geographic location. For example, timber is more susceptible to humidity and temperature variations. Loading from continuous traffic and heavy trucks can lead to vibrations and excessive loads on foundations because the streets and other structures in historical cities are not designed for modern day traffic. Similarly, the use or occupancy of the structure may change and create larger unexpected loads. Parts of older structures are sometimes used as storage, in which the magnitudes of loads are usually much higher. Other local and environmental effects, such as acid rains, may adversely affect construction materials.

Most of the factors presented here result in gradual deterioration of materials or the load carrying structural system, which can be prevented as the damage progresses and becomes visible in many cases. On the other hand, structural damage, failure or total structural collapse occurs suddenly during moderate or strong earthquakes. Thus, it is essential to evaluate the capacity of existing historical structures and to retrofit them before an expected earthquake strikes.

2.2. Construction Practice And Techniques

Overall plan dimension of a typical mosque generally varies between 12 and 25 meters. Depending on the plan dimensions, height ranges from 7 m to 15 m not including the dome. Height of the dome, a half sphere, is typically half of the plan dimension (Figure 2). Lateral seismic loads are typically resisted by relatively thick unreinforced stone masonry walls in historical mosques. Addition to load bearing masonry walls, most mosques include a few columns typically carrying the gravity loads.

Minarets can be separate or contiguous and integral with the mosque structure, and are typically built using stone, brick, wood or reinfroced concrete. They typically include cylindrical or polygonal body/shafts, one or two balconies, and a conical roof or spire (Figure 2). In masonry minarets or slender tower structures, rather small tensile strength of mortar placed between the masonry blocks presents a major problem in regions of high seismicity. The brick or stone blocks have fairly large compressive strength, however unreinforced masonry lacks tensile strength required to resist bending moments imposed by the lateral earthquake loads. Older masonry minarets were typically constructed using stone blocks or solid clay bricks or a combination of two, whereas unreinforced lightweight stone blocks are preferred in new construction.

FIGURE 2. Typical mosque and its minarets in Turkey

After a major earthquake in 1509, Ottoman architects tackled the problem of constructing tall earthquake resistant minarets (Oğuzmert, 2002). They started to use a special technique for linking adjacent stone blocks with iron bars and clamps in the vertical and horizontal directions as shown in Figure 3 (Doğangün et al. 2007). Use of iron clamps in the two perpendicular directions (transverse and vertical) has improved the lateral load carrying capacity of slender masonry minarets significantly under earthquake loads. The clamps and vertical bars were placed inside anchorage holes in the stone blocks, and melted lead was poured inside the hole to provide bond between the stone and iron clamp or vertical bars. Depending on the properties and dimensions of the stone units, different clamps were developed. For example, as shown in Figure 4a, curved clamps were used within the circular stone wall on the minaret perimeter. Sharper and thinner clamps in thin stone blocks (Figure 4b) and shorter clamps were used if low tensile stresses are expected (Figure 4c). Approximately 2000 kilograms of this heavy metal, lead was used for the construction of a typical masonry minaret. Lead performs as intended for a very long time because it does not corrode or is hardly ever influenced by the adverse environmental conditions.

3. POST-EARTHQUAKE SURVEYS

The data presented here are based on observations from two surveys conducted after the August 17 and November 12, 1999 earthquakes. Damage to historical and recently constructed mosques and minarets was documented to investigate the seismic performance of these structures. The main objectives of these surveys were to provide detailed information about the characteristics of the observed damage, and to study the relative vulnerability of these historical and modern structures to strong ground motions. The parameters considered in the survey included: type of construction material, minaret height, location and description of damage, and location (coordinates, if available). Vast majority of the surveyed minarets and mosques were located in the cities of Düzce and Bolu. The peak ground accelerations recorded in Düzce during both earthquakes were larger than 0.30g (Figure 1), whereas that recorded in Bolu was reported as 0.82g after the November 12, 1999 earthquake.

Structural performance levels and corresponding representative sample damage descriptions for mosques and minarets are shown in Table 1. It should be noted that non-structural damage is irrelevant for minarets. In general, damage to the non-structural components of the mosques was insignificant as compared to the total structural damage.

A total of 59 sites were visited after the October 12, 1999 Duzce earthquake (second earthquake). The name, location, construction date, and observed damage levels are provided in Tables 2 and 3. The first 22 mosques listed in Table 2 were located in the city of Duzce and neighboring town of Kaynasli. Mosques numbered 23 through 44 were located in the city of Bolu.

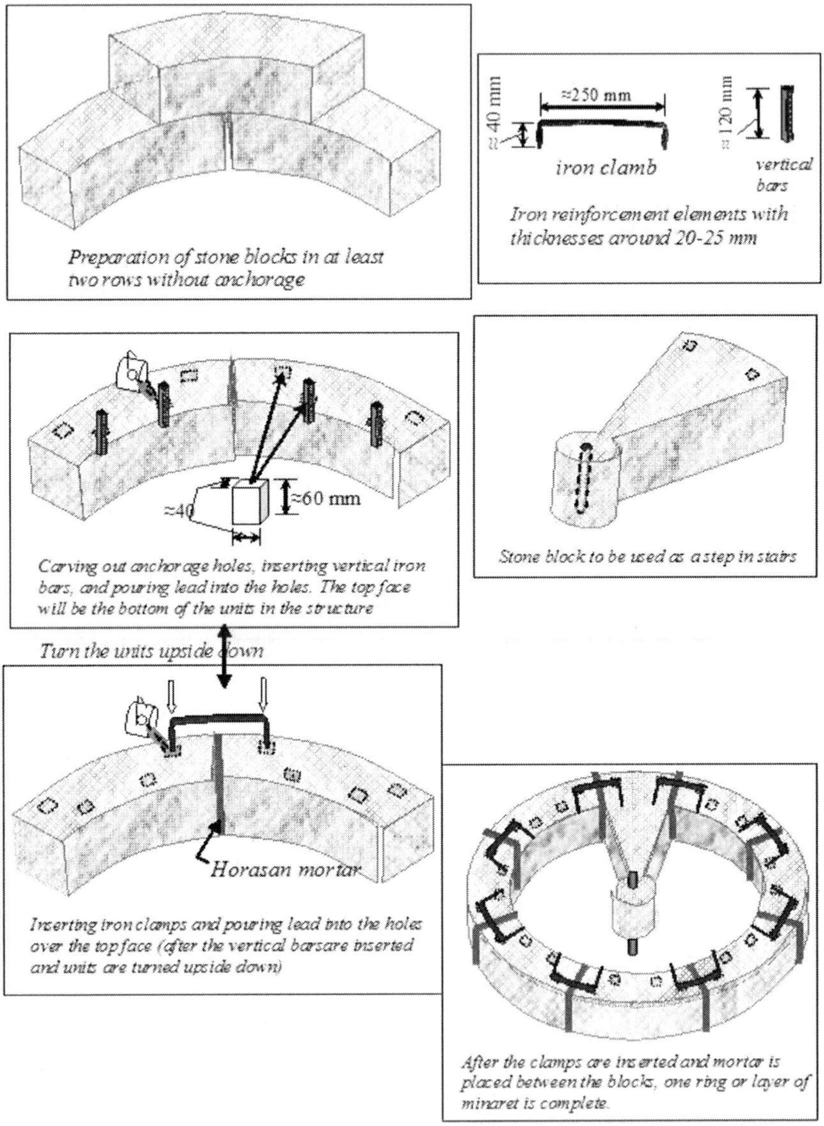

FIGURE 3. Construction of traditional Turkish minarets using stone blocks reinforced and anchored with iron bars and clamps (Dogangun et al. 2007)

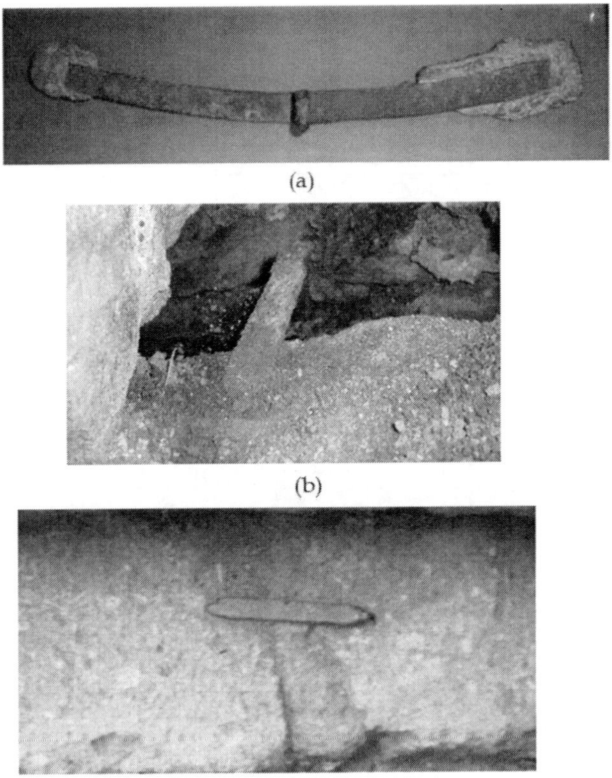

(a)

(b)

(c)

FIGURE 4. Variations in iron clamps used in historical stone masonry walls

TABLE 1. Structural damage description and classification for mosques and minarets

Performance level	Damage classification	Sample damage description (mosque)	Sample damage description (minaret)
I	None	Negligible	Negligible
II	Light	Minor cracks in primary structural components	Minor cracks in masonry or RC minarets
III	Moderate	Significant cracks in RC members or masonry walls	Significant cracks especially around the minaret base
IV	Major	Hinge formation and wide cracks in primary RC members Infill wall collapse	Permanent visible drift Wide cracks and concrete spalling
V	Collapse	Partial or total collapse	Collapse

TABLE 2. Damage to the minarets and mosques surveyed in Duzce and Bolu

Name	Location (coordinates)		Construction date	Damage level	
	North	East		Mosque	Minaret
1. Sirali Koyu	40° 49.246′	31° 11.544′	1987	Light	Collapse
2. Kocyazi Koyu	40° 50.569′	31° 10.249′	1977-79	Light	Collapse
3. Karaca	40° 50.733′	31° 09.950′	1970s	Collapse	Collapse
4. Hamidiye	40° 50.838′	31° 09.518′	1980s	Light	Light
5. Rumelipalas	40° 50.877′	31° 08.444′	1972	None	None
6. Uzun Mustafa	40° 50.755′	31° 08.736′	1989	None	Collapse
7. Kultur Mahallesi	40° 50.623′	31° 09.257′	-	None	Moderate
8. Otopark	40° 50.446′	31° 09.119′	1971-73	None	Moderate
9. Aydinpinar	40° 49.796′	31° 09.190′	1994	Light	Collapse
10. Yesil	40° 49.966′	31° 09.480′	1990	None	Collapse
11. Asar	40° 50.012′	31° 09.650′	1977	None	Moderate
12. AzmimilliYeni	40° 49.897′	31° 10.087′	1988	None	Collapse
13. Mimar Sinan Nur	40° 50.006′	31° 10.236′	1988-91	None	Collapse
14. Maresal F. Cakmak	40° 49.928′	31° 10.525′	1990	Light	Collapse
15. Huzur	40° 49.680′	31° 11.324′	1986	None	None
16. Topalakli	40° 49.606′	31° 11.539′	1951	Light	None
17. Kirazli Koyu	40° 48.918′	31° 12.844′	1965	None	None
18. Doganli Koyu	40° 48.182′	31° 14.190′	1981-94	None	None
19. Uckopru Merkez	40° 47.662′	31° 15.018′	1985	None	Light
20. Yesiltepe	40° 46.690′	31° 17.467′	1986-90	Collapse	Collapse
21. Karacaali	40° 46.496′	31° 18.204′	1992-94	None	Moderate
22. Kaynasli merkez	40° 46.406′	31° 19.138′	-	Collapse	Collapse
23. Sanayi	40° 44.280′	31° 37.562′	1988	None	None
24. Kultur	40° 44.507′	31° 36.340′	1990	None	Light
25. Oksuztekke	40° 44.488′	31° 35.851′	1993	Major	Collapse
26. Ozayan	40° 44.638′	31° 35.405′	1996	None	Moderate
27. Beskonaklar Yeni	40° 44.263′	31° 35.599′	1997	Moderate	-
28. Pasakoy Berberler	40° 43.942′	31° 34.096′	1987	None	None
29. Pasakoy Eniste	40° 43.803′	31° 34.689′	1997	None	Light
30. Sumer	40° 43.575′	31° 35.587′	1992-95	None	Light
31. Sumer Mah. Yeni	40° 43.468′	31° 35.297′	1983	None	None
32. Karacayir Siteler	40° 43.577′	31° 36.033′	1998-99	None	-
33. Semsi AhmetPasa	40° 43.852′	31° 36.635′	14th cent	Major	Collapse
34. Sarachane	40° 43.935′	31° 36.513′	17th cent	Light	None
35. Yildirim Bayezid	40° 44.040′	31° 36.576′	1804	Major	None
36. Kadi	40° 43.901′	31° 36.459′	1499	Major	Collapse
37. Aslahaddin	40° 43.981′	31° 36.732′	1978	None	None
38. Balci	40° 43.885′	31° 37.111′	1990	None	None
39. Aktas	40° 43.783′	31° 36.650′	1900s	Major	None

Name	Location (coordinates)		Construction date	Damage level	
	North	East		Mosque	Minaret
40. Karacayir	40°43.825′	31°36.515′	1946	None	None
41. Kabaklar	40°44.859′	31°36.111′	1981	None	None
42. Camli	40°44.243′	31°35.841′	1980s	Major	None
43. Sultanzade	40°44.249′	31°35.817′	1930s	Major	Major
44. Yesil	40°44.170′	31°36.170′	1966	Major	None

TABLE 3. Damage to the other mosques and minarets visited (coordinates not available)

No.	Name	Location	Mosque	Minaret
45	Cumhuriyet Mah.	Duzce, downtown	None	Collapse
46	Merkez	Duzce, downtown	Major	Collapse
47	Cedidiye Merkez	Duzce, downtown	Light	Collapse
48	Yuvacik	Yuvacik village, near Golcuk	Major	-
49	Asagi Yuvacik	Yuvacik village, near Golcuk	None	None
50	Yeni	Adapazari	None	Collapse
51	Yalova	Yalova, downtown	Light	None
52	Izmit (1)	Izmit, next to highway E5	Light	Collapse
53	Izmit (2)	Izmit, next to highway TEM	None	Light
54	Izmit (3)	Izmit, downtown	Major	Light
55	Golyaka	Golyaka, downtown	Major	Collapse
56	Suleymanbey	5 km east of Golyaka	Collapse	-
57	Golcuk (1)	4 km west of downtown	Major	-
58	Golcuk (2)	Near Ford plant, west of city	Collapse	Moderate
59	Dariyeri Hasanbeyi	East of Duzce (a village)	Major	Light

Before mid-1960s, major construction materials were wood and stone or brick masonry. Most of the recently constructed mosques and minarets were reinforced concrete. All of the reinforced concrete mosques were built after 1965. In older mosques, solid bricks were used in infill walls. Hollow clay tiles were used as infill material in the mosques built after late 1970s. As shown in Figure 5, 84 percent of the minarets surveyed in Duzce and Kaynasli were reinforced concrete, whereas only 46 percent of the minarets were reinforced concrete in Bolu. Unreinforced stone masonry was commonly used in old minarets as well as in the minarets constructed in recent years.

4. OBSERVED EARTHQUAKE DAMAGE

The Figure 6 shows the damage distribution for the mosques and minarets surveyed. Damage distribution for RC mosques and minarets are presented in Figures 6c and 6d separately.

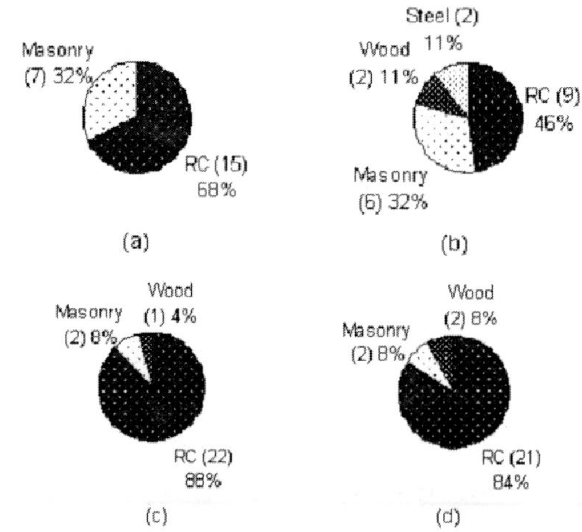

FIGURE 5. Type of construction for (a) mosques, (b) minarets in Bolu; and (c) mosques, and (d) minarets in Duzce and Kaynasli (number of mosques/minarets is given in parenthesis)

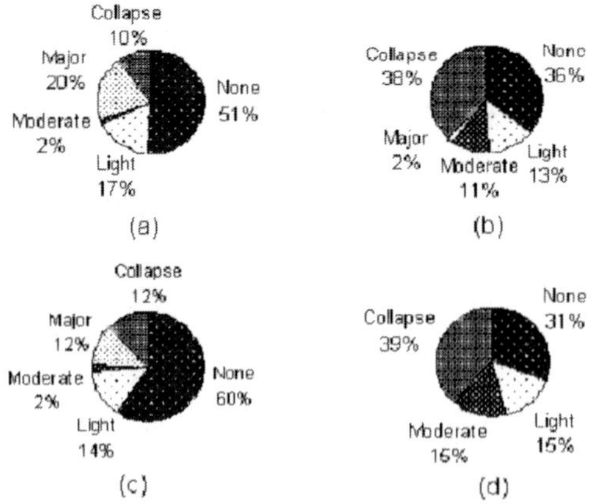

FIGURE 6. Damage distribution for (a) all mosques and (b) all minarets, and damage distribution for (c) RC mosques and (d) RC minarets

Comparison of Figures 6a and 6c indicates that the damage in RC mosques was less as compared to other structural systems. More than ten percent of the mosques surveyed collapsed. Three of the 26 mosques surveyed in Duzce and Kaynasli region collapsed. There was no collapsed mosque in Bolu, but five of the 21 mosques surveyed had to be closed after the October 12, 1999 earthquake. Closed Mosques in Bolu were Semsi Ahmet Pasa (Imaret), Yildirim Bayezid, Kadi, Camli and Yesil. Note that the first three of these mosques are at least 200 years old.

As illustrated in Figures 6b and 6d, percentage of all damaged minarets and only RC minarets are similar. Almost forty percent of the minarets collapsed, and approximately one third of the minarets were undamaged. Failure plane for almost all collapsed RC minarets was within 1.5 meter long region above the minaret base or pyramid-shaped transition segment (Figure 7a), where the longitudinal reinforcing bars were usually spliced. Horizontal circumferential cracks and spalling of concrete were commonly observed at the bottom of the cylindrical body of RC minarets (Figure 7a). Frequently, flexural cracks, concrete crushing or spalling was observed in this region of the damaged RC minarets. As shown in Figure 7, less frequently collapse or damage occurred within the transition segment and middle of the cylindrical body or near the top of the minaret.

(a) (b) (c)

FIGURE 7. Collapse of a minaret near the bottom of cylinder (left), minaret damage within transition segment (middle), and collapsed minaret at mid-height of cylinder (right)

The ratio of collapsed or damaged unreinforced solid brick or stone masonry minarets was much larger than that of RC minarets. As listed in Table 4, majority of the visited masonry minarets collapsed (Sezen et al. 2003, and Firat 1999). Note that most of these minarets were either very new or few hundred years old (Tables 2 and 3). A minaret may either have an independent foundation (referred to as Type I minaret hereafter) or the base of the minaret may be attached to the roof of the mosque (referred to as Type II minaret). The minarets in Figures 2 and 7 are Type I and II minarets, respectively. In Type II minarets, the minaret and mosque structure respond independently. Large deformatiosn or failure in one structure does not affect the other one.

Lateral laod resisting mechanism of minarets are quite different from that of other structures. The height of center of minaret's mass can be very high above ground, resulting in large bending moments and shear forces. Masonry minarets without reinforcement or clamps (Figures 3 and 4) and with weak mortat are the most vulnerable against earthquakes. Masonry minarets typically either collapsed or suffered minor or no damage. This suggests that these minarets had no ductility or very limited deformation capacity. If the demand due to lateral earthquake forces is less than minaret's ultimate strength, minaret behaves elastically and suffers no damage. However, as soon as lateral demand exceeds the elastic capacity, the minaret collapses. Unlike RC minarets, masonry minarets failed at different locations along the height of the minaret inconsistently (Figure 8). The Imaret and Kadi mosques and possibly their minarets were 600 and 500 year old, respectively. As shown in Figures 8a and 8b, their minarets collapsed at the bottom of the cylindrical body. On the other hand, the Oksuztekke minaret collapsed at its mid-height (Figure 8c). Oksuztekke mosque was constructed only six years before the 1999 earthquakes (Table 2).

TABLE 4. Masonry minarets surveyed after the 1999 earthquakes

Name	City	Location or coordinates	Type	Observed damage
Aziziye Merkez	Düzce	40.50N-31.08E	II	Failed at the bottom of cylinders and collapsed
Cedidiye Merkez	Düzce	40.50N-31.09E	I	Failed at the bottom of cylinders and collapsed
Oksuztekke	Bolu	40.44.488N-31.35.851E	I	Minaret segment above the 2nd balcony level collapsed
Semsi Ahmet Pasa (Imaret)	Bolu	40.43.852N-31.36.635E	I	Failed near the bottom of the cylinder and collapsed
Sarachane	Bolu	40.43.935N-31.36.513E	I	Cracks in stone blocks in the mosque – no observed minaret damage
Yildirim Bayezid	Bolu	40.44.040N-31.36.576E	II	Dislocation of stone blocks in the mosque – no obserevd damage in minarets
Kadi	Bolu	40.43.901N-31.36.495E	I	Severe damage to mosque, minaret collapsed

4.1. Earthquake Damage In Historical Mosques

Among the historical mosques surveyed, Düzce Merkez, 600-year-old Imaret, 500-year-old Kadi, 300-year-old Sarachane, and 200-year-old Yildirim Bayezid mosques observed to suffer significant structural damage after the 1999 earthquakes (Dognagun and Sezen, 2012). The Yıldırım Bayezid mosque was originally built in 1382 and was burned down in the 19th century. A new structure was constructed after the fire, and it was severely damaged during a 7.3 magnitude earthquake in 1944. The mosque was damaged during the 1999 earthquakes and was closed for a period of time. On the south side of the mosque, portion of the walls above and below the windows were subjected to larger shear stresses (compared to solid wall sections) during the strong ground shaking. Higher shear demand in those parts of the relatively thick walls created serious cracks and openings between the stone blocks (Figure 9). As shown in Figure 9 (middle photo), portion of the walls above and below the windows act, in a sense, like short columns during the earthquake. Larger shear demand in those parts of the relatively thick walls created wide cracks between the stones.

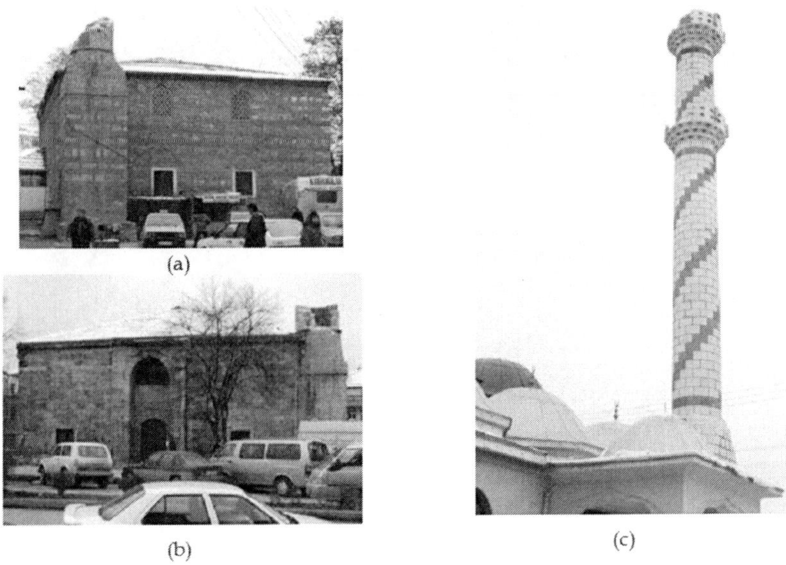

FIGURE 8. Failures near the bottom and mid-height of the masonry cylindrical minaret body: (a) Imaret, (b) Kadi, and c) Oksuztekke mosques in Bolu

CITRON

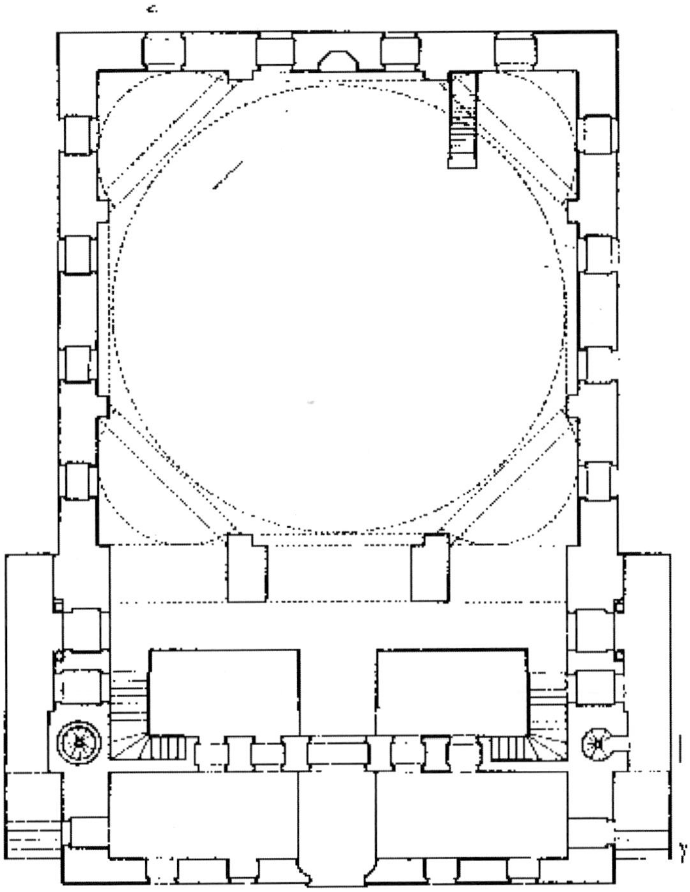

Bolu Yıldırım Camii Plânı

FIGURE 9. Wall damage and plan view of historical Yildirim Bayezid mosque in Bolu

Another historical mosque, Kadi (Figures 8b and 10a) sustained substantial damage including large cracks around historical entrance door, severe cracks and stone dislocations in the 1.5 m wide unreinforced stone masonry perimeter walls. Damage was mostly concentrated below or above the windows. The missing two key keystones and dislocated stones on top of the upper window are visible in Figure 10a. The reduced wall area along the vertical section through the windows was stressed more, causing considerable damage in those wall areas. The stone masonry minaret collapsed right above its base because the minaret base was integral with minaret walls (Figure 8b) and was quite stiff compared to the cylindrical minaret body.

The Imaret mosque, one of the oldest structures in the region, and its minaret were built using stones and small solid bricks bounded by a thick layer of mortar. The mortar between the bricks is typically as thick as the bricks. Old brick masonry minaret collapsed and Imaret mosque was closed after the 1999 earthquakes due to cracks in the walls (Sezen et al. 2003). The 1999 earthquakes caused very limited damage to the 300-year-old Sarachane mosque and there was no visible damage to its minaret. Some large cracks were observed in the walls at locations similar to those observed in other mosques discussed in this paper. Figure 10b and 10c show examples of couple such cracks; one immediately above a window and another in a corner near the roof.

(a) (b) (c)

FIGURE 10. a) cracks developed in the walls and above windows of Kadi; and b, c) damage in in the walls of Sarachane mosque

5. DYNAMIC ANALYSIS OF REPRESENTATIVE MASONRY MINARETS

Although numerous collapses and structural damage to minarets and mosques were documented after strong earthquakes, only few researchers investigated the seismic behavior and performance of minarets (Sezen et al. 2008, Portioli et al. 2011, Turk and Cosgun 2012, Altunisik 2012, Oliveira et al. 2012). It is essential to understand the dynamic behavior of these structures to improve the life safety and to preserve and strengthen the historical monumental structures. In this research, as an example, dynamic modeling and seismic analysis of unreinforced masonry minarets is presented.

Recently, a study was conducted by the authors to investigate the seismic response of reinforced concrete (RC) and masonry minarets (Dogangun et al. 2008, and Sezen et al. 2008). Modal analysis and dynamic time history analyses were conducted. Strength and deformation capacities of the RC and unreinforced masonry models are influenced by unique differences in material properties, including weight, modulus elasticity, and Poisson's ratio. For example, the flexural strength of an unreinforced masonry model is significantly lower as the reinforcing steel in a similar RC model provides tensile resistance. While Sezen et al. 2008 investigates distinct parameters affecting the behavior of RC minarets using a single model, the example presented below examines the effect of masonry material properties and minaret height on the minaret response using three different models. Detailed modeling properties and analysis results are presented in Dogangun et al. (2008).

Although the height of a minaret varies greatly depending on many factors such as the location or construction materials used, the height of a typical minaret with a single balcony over its height is about 20 to 25 m. In this research, two generic minarets with a height of 25 and 30 m and double balconies (Minaret I and Minaret II) and another 20 m tall minaret with a single balcony (Minaret III) were modeled and analyzed. As shown in Figures 11 and 12, only the height of the cylindrical minaret body was varied while the geometrical properties of the base or boot, transition segment, balconies and conical cap were kept the same in all three models. It should be noted that not all conical caps at the top of the masonry minarets are constructed using masonry materials. For uniformity, all minarets including the conical caps are modeled with the same masonry material in this study.

The finite element models of the three minarets are shown in Figure 12, and significant mode shapes of a typical minaret are shown in Figure 13 (Sezen et al. 2008). The computer program, ANSYS was used to model and analyze the minarets (see Dogangun et al. 2008 for details). In the models, the modulus of elasticity of uncracked section, Poisson's ratio and unit weight of masonry material (ordinary limestone) are taken as 3000 MPa, 0.2, and 20 kN/m^3, respectively. In the models, linear elastic material models and five percent damping ratio is used in all dynamic time history analyses. It is assumed that the minaret is located in a high seismic region Zone 1 in Turkish Earthquake Code (TEC, 2007). Considering the TEC, a structural behavior factor, R of 3 and an importance factor, I of 1.2 can be used for such structures.

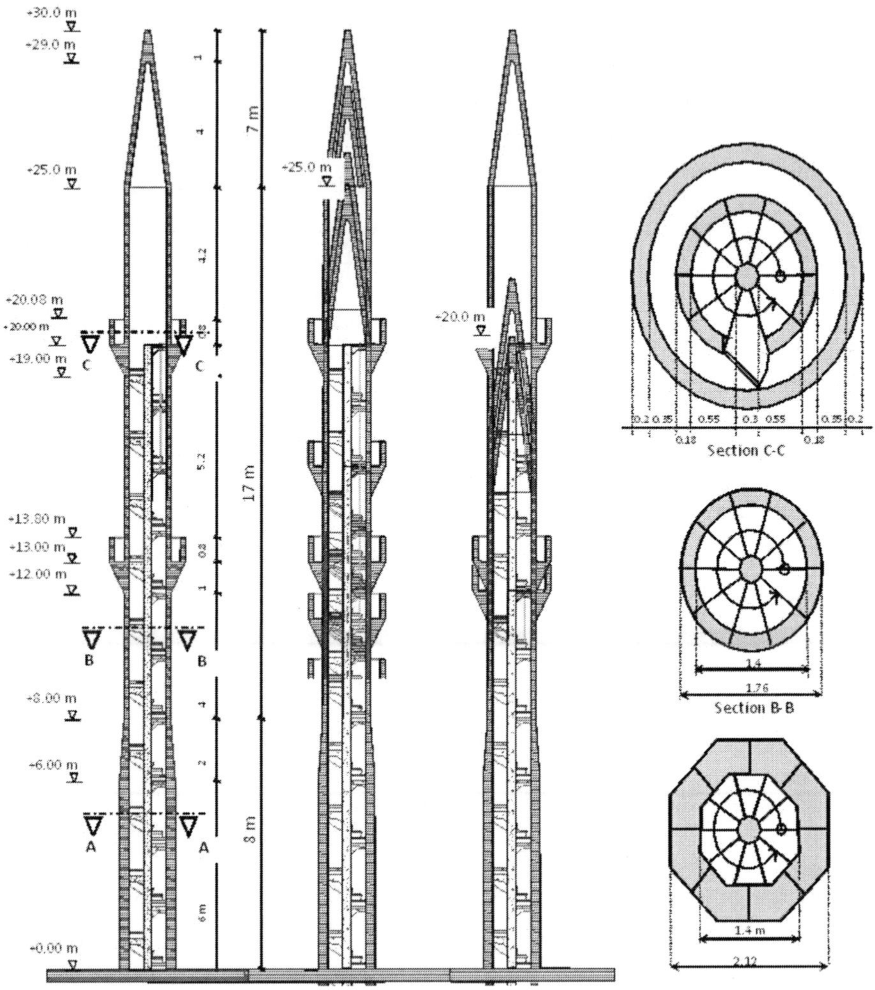

FIGURE 11. Three minaret models and their geometrical and cross-sectional properties

Dynamic time history analyses are carried out for each minaret model using two ground motions recorded during the 12 November, 1999 Duzce and 17 August, 1999 Kocaeli earthquakes (see Dogangun et al. 2008 for details). The input ground motion was applied only in one horizontal direction. The modal periods of the minaret models (calculated from the modal analysis) and their contribution to the total dynamic response are calculated. The modal periods are greatly affected by the height of the minaret which affects the total mass and flexural stiffness. The calculated first or fundamental periods were 1.41, 0.90, and 0.51 seconds for the minaret models I, II, and III, respectively. The calculated modal response quantities indicate that the contribution of higher mode effects to total dynamic response is significant. The first mode contribution to the total response is about 47 or 49 percent. The torsional or the fifth mode has virtually no effect on the total response of the almost symmetrical minaret structures.

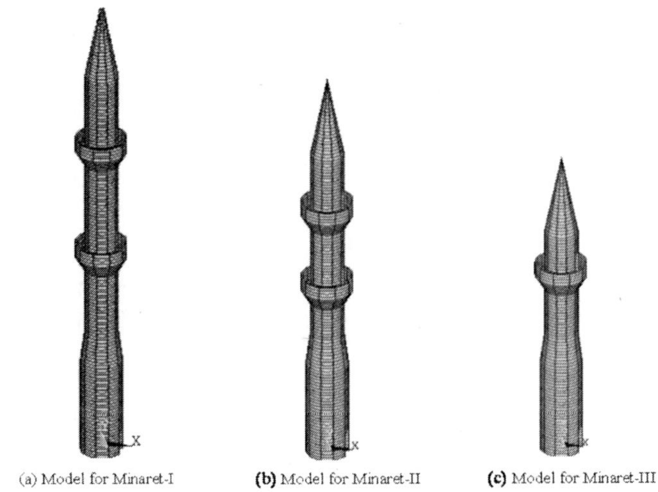

FIGURE 12. Finite element models of the three minarets

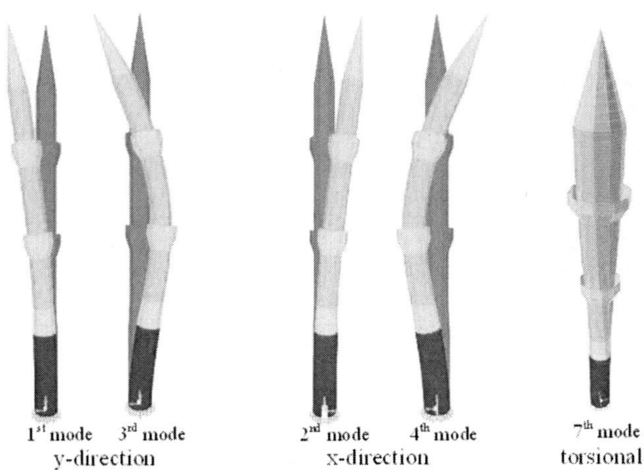

FIGURE 13. First two translational and the seventh mode shapes for a typical model

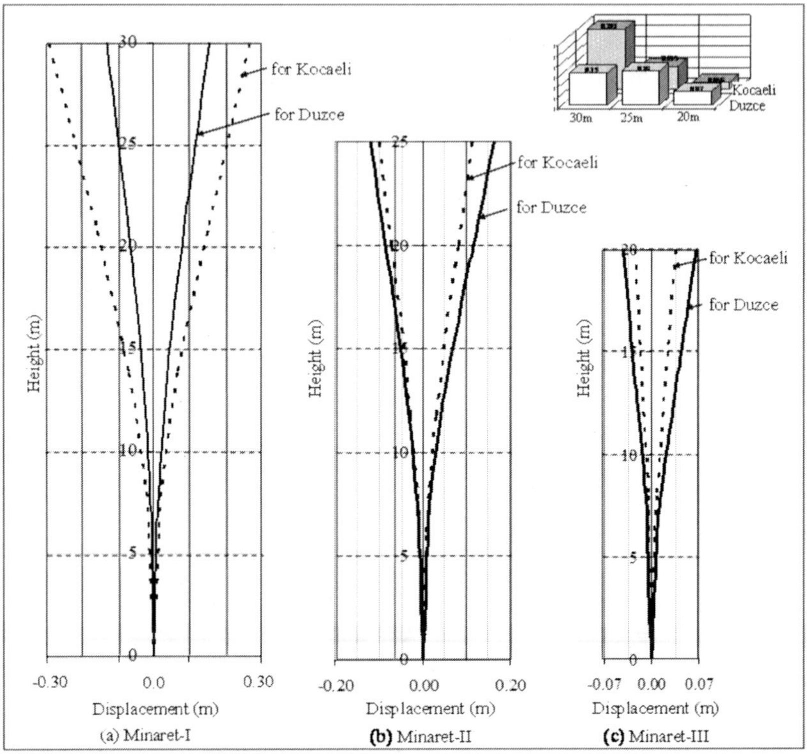

FIGURE 14. Lateral displacement distribution over the height of minarets for Duzce and Kocaeli ground motions

Figure 14 shows the calculated maximum elastic lateral displacement distribution along the length of three minaret models subjected to 1999 Duzce and Kocaeli earthquake ground motions. As shown in Figure 14 the deflected shapes of the minarets appear to show flexure dominated response with the largest displacements calculated at the top. Although the minarets act as cantilevers, the deformations are much smaller over the height of relatively stiff 6 m high base or boot. The displacements start to increase above the transition segment at about 8 m from ground level. The maximum dynamic displacement of Minaret I is 88% larger for the Kocaeli input motion (0.291 m) than that for the Duzce ground motion (0.154 m), although the maximum acceleration of the Duzce motion was larger. For the other two shorter minarets Minaret I and II, the maximum calculated displacements are larger for the Duzce ground motion. The maximum dynamic lateral displacement of 0.29 m calculated for the 30 m tall minaret under Kocaeli earthquake exceeds the TEC (2007) code limit of 0.20 m for the Minaret I model (Dogangun et al. 2008).

The most critical axial stress contours calculated during the dynamic analysis of minaret models are shown in Figure 15. The maximum tensile and compressive stresses occur at a height of 8 m, immediately above the transition zone (Figure 12). This is consistent with the most common minaret failure mode observed in the field after the earthquakes (Figures 7a, 8a and 8b). The minaret models used here were stand-alone fixed-ended cantilevers, which did have any lateral support over their height. Nevertheless, in some cases one side of the minaret boot or base (bottom 6 m of the minaret models) are supported by or attached to the mosque structure (Type II in Table 4, e.g., Figures 8a and 8b). Such a partial lateral support, even on one side, provides added stiffness near the bottom of the minaret. This will further increase failure potential at the top of the transition zone as shown in Figures 7a, 8a, 8b and 14.

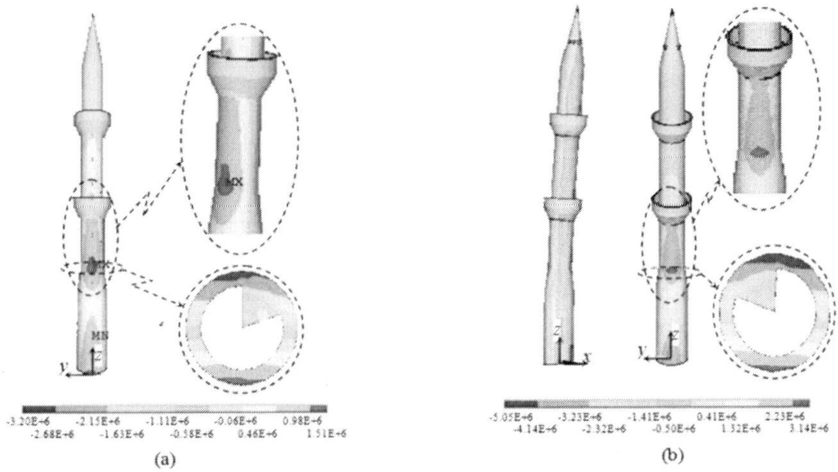

FIGURE 15. Maximum stress contours for Minaret-I under: a) Duzce, and b) Kocaeli records

6. SUMMARY AND CONCLUSIONS

Structural damage and failures observed in 59 historical and monumental structures (mosques and/or minarets) were documented after the August 17 and November 12, 1999 earthquakes. The reported damage distribution in minarets in the cities of Düzce and Bolu gives an indication of the extent of damage. Ten percent of all visited mosques collapsed while 20% suffered severe damage. Of all visited minarets, 38 percent collapsed (Figure 6). Five out of seven masonry minarets (approximately 70%) in the cities of Düzce and Bolu reporeted in this chapter collapsed (Table 4).

Historical structures with heavy and stiff walls are subjected to larger lateral earthquake forces due to their small periods of vibration. The missing keystones in the upper-floor windows and wide cracks between the stones in the walls above and below window openings were the most common damage types.

The location of the failure in the minarets that collapsed during the 1999 earthquakes was generally at the region near the bottom of the cylinder where a transition was made from a square to a circular section. At the bottom of the cylinder body the lateral stiffness and strength are smaller compared with those of the transition region or the minaret base. Consistent with the observed damage, dynamic analysis results from three generic finite element minaret models showed the largest bending stresses in the same region of the minaret models. For minarets with the base or boot attached to the mosque structure (Type II), additional rigidity and stiffness provided by the mosque prevents deformation and damage within the base. In such cases the failure mostly occurs just above the transition region or base of the cylindrical body.

REFERENCES

1. Turkish Earthquake Code.1998and 2007).Specification for structures to be built in disaster areas; Part III-earthquake disaster prevention. Ministry of Public Works and Settlement, Government of Republic of Turkey.
2. Turkish Building Code. TS-5001985Building Code Requirements for Reinforced Concrete. Turkish Standards Institute, Ankara, Turkey (in Turkish).
3. H. Sezen, A. S. Whittaker, K. J. Elwood, K. M. Mosalam, 2003Performance of Reinforced Concrete and Wall Buildings during the August 17, 1999 Kocaeli, Turkey Earthquake, and Seismic Design and Construction Practice in Turkey. Engineering Structures, 25103114
4. Uniform Building Code.1997Structural Engineering Design Provisions, 2International Conference of Building Officials, ICBO. Whittier, California.
5. N. Çamlıbel, 1998Sinan Mimarlığında Yapı Strüktürünün Analitik İncelenmesi, Doçentlik Tezi, İTÜ, Fen Bilimleri Enstitüsü, İstanbul (in Turkish).
6. A. Ural, A. Dogangun, H. Sezen, S. Angin, 2012Seismic Performance of Masonry Buildings during the 2007 Bala, Turkey Earthquakes. Natural Hazards, 60310131026
7. H. Sezen, A. Doğangün, 2009Vulnerability and conservation of historical masonry structures. Prohitech 2009, Protection of Historical Buildings, Rome, Italy, June 2124
8. M. Oğuzmert, 2002Dynamic behaviour of masonry minarets, M.Sc. Thesis, Istanbul Technical University, Turkey.

9. A. Dogangun, H. Sezen, O. I. Tuluk, R. Livaoglu, R. Acar, 2007Traditional Turkish Monumental Structures and their Earthquake Response. International Journal of Architectural Heritage, 13251271
10. H. Sezen, G. Y. Firat, M. A. Sozen, 2003Investigation of the Performance of Monumental Structures during the 1999 Kocaeli and Duzce Earthquakes. Fifth National Conference on Earthquake Engineering, Istanbul, Turkey, 2630May
11. Y. G. Firat, 2001A Study of the Structural Response of Minarets in the 1999 Anatolian Earthquakes. M.S. Thesis. Purdue University. West Lafayette, Indiana.
12. A. Dogangun, H. Sezen, 2012Seismic Vulnerability and Preservation of Historical Masonry Monumental Structures. Earthquakes and Structures, 318395
13. H. Sezen, R. Acar, A. Dogangun, R. Livaoglu, 2008Dynamic Analysis and Seismic Performance of Reinforced Concrete Minarets." Engineering Structures, 30822532264
14. F. Portioli, O. Mammana, R. Landolfo, F. M. Mazzolani, L. Krstevska, L. Tashkov, K. Gramatikov, 2011Seismic Retrofitting of Mustafa Pasha Mosque in Skopje: Finite Element Analysis. Journal of Earthquake Engineering, 154620639
15. A. M. Turk, C. Cosgun, 2012Seismic Behaviour and Retrofit of Historic Masonry Minaret. Gradevinar, 641
16. A. C. Altunisik, 2012Comparison of earthquake behavior of reinforced concrete minarets using fiber-reinforced polymer composite. The Structural Design of Tall and Special Buildings,
17. C. S. Oliveira, E. Çaktı, D. Stengel, M. Branco, 2012Minaret behavior under earthquake loading: The case of historical Istanbul. Earthquake Engineering and Structural Dynamics, 411939
18. A. Dogangun, R. Acar, H. Sezen, R. Livaoglu, 2008Investigation of Dynamic Response of Masonry Minaret Structures. Bulletin of Earthquake Engineering, 63505517

CHAPTER 17

Crustal Structure and Seismogenic Zone of Cameroon: Integrated Seismic, Geological and Geophysical Data

Eloumala Onana Parfait Noel[1], Mouzong Pemi Marcelin[2], Ateba Bekoa[3]

[1]*Civil Engineering Department, University Institute of Technology (IUT), Douala, Cameroon*

[2]*Physics Department, Sciences Faculty, University of Yaoundé, Yaoundé, Cameroon*

[3]*Antenna of Geophysics and Volcanology Researches (ARGV), Buea, Cameroon*

ABSTRACT

This study describes the seismo-tectonic evolution of Cameroon. It is noticed that seismic activity is largely related to the Cameroon volcanic line (CVL) and many of the tectonic phenomena follow the previous structural lines. Although the area is that of a relatively low seismicity, the effects of a given earthquake could be considerable. This study combines seismic records, from 1952 to 2002 and from 2005 to 2007, with the integration of existing structural information to better define the seismogenic zones. An important identified cluster of epicenters in S-W defines the first seismic source region where events' characteristics show a weak seismicity related to volcanic activity. The second seismogenic zone, in the north of Mount Cameroon volcano, illustrates the activity of Central Cameroon Shear Zone

(CCSZ) faults; its seismicity is considered weak to moderate with maximum magnitude recorded which is 5.1 Mb. The area of "Sanaga Shear Zone" (SSZ) constitutes the third seismic source region with moderate seismicity in which maximum magnitude recorded is 5.8 Mb; Depth of its faults segments is evaluated at 33 km. The fourth Source Region follows the North boundary of Congo Craton; a characteristic event with magnitude of 6 M allows evaluating fault depth at 33 km. Also three zones define crustal structure: the South Domain with thick crust, the Center Domain characterized by thin crust, active tectonic and volcanic structures and the north domain with shallow Moho.

Keywords: Seismogenic Zones, Source, Shear Zone, Crustal Structure

1. INTRODUCTION

Cameroon spans the area between latitude 2°N to 13°N and longitude 08°E to 16°E. It corresponds to a complex which includes the Congo Craton in the South and the major geological feature in Central Africa called "Cameroon Volcanic Line", uprising from the West in the Gulf of Guinea to the North (Figure 1).

The edification of Cameroon has developed grabens, horsts, and tectonic lineaments which are superimposed on older structures. Fault structures can be considered in three main groups according to the direction of lineaments: ENE-WSW, NNE-SSW and ESE-WNW (Figure 2). As a consequence, Cameroon is characterized by heterogeneous seismicity where earthquakes that affect most parts are located around the mount Cameroon. The 6.0 Ms earthquake on September 12, 1945 (IRGM, 2002) that hit the south region is one of the largest earthquakes in the country in the historical era of earthquake feeling.

The Moho (Mohorovičić discontinuity) that separates the Earth's crust from the underlying mantle is characterized by major changes in seismic velocity, chemical composition, and rheology [1] . Depth to Moho also varies according to the geo-tectonic evolution of each area.

Although the structures have been the subject of several previous papers, till date, little has been done to identify the seismogenic zone and to confirm the vulnerability of the study area. This study characterizes the recent seismicity of the Cameroon and examines the influence of the structure on the overall seismicity of the region. It also provides information about earthquakes prone areas, seismic zoning maps and crustal thickness for Cameroon. The correlation of seismic, geological and geophysical data allows for a better seismo-tectonic analysis and understanding of the implications for seismic hazard. This work shows three main parts of analysis including the crustal structure, the spatial distribution of events and the "Seismic Source Regions". It identifies structures generating events, characterizes density of seismicity and allows the establishment of the relationship between seismicity and tectonic structures.

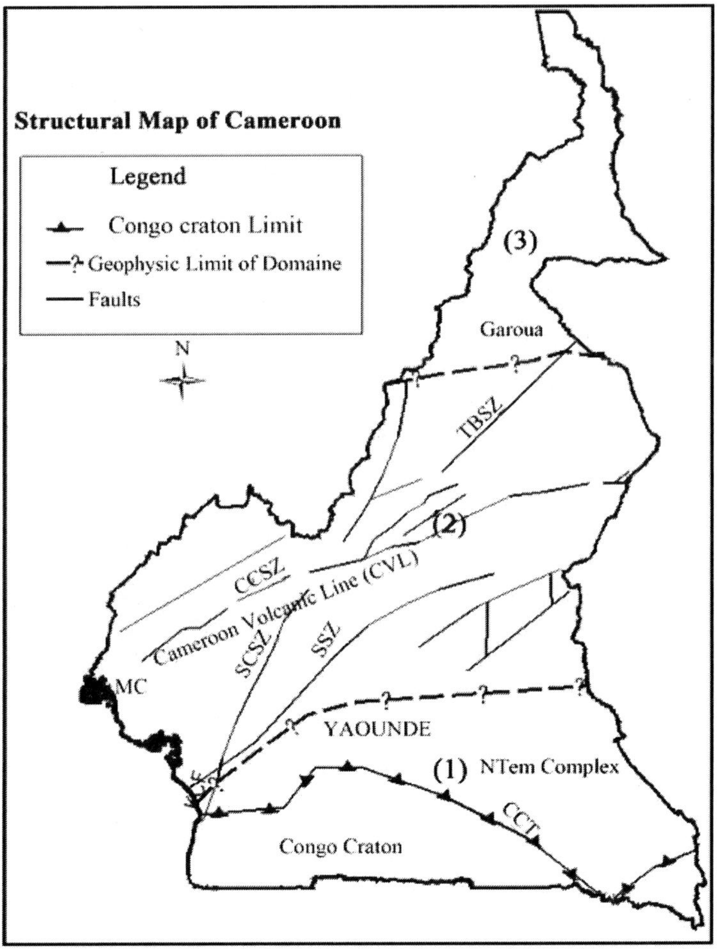

Figure 1. Structural map of Cameroon. CCSZ: Cameroon Shear Zone; SSZ: Sanaga Shear Zone; SCSZ: South Cameroon Shear Zone; CCL: Congo Craton Limit; TBSZ: Tchollié-Banyo Shear Zone; KCF: Kribi-Campo Fault; MC: Mont Cameroon.

2. GEODYNAMIC SETTING

Cameroon has known three orogenic cycles:

1) The Liberian cycle is illustrated by the Ntem complex in South Cameroon. It dates from the Archaean era (about 2.5 billion years old) and is bounded by major thrusts that separate it from the Nyong unit in the North West and the Pan-African Yaoundé Group in the North [2] .

2) The Eburnean or Transamazonian cycle dates from the Palaeoproterozoic period (2.5 - 1.8 billion years ago). It corresponds to the Nyong and Ayna formations, respectively in the West border of Congo Craton and in the East of Ntem complex.

3) The Pan-African cycle, which comprises formations from the Neoproterozoic era 1000 - 600 million years (Ma) ago, includes three main groups (Figure 1) [3] [4] : the southern Domain a) consists of the Yaoundé Group which is limited to the west by the Kribi-Campo fault; The central domain b) (Adamaoua-Yaoundé area) lies from the north of the Sanaga fault to Tchollirć Banyo shear zone; The northern domain c) which is located north of the Tchollirć-Banyo shear zone and extends along the western edge of Cameroon. It continues in the regions of Poli.

3. GEOLOGY SETTING

The geological history of Cameroon begins with the Archaean era between 3.5 and 2.5 billion years (Ga) ago. Its different phases of development are illustrated by geological masses formed during successive orogenic cycles. It is characterized by formation of craton and mountain ranges, and subsequent extension phases by the splitting of the continental crust. From the South to the North we have (Figure 1):

1) The Cameroonian southern domain consists of Pan-African meta-sedimentary units such as the Ntui-Be- tamba, Yaoundé and Ayos-Mbalmayo-Benbis units. The layers were plunged onto the Archaean Congo Craton towards the South [3] from this domain, which were affected by four stages of ductile deformation, corresponding to alternating phases of E-W to NW-SE contraction (D1, D3) and N-S to NE-SW extension D2 [5] .

2) The Cameroonian central domain, broad area that extends between the fault of the Sanaga to the South and the Tibati-Banyo fault to the North. It consists of Archaean to Palaeoproterozoic high-grade gneisses intruded by widespread Pan-African syntectonic plutonic rocks of high-

K calc-alkaline affinities [6] . Major setbacks in this area seem to have guided the implementation of plutonism having an orthogneissification with variable intensity. This area is subjected to an advanced general metamorphism where banking is composed of gneiss and amphibolite.

3) The Cameroonian northern domain is characterized by subordinate 830 Ma old meta-volcanic rocks of tholeiitic and alkaline affinities associated with meta-sediments known as the Poli series. This domain is characterized by three stages of deformation: an early phase D1 is associated to the metamorphism with granulite facies, medium pressure [7] . A phase D2 dated 600 - 580 Ma is synchronous with an intense migmatization [7] and a granitization associated to a metamorphism with amphibolite facies (600° 5 - 7 Kb) and greenschistfacies (550° 5 Kb). And a phase D3 is liable to the dextral setbacks E-W and drive folds N-S to E-W.

4. MAJOR TECTONIC STRUCTURES

The region of Cameroon corresponds to a complex which includes the Congo Craton, major geological feature in Central Africa called "Cameroon Volcanic Line" and diverse tectonic structures (Figure 1). Its edification developed grabens, horsts, and tectonic lineaments which were superimposed on older structures. The fault systems which influence the geodynamic evolution can be considered in different main groups [4] [8] :

1) The ENE-WSW (30°N - 60°N) fault set, the most frequent of the structural trends, was developed at the end of the Cretaceous. It is made up of: Cameroon Shear Zone (CCSZ), Sanaga Shear Zone (SSZ) and Tchollíré Shear Zone (TBSZ).

Central Cameroon Shear Zone (CCSZ) is a fault zone in Cameroon which is part of Central Africa Shear zone. It extends from the Trans-Brazilian Lineament to Sudan. The zone was rejuvenated several times, usually with a dextral movement, before and during the opening of the South Atlantic in the Cretaceous period [9] . The CCSZ underlies a chain of extinct volcanoes, called "Cameroon Volcanic Line" which includes the active volcanos as Mount Cameroon, Nyos and Monoun crater lakes and non-active others.

Sanaga Shear Zone (SSZ), Precambrian Central African Shear Zone, had a prior dextrial movement probably during the end of the Pan-African orogeny, and a later N-S shortening. It is expected to be implicated in a dextrial movement of the northern part of the Mobile Zone [4] .

TheTchollire-Banyo Shear Zone (TBSZ) is thePrecambrian Central African Shear Zone which extends across the African continent from the Darfur in

Sudan, through Central Africa, to the Adamawa Plateau in Cameroon [10] . The Pernambuco lineament in Brazil is probably the extension of the shear zone on the conjugate Atlantic-rifted margin [9] .

Geo-tectonic studies along the Central African Shear Zone estimate ≥40 km of dextral displacement [11] .

2) The NNE-SSW (00°N - 15°N) lineament principally includes South Cameroun Shear Zone (SCSZ). Many segments are identified as the Kribi-Campo Fault (KCF) trends almost in a north-south direction and may correspond to the north-western margin of the Congo Craton.

3) The ESE-WNW (095°N - 110°N) lineament marks out the limit between the Congo Craton with Archaen rocks in the South and the mobile zone with reworked Paleoproterozoic material in the North. This lineament of faults called Congo Craton Limit (CCL) is located around the parallel of latitude 4°N. Aeromagnetic studies have shown that the limit of the Congo Craton is characterized by a series of buried and deep seated fault [12] . The recent studies [13] show that the crust thickness in the Craton complex is around 43 km to 45.5 km and beyond.

The Cenozoic Cameroon Volcanic Line (CVL) trends 30°N and extends from Pagalu Island to northernmost Cameroon. The lithosphere-asthenosphere boundary is <100 km deep beneath the CVL and the mantle is characterized by slow seismic wave speeds [14] . A continuous low-velocity zone (δVS = −2% to −3%) underlies the entire CVL to a depth of at least 300 km and was attributed to a thermal anomaly of ≥280 K [15] . Bulk crustal Vp/Vs ratios along the CVL are markedly low with ~1.74 [14] .

5. SEISMICITY KNOWLEDGE

Studies of continent-scale surface wave show that mantle structure beneath Cameroon is seismically slow [16] [17] . Seismic anisotropy in Cameroon which has been studied by [18] through analysis of SKS splitting allows to identify four regions of distinct anisotropy: moderately strong NE-SW oriented fast polarization directions ($\delta t \approx 1.0$ s) beneath two region: the Congo Craton in the south and the Garoua rift in the north; weak anisotropy ($\delta t \approx 0.3$ s) between the Congo Craton and the CVL; N-S oriented fast polarization directions within the CVL, with $\delta t \approx 0.7$ s. Koch et al. [18] interpreted these observations as evidence for small-scale mantle convection beneath Cameroon.

Tadjou et al. [19] help to constrain the lithospheric subdivisions of the region. For example, thinned crust (~25 km) is imaged beneath the Garoua Rift (Figure 2), while the transition from the CVL to the Congo Craton is characterized by a transition from ~36 km thick crust to 43 - 48 km thick crust [13] [20] . Average crustal shear wave velocity also increases from the CVL with ~3.7 km·s^{-1} into the craton with ~3.9 km·s^{-1} [13] .

6. SEISMIC DATA

For this study, the seismic data we used are extracted from a variety of sources including: the Cameroonian catalogue established by Mining and Geological Research Institute [21] which covers the period from 1852 to 2002, the catalogue established from seismic data recorded by the temporary seismic network installed from 2005 to 2007 (Figure 2) and some events have been taken from international catalog as IRIS and ISC. We have also used crustal structure fund by [13] [17] .

7. ANALYSIS AND INTERPRETATION

7.1. Evaluation of Crustal Thickness

Using joint inversion analysis, [13] presented a crustal thickness underneath 30 stations installed in Cameroon and [16] presented a crustal thickness and mantle velocities maps for the African continent map using analysis of group velocity dispersion. From these data, we execute the interpolation on ArcGIS to create an outer and inner ellipse in order to obtain the crust thickness map. Thus, three main groups are observed:

Along the Cameroon volcanic line (CVL), the interpolation map shows that the crust thickness, less than 37 km depth, is low compared to both sides (Figure 3). This thickness varies from 37 km to 45 km in the southern side of the Congo Craton and also grows in the Northern side of CVL. Although the CVL is structured and up, the average Moho shows the compression conditions where lithosphere is deformed by uprising accompanied with thinning of the crust. The Moho and upper mantle are also uprising, there is no isostacy compensation root in such condition; rather the crust thickness decreases as a layer undergoing compressional stress.

Laterally of the Atlantic coastline, it appears that crustal thickness considerably decreases from inland to the Ocean. The sharp portion of Moho at about 25 km depth is observed in the area of Mount Cameroon. Thus, the Moho under this area, shallow and gradational, probably reflects a transitional zone between continental and oceanic crust. Shallow Moho is consistent with geologic and isostasy normal conditions, which suggest continental and oceanic Moho at 35 and 15 km respectively.

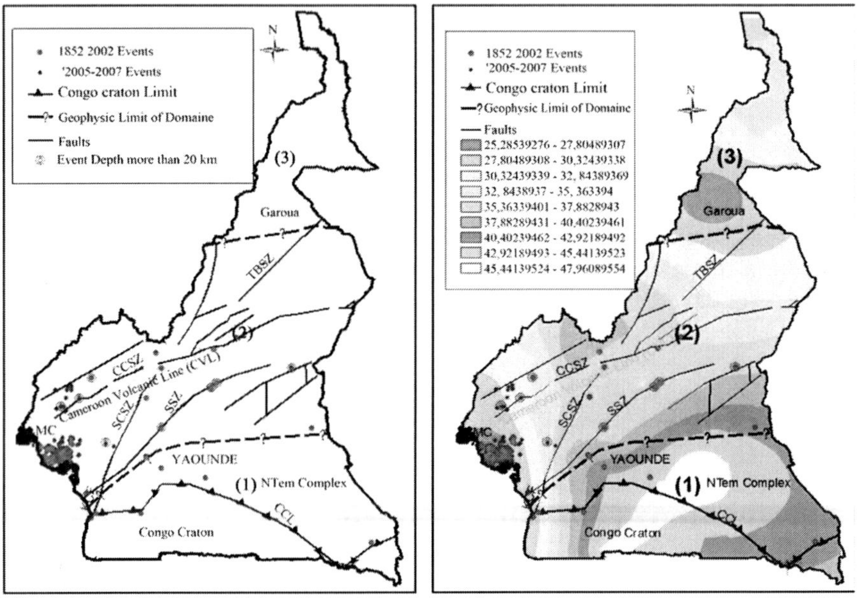

Figure 2. Seismicity map of Cameroun.

The Domain (3) is also characterized by shallow Moho where crust thickness decreases until 25 km. This region corresponding to Garoua Rift that [18] has also identified through analysis of SKS splitting.

7.2. Spatial Distribution of Seismicity and Tectonic Structures

Cameroon has been the subject of several structural studies [13] -[15] . According to these studies, the available geotectonic data, combined with historical and recent seismic data, have allowed an improvement of the know- ledge of the seismo-tectonics in the study area.

From the three main domains, the seismic activity is distributed mostly over the Central Domain, decreasing from the West to the East. A very dense concentration is observed in the Mount Cameroon region whereas very few events are located in the vicinity of strike-slip and thrust faults (Figure 3). The Southern Domain has also developed seismicity along the Northern boundary of Congo Craton.

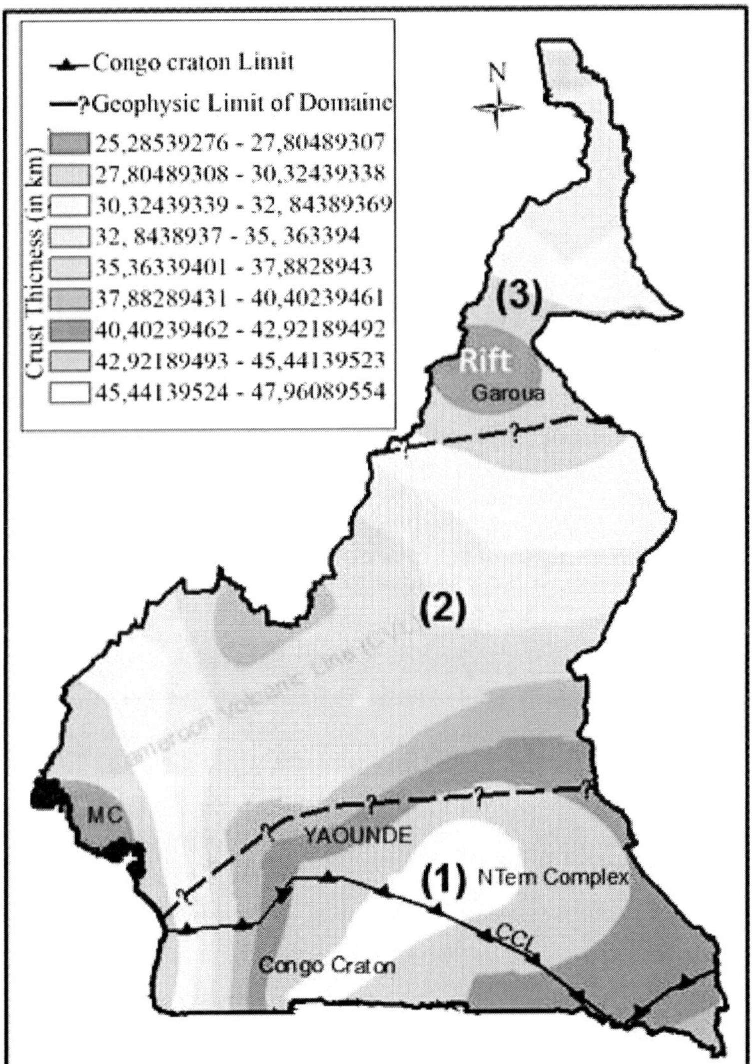

Figure 3. Crustal thickness map of Cameroon. MC: area of Mount Cameroon; CCL: Congo Craton Limit. The thickness varies from 25 to 47 km.

Figure 4 shows the seismic events based on their focal depth. It can be seen that 28.6% of events occur at depths inferior to 1 km, 11.1% at depths between 1 and 5 km, 40.4% at depths between 5 and 10 km, 10.1% at depths between 10 and 20 km and 9.8% at depths between 20 and 40 km. 79.5% of the events are at very shallow depths inferior to 10 km. Importantly, events occurring on the Sanaga Shear Zone (Seismic Source

Region 3) have their focal depth at 33 km. We conclude that Sanaga Shear Zone faults have average depth of 33 km. This observation is consistent with our results which found crust thickness at about 35 km along the SSZ (Figure 2(b)).

7.3. Seismic Source Region in Cameroon

Source region model is built by delineating area where a given seismo-tectonic structure is active. It is done based on detailed geological studies of the area together with consistent and comprehensive historical and instrumental data of seismic events to substantiate the findings [22] .

Several types of source modelling have been identified. Each source region model is adopted and assumed to have homogeneous seismicity, because seismic activities are scattered over a large area without a well identifiable point or specific fault location to warrant a point or a plane seismic source [23] [24] . Here, tectonic characteristics of each area and density of spatial distribution of seismic events are used to delineate Cameroon into four main Seismic Source Regions (Figure 5).

Seismic Source Region I: It corresponds to the area of "Mount Cameroon" volcano, in South-West Cameroon. From data recorded by temporary seismic network, 93.4 % of events are located in this Region. Mostly

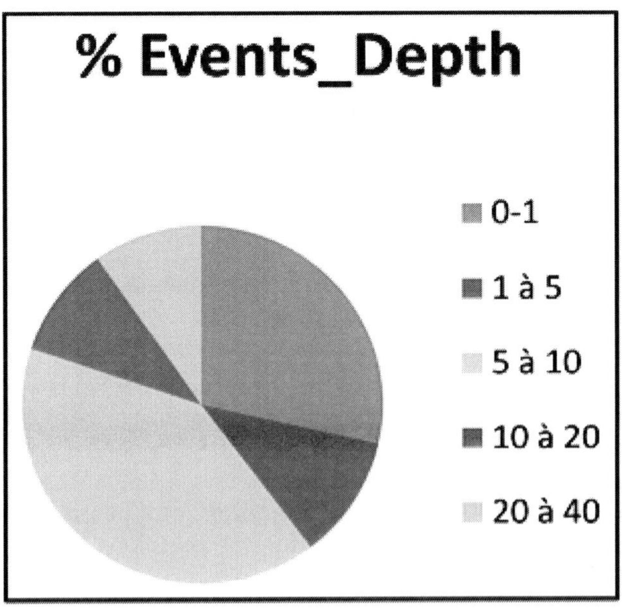

Figure 4. Percentage of events accord- ing to focal depth.

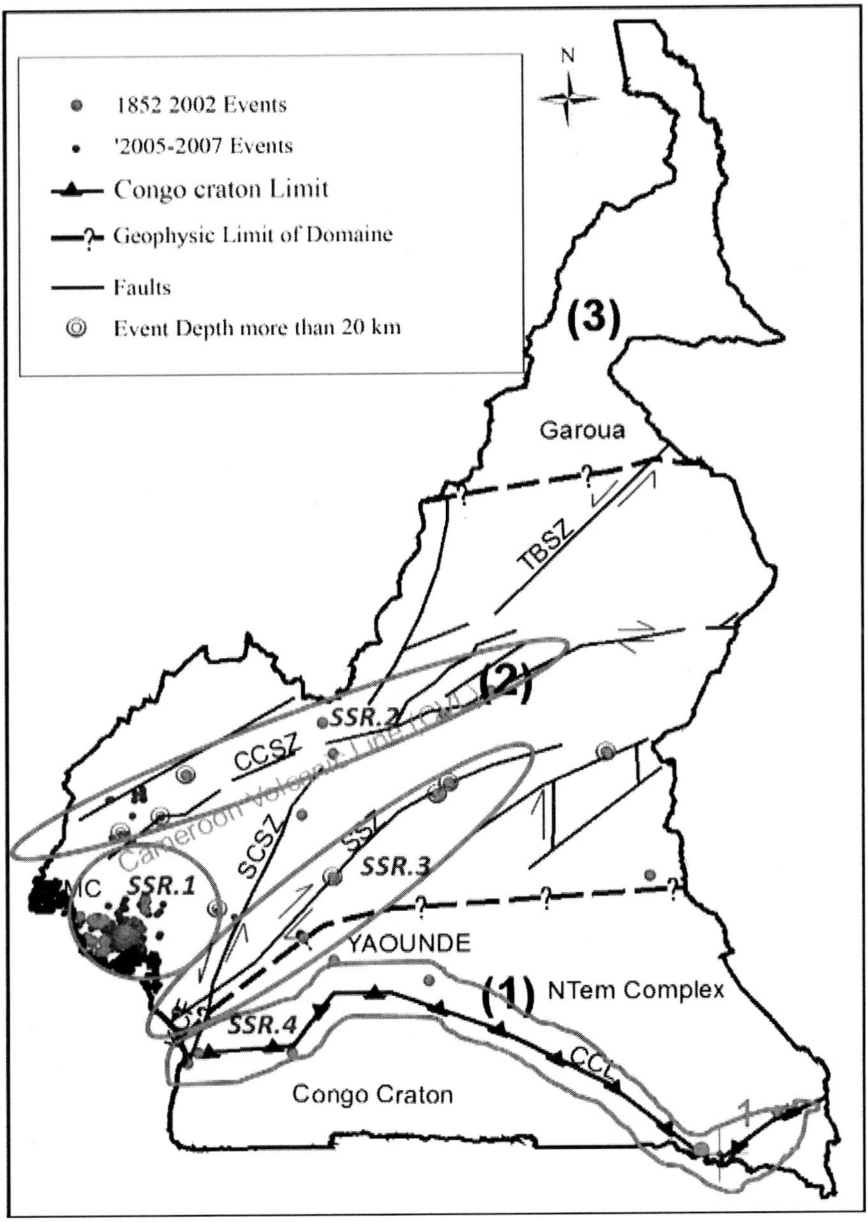

Figure 5. Seismic Source Regions (SSR); SSR. 1: Mount Cameroon Seismic Source; SSR. 2: Central Cameroon Seismic Source; SSR. 3: Sanaga Seismic Source; SSR. 4: Congo Craton Seismic Source.

events are shallow with depth inferior to 25 km. However some historical (Table 1) and Instrumental (Table 2) events have their focal depth more than 25 km depth which correspond to the thickness of the crust in this region. Importantly, according to geotectonic data available, there are no faults identified in this area. We conclude that seismic activity in the "Mount Cameroon Source Region" is related to the magmatic activity, probably related to small-scale mantle convection mentioned by [18] . The maximum magnitude recorded in the source region is 4.4 Mb; this suggests a weak seismicity.

Seismic Source Region II: It is also located in the South West region, in the North of mount Cameroon. In contrary to the "Seismic Source Region I", "Source Region II" is affected by southern segments faults of "Central Cameroon Shear Zone" (CCSZ). Shallow and very shallow events are found, but clusters have few events compared to Source Region I. The maximum magnitude recorded is 5.1 Mb; this might suggest a weak to moderate seismicity.

Seismic Source Region III: Central Cameroon, along Fault called "Sanaga Shear Zone" (SSZ). Despite the seeming diffuseness in seismicity, the epicenters of some events appear to be aligned along the Sanaga Shear Zone. Events occurring on the Sanaga Shear Zone (Seismic Source Region III) have their focal depth at 33 km. We conclude that Sanaga Shear Zone faults have average depth of 33 km. This observation is consistent with our results which found crust thickness at about 35 km along the SSZ (Figure 2). The maximum magnitude recorded in this source region was 5.8 Mb. Although this might suggest a moderate seismicity, the parameter of SSZ with average depth of 33 km and length of 900 km, convinced us that it is an important seismogenic area.

Table 1. Historical events with focal depth between 20 and 40 km. SSR: Seismic Source Region.

Latitude	Longitude	Depth	Magnitude	SSR
2	15	35	6	IV
6	14	33	5.1	III
3	8	33	5.4	
5.8	10	33	5.1	II
4.8	11.4	33	5.1	III
5.6	12.4	33	5.8	III
5.7	12.5	33	5.5	III

Table 2. Instrumental events with focal depth between 20 and 40 km.

Latitude	Longitude	Depth	Magnitude	SSR
4.5	10.29	39.83	4.3	V
4.04	9.13	34.41	4.1	I
4.33	9.51	33.15	4.8	I
4.16	9.44	29.78	4.3	I
4.1	9.07	29.31	4.9	I
5.24	9.38	26.94	4.0	II
4.26	9.41	26.93	4.2	I
4.13	9.5	25.24	4.5	I
5.4	9.76	20.49	4.6	II

Seismic Source Region IV follows the northern boundary of Congo Craton. One characteristic event has focal depth at 33 km and magnitude of 6 M. Although most events are shallow and have weak magnitude, this characteristic event shows that segment faults of this source region can generate large earthquakes.

Spatial distribution of the epicenters shows an activity on the whole area of Central Domain (2). It is not easy to correlate activity of each event to a given seismo-tectonic structure. Some events are isolated and the others can't give coherent information about its location area.

7.4. Impact on the Hazard Management in Cameroon

The result of our study shows that, over the last few decades many parts of Cameroon have shown the potential to generate earthquakes with a magnitude of 6 or more. Although the recorded seismic activity is not very significant, the results of the present study cannot preclude the possibility of destructive seismic events occurring. In addition to its 20,000,000 inhabitants, many buildings with systemic and economic impacts are exposed.

The presence of buildings of all types from mud buildings to reinforced cement concrete framed structures and steel construction and evidence of improper and low quality construction practices makes Cameroon vulnerable even against average earthquakes. This study is a first step toward a comprehensive evaluation of the seismic hazard and risk analysis. The appropriate seismic safety assessments can be properly taken into account in Cameroon Code for Para-seismic Construction.

8. CONCLUSIONS

Cameroon is affected by many local and regional tectonic structures. This work, which integrates the most recent seismic data, was undertaken in an attempt to analyze and interpret crustal structure and seismic activity.

Crustal structure shows three main domains which are consistent with their contrasting tectonic environments. Very thin crust in S-W Cameroon results principally from area's location in a transition zone between continental crust to the East and oceanic crust of the Atlantic margin to the West. The Ntem complex which has the highest crust thickness is the equilibrium created by the uprising of Cameroon Volcanic Line. Besides, according to the crustal thickness map, both Ntem complex and CVL appear to be parallel (Figure 3).

It has been shown that the seismicity is heterogeneous. Shallow and weak magnitude earthquakes are pre- dominant. Four Seismic Source regions are identified; three in the Central Domain and one in the South domain. The distribution of seismic activity is particularly denser in the area of Mount Cameroon and related to the magmatic activity. Elsewhere the seismicity, as induced by the tectonic activity, is weak to moderate. Sanaga Source Region is characterized by events with 33 km depth for 35 km of Crust thickness in this area; that allows one to think that the faults segments are deep for about 33 km. An event with 33 km depth in Congo Craton Source Region allows also predicting deep faults.

Eloumala Onana Parfait Noel, Mouzong Pemi Marcelin, Ateba Bekoa For the whole study area of Cameroon, the seismicity is weak to moderate. It is related to the magmatic activity of Mount Cameroon on one hand, and on the other hand to the tectonic structures of CVL and the boundary of Congo Craton.

REFERENCES

1. Zhu, L. and Kanamori, H. (2000) Moho Depth Variation in Southern California from Teleseismic Receiver Functions. Journal of Geophysical Research, 105, 2969-2980.
2. Takam, T., Mokoto, A., Kokonyangi, J., Dunkley, D.J. and Nsifa, E.N. (2009) Paleo-Archaeancharnockite in the Ntem Complex, Congo Craton, Camroon: Insights from SHRIMP Zircon U-Pb Ages. Journal of Mineralogical and Petrological Sciences, 104, P1-P11.
3. Nedelec, A., Macaudière, J., Nzenti, J.P. and Barbey, P. (1986) Evolution structurale et métamorphique de la série de Mbalmayo (Cameroun): Implications pour la structure de la zone mobile panafricaine au contact du craton du Congo. Acad des Scies Paris, 303, 75-80.

4. Dumont, J.F. (1986) Identification par télédétection de l'accident de la Sanaga (Cameroun). Sa position dans le contexte des grands accidents d'Afrique centrale et de la limite nord du craton congolais. Géodynamique, 1, 13-19.
5. Mvondo, H., Owona, S., Mvondo-Ondoa, J. and Essono, J. (2007) Tectonic Evolution of the Yaoundé Segment of the Neoproterozoic Central African Orogenic Belt in Southern Cameroon. Canadian Journal of Earth Sciences, 44, 433- 444.
6. Ganwa, A.A., Frish, W., Siebel, W., Ekodeck, G.A., Shang, C.K. and Ngako, V. (2008) Archean Inheritances in the Pyroxene-Amphibole Bearing Gneiss of the Méiganga Area (Central North Cameroun): Geochemical and $^{207}Pb/^{206}Pb$ Age Imprints. Comptes Rendus Geoscience, 340, 211-222.
7. Toteu, S.F., Van Schmus, W.R., Penaye, J. and Michard, A. (2001) New U-Pb and Sm-Nd Data from North-Central Cameroon and Its Bearing on the Pre-Pan-African History of Central Africa. Precambrian Research, 108, 45-73.
8. Toteu, S. F., Penaye, J. and Poudjom, D.Y. (2004) Geodynamic Evolution of the Pan-African Belt in the Central Africa with Special Reference to Cameroon. Canadian Journal of Earth Sciences, 41, 73-85.
9. Dorbath, C., Dorbath, L., Fairhead, D. and Stuart, G.W. (1986) A Teleseismic Delay Time Study across the Central African Shear Zone in the Adamawa Region of Cameroon, West Africa. Geophysical Journal International, 86, 751- 766.
10. Castaing, C., Feybesse, J.L., Thieblemont, D., Triboulet, C. and Chevremont, P. (1994) Palaeogeographical Recon structions of the Pan-African/Brasiliano Orogen: Closure of an Oceanic Domain or Intracontinental Convergence between Major Blocks? Precambrian Research, 67, 327-344.
11. Daly, M., Chorowicz, J. and Fairhead, J. (1989) Rift Basin Evolution in Africa: The Influence of Reactivated Steep Basement Shear Zones. In: Cooper, M.A. and Williams, G.D., Eds., Inversion Tectonics, Vol. 44, Geological Society, Special Publications, London, 309-334.
12. Ndougsa-Mbarga, T., Yufenyiu, L.D., Yene-Atangana, J.Q. and Tabod, C. (2014) Analysis and 2.5D Delineation of the Northern Limit of the Congo Craton Based on Spectral Modeling of Aeromagnetic Data in Akonolinga-Mbama Area, Cameroon. Geofísica Internacional, 53, 5-16.
13. Tokam, A.P.K., Tabod, T.C., Nyblade, A., Julià, J., Wiens, D.A. and Pasyanos, M.E. (2010) Structure of the Crust beneath Cameroon, West Africa, from the Joint Inversion of Rayleigh Wave Group Velocities and Receiver Functions. Geophysical Journal International, 183, 1061-1076.
14. DePlaen, R.M., Bastow, I.D., Chambers, E.L., Keir, D., Gallacher, R.J. and Keane, J. (2014) The Development of Magmatism along the Cameroon Volcanic Line: Evidence from Seismicity and Seismic Anisotropy. Journal of Geophysical Research: Solid Earth, 119, 4233-4252.
15. Reusch, A., Nyblade, A., Wiens, D., Shore, P., Ateba, B., Tabod, C. and Nnange, J. (2010) Upper Mantle Structure beneath Cameroon from Body Wave Tomography and the Origin of the CVL. Geochemistry, Geophysics, Geosystems, 11, Q10W07.
16. Pasyanos, M.E. and Nyblade, A.A. (2007) A Top to Bottom Lithospheric Study of Africa and Arabia. Tectonophysics, 444, 27-44.
17. Fishwick, S. (2010) Surface Wave Tomography: Imaging of the Lithosphere-Asthenosphere Boundary beneath Central and Southern Africa. Lithos, 120, 63-73.
18. Koch, F., Wiens, D., Nyblade, A., Shore, P., Tibi, R., Ateba, B., Tabod, C. and Nnange, J. (2012) Upper Mantle Anisotropy beneath the Cameroon Volcanic Line and Congo Craton from Shear Wave Splitting Measurements. Geophysical Journal International, 190, 75-86.
19. Tadjou, J., Nouayou, R., Kamguia, J., Kande, H. and Manguelle-Dicoum, E. (2009) Gravity Analysis of the Boundary between the Congo Craton and the Pan-African Belt of Cameroon. Austrian Journal of Earth Sciences, 102, 71-79.
20. Gallacher, R. and Bastow, I. (2012) The Development of Magmatism along the Cameroon Volcanic Line: Evidence from Teleseismic Receiver Functions. Tectonics, 31, TC3018.

21. IRGM: Institut National de Rechechrche Géologique et Minière (2002) Catalogue sismique du Cameroon. Yaoundé.
22. Karnik, V. and Algermissen, S.T. (1977) Seimic Zoning UNESCO Report.
23. Rizzo, P.C., Vaidya, N.R., Bazan, E. and Heberling, C.F. (1995) Seismic Hazard Assessment in the Southeastern United States. Earthquake Spectra, 11, 129-160.
24. Abdalla, J.A., Mohamedzein, Y.E.A. and Wahab, A.A. (2001) Probabilistic Seismic Hazard Assessment of Sudan and Its Vicinity. Earthquake Spectra, 17, 399-415.

CHAPTER 18

Seismicity in the Antarctic Continent and Surrounding Ocean

Masaki Kanao

National Institute of Polar Research, Midori-cho, Tachikawa-shi, Tokyo, Japan

ABSTRACT

Seismicity in the Antarctic and surrounding ocean is evaluated based on the compiling data by the International Seismological Centre (ISC). The Antarctic continent and surrounding ocean have been believed to be one of the aseismic regions of the Earth for many decades. However, according to the development of Global Seismic Networks and local seismic arrays, the number of tectonic earthquakes detected in and around the Antarctic continent has been increased. A total of 13 seismicity areas are classified into the Antarctic continent (3 areas) and oceanic regions within the Antarctic Plate (10 areas). In general, seismic activity in the continental areas is very low in Antarctica. However, a few small earthquakes are identified. Wilkes Land in East Antarctica is the most tectonically active area in the continent, with several small earthquake events having been detected during the last four decades. In the oceanic region, in contrast, seismic activity in the area of 120°-60° W sector is three times higher than that in the other oceanic areas. This may be considered to be involved in a tectonic stress concentration toward the Easter Island Triple Junction between the Antarctic Plate, the Pacific Plate and the Nazuca micro-Plate. Three volcanic areas, moreover, the Deception Island, the Mts. Erebus and Melbourn, are also found to be high seismic activities in contrast with surrounding vicinity areas.

Keywords: Seismicity; Antarctic Plate; Antarctic Continent; Southern Ocean; Tectonic Stress

1. INTRODUCTION

The Antarctic Plate has a unique tectonic setting that it is surrounded by almost completely divergent margins, with a very small amount of convergent or transformed margins. The divergent margins are characterized by the circum Antarctic seismic zones and cover about 92% of the surroundings of the Antarctic Plate. The convergent margins locate in the northwestern part of the South Shetland Islands and are less than 2% of the plate boundary. The transformed margins cover less than 7% along the boundary of the Scotia micro-Plate (Figure 1).

Seismicity in the Antarctic Plate is not so high compared with other plates. However, the seismicity becomes gradually clear in the last few decades by the development of Global Seismic Networks as well as the local seismic stations deployed in the Antarctic. It was a general knowledge by 1957, when the International Geophysical Year (IGY) started, that the seismic activity in the Antarctic continent was very low and only minor activity in the vicinity of active volcano, such as the Mount Erebus, was to be expected [1,2]. One of the other active volcanoes, the Deception Island in the South Shetland Islands was already known since the 19th century, but no seismic observation was carried out in the Island until the IGY. During the IGY period, ten seismic stations were established in the Antarctic and arrival time information on seismic phases recorded at the deployed stations had been reported to the United States Coastal Geological Survey (USGS), significantly contributed to the determination of earthquake locations globally. However, no earthquakes were determined by their hypocenters in the Antarctic at the time. USCGS had published "Antarctic Seismological Bulletin" based on travel time information received telegraphically from the Antarctic seismic stations during 15 years since IGY.

As the number of seismic stations in the Antarctic has been increased, local and regional earthquakes came to be detected within the Antarctic continent, even though their seismic activities were very low [3,4]. It became clear that an earthquake with magnitude around 4 occurred every few years in the Antarctic continent. However, no earthquakes with magnitude larger than 5 have occurred during the last four decades. [5] proposed an explanation for the lack of seismic activity of the Antarctic continent in terms of loading effects by overlying continental ice sheet. As the number of seismic stations of the globe increased, several tectonic earthquakes have been detected in the Antarctic. Inside the Southern Ocean, moreover, the intra-plate seismicity has been very low. However, one great earthquake with Mw 8.1 struck the Balley Islands region on March 25, 1998. This intra-plate earthquake is the largest one ever recorded in the Antarctic Plate [6,7].

[8] compiled the seismicity in and around the Antarctic continent for the first time, and also mentioned the existence of the ice related seismic signals (i.e., icequakes, ice shocks) recorded at several seismological stations. [9] also evaluated the seismicity in the Antarctic from the viewpoint of seismotectonics and neotectonics, by dividing the continent into several tectonic brocks. [10] presented a hypocentral distribution map of Antarctic intra-plate earthquakes in the period of 1900-99 including the records by Global Seismic Network, combined with those by permanent Antarctic stations in addition to temporary field stations.

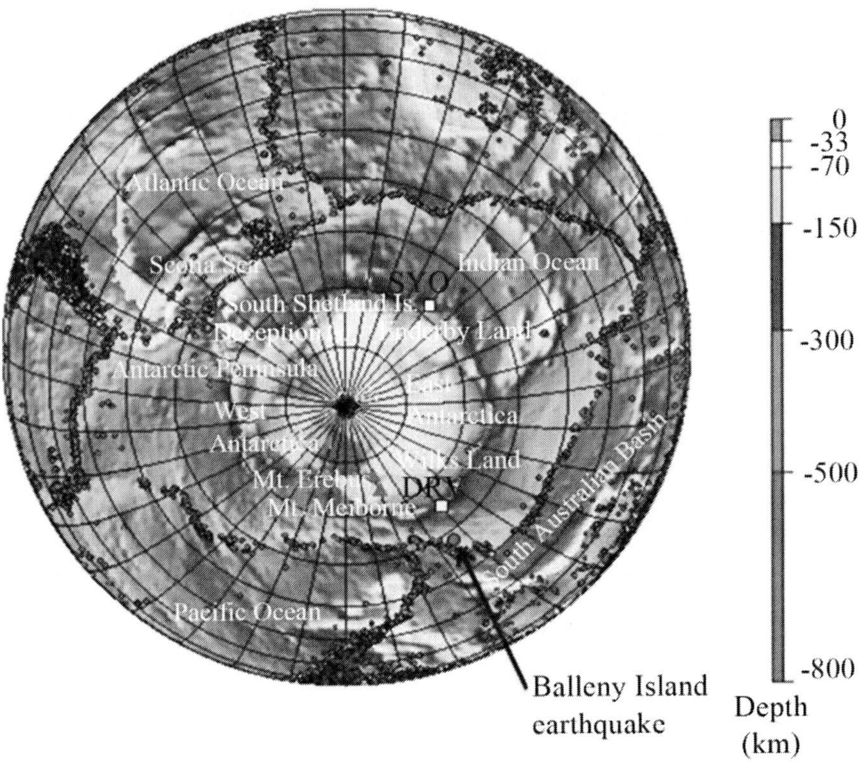

Figure 1. Seismotectonic regions discussed in this paper. Earthquake distribution in and around the Antarctic plate was compiled by NEIC (National Earthquake Information Center) of USGS. Three volcanoes and South Shetland Islands are also shown by solid (red) triangles, and the Balleny Island earthquake is shown by solid (red) circle, respectively.

In this paper, seismicity in the Antarctic continent and surrounding ocean is overviewed by using the data from the International Seismological Centre (ISC). Seismicity in terms of each tectonic setting is discussed by being divided into the following five regions: intra-plate low-seismic regions; high-seismic region around the Antarctic Peninsula; the Antarctic Continent aseismic region; lowseismic region at the edge of the Continent; and the volcanic regions.

2. ANTARCTIC CONTINENT

A general seismicity map in and around the Antarctic Plate is shown in Figure 1. The earthquake locations are compiled by the National Earthquake Information Center (NEIC) of USGS. The Antarctic continent is roughly divided into two large tectonic provinces; East Antarctica (eastern hemisphere part from the Transantarctic Mountains) and the other younger province; West Antarctica. East Antarctica is characterized by a fragment of Gondwana super-continent, one of the members of adjacent Pre-Cambrian terrains in southern hemisphere (South America, Africa, India and Australia; Figure 1). West Antarctica, on the contrary, is attributed by a chain of many islands beneath the ice sheet of Cenozoic era, including several active volcanoes.

Four earthquakes inside the continental area of Antarctica were reported to the ISC catalog in early stage of Antarctic scientific research. However, except only one event out of the four, no significant earthquakes were located in the continent before the IGY, because neither their locations nor their magnitudes were accurately determined [11]. One event which occurred on June 26, 1968 was located in Coats Land, 20°W, 80°S using the initial phase readings of five seismic stations on the Antarctic continent [3]. The magnitude of the event was 4.3 determined by the waveform amplitude of a three-component seismograph at South Pole Station (SPA). This was the first earthquake located instrumentally in the continent using the data of only seismic stations in Antarctica. However, this event was not listed in the ISC catalog. [3] used the data of five stations for hypocenter determination and two of the five stations were Byrd Seismic Array. As the focal depth was determined 1 km beneath the sea level, the event was considered to be an earthquake in the crust, not an ice-quake.

On the other hand, [4] reported an earthquake near the coast of Oates Land at 70.5°S and 161.3°E on October 15, 1974. This event was the only shock to be located on the Antarctic continent by international agency by that time. Magnitude of the event was estimated to be 5 and he concluded that the event might be an ice-quake associated with ice movement or cracking. Therefore, there are ambiguity to identify the very shallow events near the surface only from the hypocentral information, whether they are tectonic events or the ice related signals.

Regarding the wide inland area of East Antarctica, seismicity in Wilkes Land ("G" in Figures 2 and 3) is found to be one order higher than that in vicinity of Syowa Station (69.0°S, 39.0°E), the Lüzow-Holm Bay (LHB), East Antarctica [9,12]. Seven earthquakes with Mb 4.0 - 4.9 were located in the coastal area of 100° - 170°E and 66° - 82°S in Wilkes Land in 1964-1996. Another nine earthquakes were located in the inland area and two were located offshore. The magnitudes of these eleven earthquakes were not accurately determined. Not only micro-seismic activity, but also small earthquake activity in Wilkes Land and surrounding coasts are higher than that in other area on the East Antarctic continent. The sub-glacial topography in Wilkes Land is characterized by a sub-glacial basin with 1000 m below sea level in minimum elevation of the bedrock [13]. The maximum thickness of the ice sheet in the area is over 4000 m and the surface elevation of the ice sheet is mostly over 2000 m. The formation, distribution and stability of these sub-glacial lakes might affect on tectonic processes involving relatively high seismicity in this area.

3. SOUTHERN OCEAN

Seismic activity inside the Antarctic Plate has been known as very few distributions of their epicenters both in ocean and continent areas. As the space distribution and time variations in seismicity around the Antarctic Plate, in particular for the Indian Ocean sector (0° - 160°E, 20° - 80°S), was previously evaluated and intraplate seismicity was discussed associated with far-field tectonic stress in the oceanic lithosphere [14]. Therefore in this study, a detail hypocentral distribution within the whole Antarctic Plate is investigated by using the data from ISC in 1964-2002. Figure 2 represents the earthquake locations compiled by ISC in the area of 20° - 80°S and every 60° of longitude. Among the individual 60° longitude sectors and South Pole area in 80° - 90°S, the earthquake activities are divided into 13 regions (from "A" to "M", Table 1). Since the seismic activities are extremely high along the plate boundaries around the Antarctic Plate between the surrounding plates, a criterion of the area for intra-plate was selected very carefully not to include the events associated with the plate boundaries.

4. CONTINENTAL MARGINS

[10] pointed out the considerable number of intra plate earthquakes in the 90° - 180°E quadrant and divided the earthquakes into two groups as poorly located earthquakes and well located ones. Inside the continental area in Antarctica, seismicitiy is almost very few in "D", "G" and "M" areas of Figure 2. However, the Wilkes Land ("M" in Figures 2 and 3) had been identified as the most active within the Antarctic continent at IGY period. Poorly located earthquakes were lined from north to south along the 140°E longitude [9]. In the earthquake locating area, the Resolution sub-glacial highland, the Adventure sub-glacial trench, and the Belgica sub-glacial highland are existed along the longitudinal direction from east to west [13]. There is a possibility that the poorly located earthquakes were ice-quakes, because the thick ice sheet and complicated sub-glacial topography must cause ice-shocks. Generally the edge of the continent, the coast area, is aseismic region.

Over the past few decades, more seismic observations in polar region have detected local seismicity by both temporary seismic networks and permanent stations. [16] found the majority of seismicity near the Scott Base (SBA, 167°E, 78°S) and Wright Valley area (VNDA, 162°E, 78°S) located along the coast, particularly near large glaciers. They suggested a few generation mechanisms for these events, distinguishable by their focal mechanism and depth: basal sliding of the continental ice sheet, movement of ice streams associated with several scales of glaciers, movement of sea-ice, and tectonic earthquakes.

[17] deployed a local seismic network around the Neumayer Station (08°W, 71°S), and determined hypocenters of local tectonic events, located along the coast and the mid area of the surrounding bay. A seismic array has been operated more than one decade at the Neumayer Station. Since the deployment of the seismic network/ array, several local events could be detected. Two seismic active regions were figured out at inland area and offshore of the continent. In addition, a broadband seismic network had been developed in the large region between Mawson and Casey stations and inland as far as 75°S by Australia [18]. The aim to establish the seismic network is to discover the" seismic structure of the continent under Antarctica (SSCUA)". Moreover, India has also been carried out seismic observation at the Maitri Station (12°E, 71°S) since 1997. The seismic data have already been contributed to earthquake locations by ISC. India also has published "Seismological Bulletin of Maitri Station, Antarctica" in every year [19].

5. LÜTZOW-HOLM BAY

Once a denser seismic network was established, small/ micro earthquake activities became gradually clear. The LHB area around Japanese Syowa Station, East Antarctica, is one of the area where seismic activity has been well studied since 1980s [20]. Since seismic observations started by a tripartite network in 1987, seismicity for relatively small events became clear in and around Syowa Station. A total of 18 local earthquakes were located during the 15 years in 1987-2003 ([12]; Figure 4).

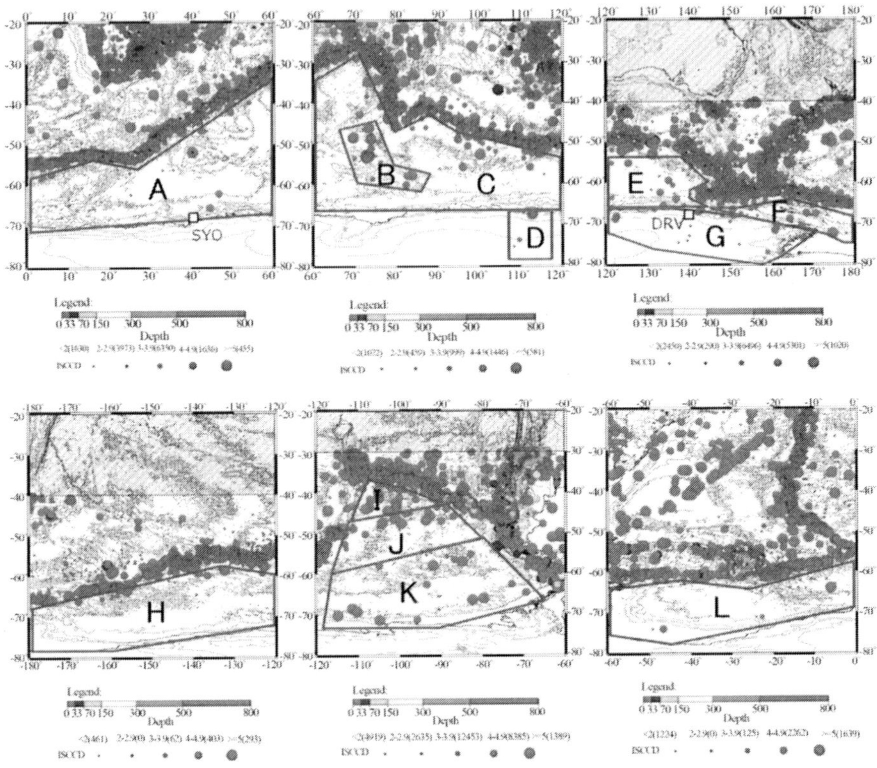

Figure 2. Earthquake locations determined by ISC in 20° - 80°S and every 60° of longitude. Surrounded areas by solid red lines indicate individual blocks ("A" - "L") discussing seismicity in this paper. These areas were classified into Antarctic continent ("D" and "G") and oceanic region within Antarctic Plate (other regions). Numerous denoted in the brackets correspond to magnitude ranges representing the number of included events for each 60° longitude sector.

Characteristic features in time variations of seismic activity are summarized as follows; Seismicity in 1987- 1989: A three-station seismic network was operated around the Syowa Station. Epicenters of ten local earthquakes were determined during these three years. Many different types of earthquakes, such as a main-shock-aftershock, twin earthquake, earthquake swarms, were detected and identified at that time. The seismic activity during this period was higher than that of the following decade.

In 1990-1996, nine local earthquakes were recorded with many different types of events. The seismic activity during this period was very low and the magnitudes of the earthquakes ranged from −0.5 to 1.4. One local event was detected in 1997, two events in 1998 and one event in 2001 and 2003, respectively. The low seismic activity continues to the present day in 2004.

The seventeen events were only detected by local seismic network deployed around the LHB, except for the September 1996 Mb = 4.6 earthquake in the southern Indian Ocean. Almost all the hypocenters were located along the coast, apart from a few on the northern edge of the continental shelf. Local earthquakes in and around Syowa Station were presumably caused by tectonic stress accumulated with crustal uplift after deglaciation [21]. The effect of ice sheet changes may have caused phenomena such as crustal deformation, earthquake occurrence, faulting systems in the shallow part of the lithosphere

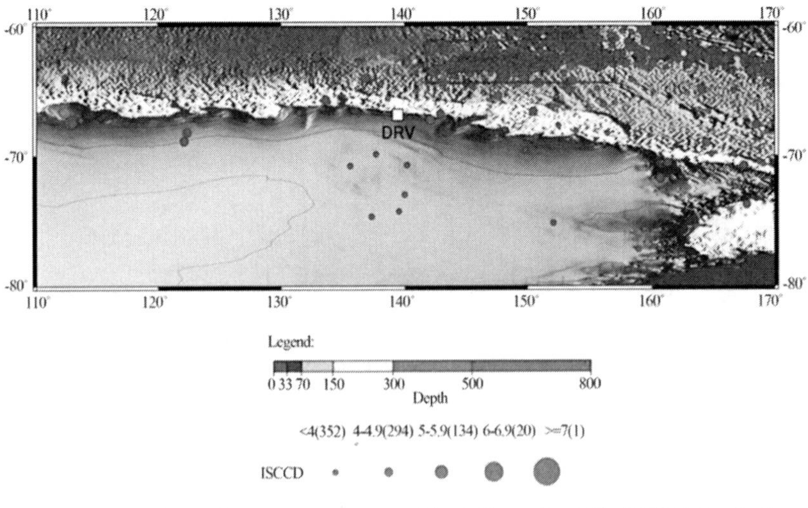

Figure 3. Earthquake locations in 60°-80°S and 110°-170°E, including both the Wilkes Land and aftershock area (red square) of the March 25, 1998 Balleny Island earthquake (Mw = 8.1, red circle). Numerous denoted in the brackets correspond to magnitude ranges representing the number of included events.

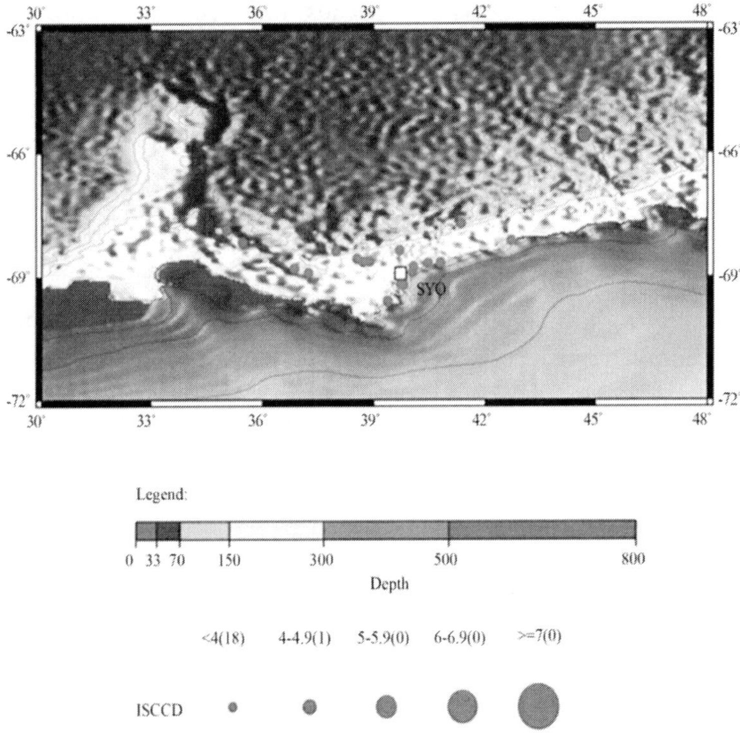

Figure 4. Earthquake locations in and around the Lützow-Holm Bay (LHB), East Antarctica detected at Syowa Station. Total number of 19 events were determined their hypocenters. Numerous denoted in the brackets correspond to magnitude ranges representing the number of included events.

beneath Antarctica [16]. Several of these events, moreover, could be large ice-quakes associated with the sea-ice dynamics around the LHB or in the Southern Ocean.

6. ANTARCTIC PENINSULA

The epicentral locations in the Antarctic Peninsula of 20° - 80°W, 50° - 70°S are shown in Figure 5, on the basis of the ISC data catalog in 1964-2002. The earthquake activity in this region is the highest in Antarctic including active volcanic areas in Deception Island, together with subduction zone in the Bransfield Strait. The focal depths of the earthquakes are mostly shallower than 40 km. During two decades in 1971-1989, only four earthquakes occurred with focal depths between 40 and 100 km [9]. One big event occurred on Feb. 8, 1971, for which Mb and Ms were determined to be 6.3 and 7.0 respectively. In addition that the event is the largest earthquake recorded in the region, the only recorded event with magnitude lager than 7.0 in Antarctica. This earthquake was the first tectonic earthquake to be felt in the Antarctic, observed at the Farady Station (65°S, 64°W) of United Kingdom [22].

Table 1. Classification of the Antarctic intra-plate earthquakes in 1964-2002 period compiled by ISC, corresponding the individual blocks in **Figure 2**.

Block	LAT	LON	Event number	Comment
A	0°-60°E	20°-80°S	33	Ocean
B	60°-120°E	20°-80°S	58	Ocean. Kerguelen Plateau
C	60°-120°E	20°-80°S	26	Ocean
D	60°-120°E	20°-80°S	5	Continent
E	120°-180°E	20°-80°S	33	Ocean
F	120°-180°E	20°-80°S	26	Ocean
G	120°-180°E	20°-80°S	23	Continent. Wilks Land
H	180°-120°W	20°-80°S	22	Ocean
I	120°-60°W	20°-80°S	19	Ocean. Easter Is. Triple Junct.
J	120°-60°W	20°-80°S	20	Ocean
K	120°-60°W	20°-80°S	17	Ocean
L	60°-0°W	20°-80°S	33	Ocean
M		80°-90°S	8	Continent. South Pole

According to the report of seismic events recorded at King Sejoung Station (60°S, 59°W) of South Korea, King George Island of the South Shetland Islands, some earthquake swarms occurred around Bridgeman Island [23]. [24] evaluate these earthquake swarms in view from physical volcanology. The earthquake swarms appeared to represent typical volcanic earthquake waveforms. The earthquake swarm might be originated around ORCA Sea Mount about 20 km southeast of King Sejoung Station. Many earthquakes are located in the back-arc area, either on large submarine volcanoes or rifting region along the center of the Bransfield Strait. Earthquakes concentrated around ORCA Sea Mount and Bridgeman Island. These earthquake locations indicate that the seismicity is associated with active volcanism rifting, non active subduction systems.

[25] deployed seven continuously recording broadband seismometers in the South Shetland Islands region during 1997 and 1998. High level of local seismicity was identified in the first 15 months. About 90 earthquakes with magnitude mb 2 - 4 were located. Many earthquakes with 10 - 30 km depth are located in the forearc region extending from the South Shetland Trench axis toward and beneath the South Shetland Islands. The earthquake locations suggest active convergence along the South Shetland subduction zone, however, there is no significant subduction zone extending into mantle.

7. VOLCANIC REGIONS

Seismic observations in volcanic regions have been carried out only three volcanoes as shown in Figure 1. Eruptions were recognized at Deception Island (63°S, 61°W) and Mount Erebus (78°S, 167°E, 3974 m), on the contrary, no eruption was recognized at Mount Melbourne (74°S, 165°E, 2732 m). The first felt shock in the Antarctic was the magnitude 4.7 earthquake accompanied by the volcanic eruptions of Deception Island on December 4, 1967 [22]. This eruption was the first recognized explosion of Deception Island by humankind. The buildings of the stations on Deception Island were destroyed by the eruptions and all members in the stations of Argentina, Chile and the UK evacuated safely after the eruption. All of the stations at Deception Island have been closed since that time.

Since 1980's Spanish established a summer stations at Deception Island and started geophysical research on Deception Volcano. [26] have carried out seismic observations with five stations during the austral summer since 1987, and observed approximately 1000 local events per month. Earthquake locations seem to be concentrated along the NE-SW direction crossing the central part of the island. In the 1991-92 survey, the seismic activity was significantly increased with total number of recorded earthquakes more than 700, an extending of the areas with geothermal anomalies and major activity of the fumaroles. The seismic activity was low in 1992-93 and 1993-94 surveys, and no other volcanic anomalies were observed. In the 1994-95 survey, a new increase of seismic activity was recorded with more than 800 seismic events [27].

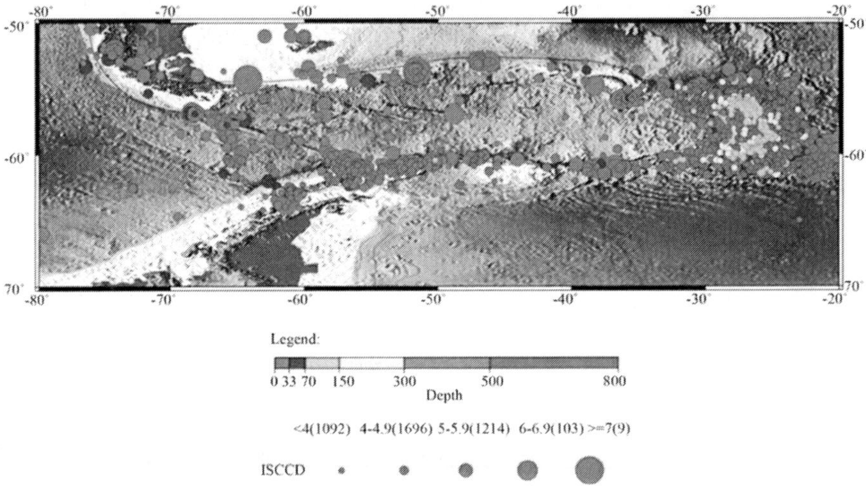

Figure 5. Earthquake location around Antarctic Peninsula. Numerous denoted in the brackets correspond to magnitude ranges representing the number of included events.

Seismic observations by radio-telemetry continued in the summit area and on the slope of Erebus Volcano on Ross Island by the international cooperation among Japan, New Zealand and US during 1981-90 [28]. Seismic activity around Mount Erebus became clear and remarkable change of seismic activity were recognized before and after the new phase of volcanic activity started on September 13, 1984. It was characterized by larger and more frequent strombolian eruptions. Significant changes in seismicity within Ross Island and Mount Erebus were recognized both before and after the increased eruptive activity. The seismic activity of Mount Erebus in 1980- 1990 is divided into the following four stages: 1) normal high activity, 2) preceding the new phase, 3) new phase in volcanic activity and 4) low seismic activity. These stages suggest a general pattern of volcanic activity and give a fundamental information concerning production of volcanic eruption.

US scientists have still been carrying on the monitor of seismic activity around Mount Erebus [29]. A research program on physical volcanology has been conducted around Mount Melbourne by Italian scientists since the end of the 1980's. Four seismic stations were installed on Mount Melbourne since 1990 and many local earthquakes have been reported together with other geophysical data [30]. Local seismicity clustered along the eastern flank of Mount Melbourne, possibly in spatial association with the local north-south magnetic trend. One event occurred on December 10, 1990 is the largest earthquake with magnitude 1.9 around Mount Melbourune [31].

8. ICE-QUAKES, GLACIAL EARTHQUAKES

Seismic signals involving ice-related phenomena are called "ice-quakes (ice-shocks for smaller ones)", and are most frequently reported in association with glacially related mass movements of ice-sheets, or with sea-ice, tidecracks and icebergs in the other polar areas [32,12]. The so-called "ice-micity" detected around the Bransfield Strait and Drake Passage by a local network of hydrophone arrays in 2006-2007 illustrate the dynamic behavior of sea ice in the Bransfield and Antarctic Peninsula regions [33].

Dynamics of sea-ice and icebergs also affect on the seismic signals. For example, a large volume of sea ice was discharged from LHB during the 1997 austral winter, and clearly imaged by the NOAA satellite. The broadband seismographs at Syowa Station recorded distinct waveforms associated with the discharge events [34]. The long-duration sea-ice tremors had very distinct spectral characteristics that distinguished them clearly from ordinary teleseismic and/or local tectonic events. Several sequences of harmonic over-toned signals, presumably associated with the merging of multiple ice volumes, appeared on the seismic signals [34]. The characteristic signals also represent the surge events that seem more closely related to the break-up process of the sea-ice mass volume. Both kinds of cryoseismic waves occurred continuously for few hours, and repeated themselves several times within a few days during late July, 1997. Identification of the exact sources that produced these characteristic signals has not yet been completed, and theoretical modeling will most likely be required to explain the physical processes.

Similar cryoseismic phenomena have also been reported around the Ross Sea region (MacAyeal et al., 2009), the marginal sea of the Antarctic Peninsula [33], as well as the continental margin of Dronning Maud Land [17]. In particular, iceberg-originated harmonic tremor emanating from tabular icebergs was observed by both seismo-acoustic and local broadband seismic signals [35]. The tremor signals consisted of extended episodes of stick-slip ice-quakes generated when the ice-cliff edges of two tabular icebergs rubbed together during glancing, "strike-slip" type iceberg collisions. Source mechanisms of such harmonic tremors might provide useful information for the study of iceberg behavior, and a possible method for remotely monitoring iceberg activity.

In addition to the short-period cryoseismic signals mentioned above, a new class of seismic events associated with melting of large ice cap was discovered in particular around Greenland [36,37]. These large events were called "glacial earthquakes", generating long-period ($T > 25$ s) surface waves equivalent in strength to those radiated by standard magnitude five earthquakes, and were observable worldwide. The glacial earthquakes radiated little high-frequency energy, which explains why they were not detected or located by traditional earthquake-monitoring systems. These events are two magnitude units larger than previously reported seismic phenomena associated with glaciers, a size difference corresponding to a factor of 1000 in a seismic moment.

9. DISCUSSION

For many decades, the Antarctic continent and surrounding area have been believed as aseismic. Several earthquakes were found to be located in the Antarctic continent since IGY, as mentioned in this paper. It became general knowledge that there are small earthquake activities detected in the Antarctic Continent even though the activity itself is not so high as other continents. However the event of which magnitude lager than 5 is not detected during last four decades. The number of tectonic earthquakes located in the Antarctic has increased according to the development of Global Seismic Networks and local seismic arrays. The most of the tectonic earthquakes along the coast of the Antarctic continent are caused by tectonic stress accumulated by crustal deformation after deglaciation.

The micro-earthquake activity around Syowa Station as mentioned the previous section might be a typical activity on the coastal area of the Antarctic continent where crustal uplift after deglaciation is now going on. As seismological observation like the local array deployment around Syowa Station have been developing at several stations in the coastal area, high activity of micro-earthquakes will be detected on the continental margin of Antarctica. If the other geophysical observations such as tide gauges, gravimeters and GPS have been carried out at coastal stations, the relation between micro-earthquake activity and crustal movement will become clear.

Around continental area in East Antarctica from 90° - 180°E, in generally, seismicity is low in an average. As already treated in Chapter 2, the Wilkes Land, in contrast, has been identified as the most tectonically active region within the Antarctic. It is also noticed that there is a significant number of sub-glacial lakes in this area, then can be considered a relation to produce the occurrence of local seismic activities. The formation, distribution and stability of sub-glacial lakes could give rise to the fundamental question how the tectonic processes control the existence of the lakes [38]. The poorly located earthquakes, accordingly, could be a kind of large ice-quakes, because complicated sub-glacial topography with overlying ice-sheet in this area may efficiently cause ice related seismic phenomena.

A complexity of the surface bedrock structure of Wilkes Land, moreover, might be influenced by tectonically weakened upper crust along with the present mobile belts, forced by the Tasmania micro plate [7,12]. The aftershock area of the large Balleny Island Earthquake in the Antarctic Oceanic Plate appears to be continued with the continental area of Wilkes Land (Figure 3). It might be supposed that the large earthquakes generate in this area, by assuming thermal stress of the young lithosphere, associated with unusual deformation due to Macquarie triple junction [39]. Recently several intra-plate earthquakes became to be located in the southern ocean. In contrast, there is a possibility that some earthquakes off of the Antarctic Plate are also caused by crustal deformation and tectonic stress involving deglaciation [6]. It is strongly suggested that the effect of volume and shape change of ice sheet causes some phenomena in the crust and upper mantle such as crustal uplift, earthquake occurrence, lithospheric deformation.

10. CONCLUSION

Seismicity in the Antarctic and surrounding ocean is evaluated based on the compiled data in 1964-2002 by International Seismological Centre. The Antarctic continent and surrounding ocean were believed to be one of the aseismic regions of the Earth for many decades. However, according to the development of Global Seismic Networks and local seismic arrays, the number of tectonic earthquakes detected in and around the Antarctic continent has increased. The total of 13 seismicity areas were classified into the Antarctic continent (three areas) and oceanic regions within Antarctic Plate (the other 10 areas). Generally, seismic activity in the continental areas is very low; few small earthquakes located. The Wilkes Land is the most active area in the Antarctic continent; several small earthquakes were detected during these four decades. In the ocean area, in contrast, the seismic activity in the area of 120° - 60°W sector is three times larger than that in the other areas. This evidence is considered to be involved in a stress concentration toward the Easter Island Triple Junction among Antarctic Plate, Pacific Plate and Nazuca micro-Plate. Three volcanic areas, Deception Island, Mt. Erebus and Mt. Melbourn, are also high seismic activity areas. Finally, there are a lot of undefined origin events which include both tectonic earthquakes involving deglaciation, as well as cryoseismic signals associated with variations in the surface environment.

ACKNOWLEDGEMENTS

Masaki Kanao (ISC), for the utilization of the compiled data.

REFERENCES

1. B. Gutenberg and C. F. Richter, "Seismicity of the Earth and Associated Phenomena," Princeton University Press, Princeton, 1954.
2. C. F. Richter, "Elementary Seismology," W. H. Freeman (Ed.), San Francisco, 1958, p. 768.
3. K. Kaminuma and M. Ishida, "Earthquake Activity in Antarctica," Antarctica Record, Vol. 42, 1971, pp. 53-60.
4. R. D. Adams, "Source Properties of the Oates Land Earthquake," In: C. Craddock, Ed., Antarctic Geoscience, The University of Wisconsin Press, Wisconsin, 1982, pp. 955-958.
5. A. C. Johonstone, "Suppression of Earthquakes by Large Continental Ice Sheets," Nature, Vol. 330, 1987, pp. 467- 469.
6. S. Tsuboi, M. Kikuchi, Y. Yamanaka and M. Kanao, "The March 25, 1998 Antarctic Earthquake: Great Earthquake Caused by Postglacial Rebound," Earth Planets Space, Vol. 52, 2000, pp. 133-136.
7. T. Himeno, M. Kanao and Y. Ogata, "Statistical Analysis of Seismicity in a Wide Region around the 1998 Mw 8.1 Balleny Islands Earthquake in the Antarctic Plate," Polar Science, Vol. 5, No. 4, 2011, pp. 421-431.
8. K. Kaminuma, "Seismic Activity in and around Antarctic Continent," Terra Antartica, Vol. 1, 1994, pp. 423-426.
9. K. Kaminuma, "A Revaluation of the Seismicity in the Antarctic," Polar Geoscience, Vol. 13, 2000, pp. 145- 157.
10. A. M. Reading, "Antarctic Seismicity and Neotectonics," In: J. A. Gamble, et al., Eds., Antarctica at the Close of a Millennium, The Royal Society of New Zealand Bullutin, Wellington, Vol. 35, 2002, pp. 479-484.
11. R. D. Adams, A. A. Hughes and B. M. Zhang, "A Confirmed Earthquake in Continental Antarctica," Geophysical Journal International, Vol. 81, No. 2, 1985, pp. 489- 492.
12. M. Kanao and K. Kaminuma, "Seismic Activity Associated with Surface Environmental Changes of the Earth System, around Syowa Station, East Antarctica," In: D. K. Futterer, et al., Eds., Antarctica: Contributions to Global Earth Sciences, Springer-Verlag, Berlin, Heidelberg, New York, 2006, pp. 361-368.
13. D. J. Drewry, "Antarctica: Glaciological and Geophysical Folio, 9 Sheets," Scottish Polar Research Institute, University of Cambridge, Cambridge, 1983.
14. M. Kanao, Y. Nogi and S. Tsuboi, "Spacial Distribution and Time Variation in Seismicity around Antarctic Plate—Indian Ocean," Polar Geoscience, Vol. 19, 2006, pp. 202-223.
15. D. L. Anderson, "Superplumes or Supercontinents?" Geology, Vol. 22, No. 1, 1994, pp. 39-42.
16. S. Bannister and B. L. N. Kennett, "Seismic Activity in the Transantarctic Mountains—Results from a Broadband Array Deployment," Terra Antarctica, Vol. 9, 2002, pp. 41-46.

17. C. Muller and A. Eckstaller, "Local Seismicity Detected by the Neumayer Seismological Network, Dronning Maud Land, Antarctica: Tectonic Earthquakes and Ice-Related Seismic Phenomena," IX International Symposium on Antarctic Earth Science Programme and Abstracts, Potsudam, 2003, p. 236.
18. A. M. Reading, "On Seismic Strain-Release within the Antarctic Plate," In: D. K. Futterer, et al., Eds., Antarctica: Contributions to Global Earth Sciences, SpringerVerlag, Berlin, Heidelberg, New York, 2006, pp. 351-356.
19. G. B. N. Chander, S. V. R. Ramachandra Rao, G. S. Srinivas, E. C. Malaimani and N. R. Kumar, "Seismological Bulletin of Maitri Station, Antarctica, 2002," National Geophysical Research Institute, 2003, pp. 1-71.
20. K. Kaminuma and J. Akamatsu, "Intermittent MicroSeismic Activity in the Vicinity of Syowa Station, East Antarctica," In: Y. Yoshida et al. (Eds.), Recent Progress in Antarctic Earth Science, Terra Science Publication, Tokyo, 1992, pp. 493-497.
21. K. Kaminuma and M. Kanao, "Local Seismicity and Crustal Uplift around Syowa Station, Antarctica," Korean Journal of Polar Research, Vol. 10, 1999, pp. 103-107.
22. K. Kaminuma, "Seismicity around the Antarctic Peninsula," Polar Geoscience, Vol. 8, 1995, pp. 35-42.
23. Y. K. Jin, D. K. Lee, S. H. Nam, Y. Kim and K. J. Kim, "Seismic Observation at King Sejong Station, Antarctic Peninsula," Terra Antartica, Vol. 5, 1998, pp. 729-736.
24. K. Kaminuma, "A Possibility of Earthquake Swarms around ORCA Sea Mount in the Bransfield Strait, the Antarctic," In: Y. Kim and B. K. Khim, Eds., Proceedings of the Joint International Seminar: Recent Interests on Antarctic Earth Sciences of Korea and Japan, 2001, pp. 23-34.
25. S. D. Robertoson, D. A Wiens, P. J. Shore, G. P. Smith and E. Vera, "Seismicity and Tectonics of the South Shetland Islands and Bransfield Strait from the SEPA Broadband Seismograph Deployment," In: J. A. Gamble et al., Eds., Antarctica at the Close of a Millennium, The Royal Society of New Zealand Bullutin, Wellington, Vol. 35, 2002, pp. 549-554.
26. J. Vila, R. Ortiz, A. M. Correig and A. Garcia, "Seismic Activity on Deception Island," In: Y. Yoshida et al., Eds., Recent Progress in Antarctic Earth Science, Terra Science Publication, Tokyo, 1992, pp. 449-456.
27. R. Ortiz, A. Garcia, A. Aparicio, I. Branco A. Felpeto, R. Rey Del, M. T. Villegas, J. M Ibanez, J. Morales, E. Pezzo Del, J. C. Olmedillas, M. Astiz, J. Vila, M. Ramos, J. G. Viramonte, C. Risso and A. Caselli, "Monitoring of the Volcanic Activity of Deception Island, South Shetland Islands, Antarctica (1986-1995)," In: C. A. Ricci, Ed., The Antarctic Region: Geological Evolution and Processes, Terra Antartica Publication, Siena, 1997, pp. 1071- 1076.
28. K. Kaminuma and R. R. Dibble, "Seismic Activity of Mount Erebus 1981-1988," Polar Geoscience, Vol. 4, 1990, pp. 142-148.
29. R. Aster, W. McIntosh, P. Kyle, R. Esser, B. Bartel, N. Dunbar, B. Johns, J. Johnson, R. Karstens, C. Kurnik, M. McGowan, S. McNamara, C. Meertens, B. Pauly, M. Richmond and M. Ruiz, "Real-Time Data Received from Mount Erebus," EOS Transactions, Vol. 85, No. 10, 2004, p. 99.
30. A. Banaccorso, S. Gambino and E. Privitera, "A Geophysical to the Dynamics of Mt. Melbourne (Northern Victoria Land, Antarctica)," In: C. A. Ricci, Ed., The Antarctic Region: Geological Evolution and Processes, Terra Antartica Publication, Siena, 1997, pp. 531-538.
31. E. Armadillo, A. Bonaccorso, E. Bozzo, G. Caneva, A. Capra, G. Falzone, F. Ferraccioli, S. Gandolfi, F. Mancini, E. Privitera and L. Vittuari, "Geophysical Features of the Mt. Melbourne Area, Antarctica, and Preliminary Results from the Integrated Network for Monitoring the Volcano," In: J. A. Gamble et al., Eds., Antarctica at the Close of a Millennium, The Royal Society of New Zealand Bulletin, Wellington, Vol. 35, 2002, pp. 571-577.
32. S. Anandakrishnan and R. B. Alley, "Tidal Forcing of Basal Seismicity of Ice Stream C, West Antarctica, Observed Far Inland," Journal of Geophysical Research, Vol. 102, No. B7, 1997, pp. 15183-15196.

33. R. P. Dziak, M. Parlk, W. S. Lee, H. Matsumoto, D. R. Bohnenstiehl and J. H. Haxel, "Tectono-Magmatic Activity and Ice Dynamics in the Bransfield Strait Back-Arc Basin, Antarctica," The 16th International Symposium on Polar Science, Incheon, 2009, pp. 59-68.
34. M. Kanao, A. Maggi, Y. Ishihara, M.-Y. Yamamoto, K. Nawa, A. Yamada, T. Wilson, T. Himeno, G. Toyokuni, S. Tsuboi, Y. Tono and K. Anderson, "Interaction on Seismic Waves between Atmosphere—Ocean—Cryosphere and Geosphere in Polar Region," In: M. Kanao et al., Eds., Seismic Waves—Research and Analysis, InTech. Publisher, Rijeka, 2012, pp. 1-20.
35. D. MacAyeal, E. Okal, R. Aster and J. Bassis, "Seismic Observations of Glaciogenic Ocean Waves (Micro-Tsunamis) on Icebergs and Ice Shelves," Journal of Glaciology, Vol. 55, No. 190, 2009, pp. 193-206.
36. G. Ekström, M. Nettles and V. C. Tsai, "Seasonality and Increasing Frequency of Greenland Glacial Earthquakes," Science, Vol. 311, No. 5768, 2006, pp. 1756-1758.
37. M. Nettles and G. Ekström, "Glacial Earthquakes in Greenland and Antarctica," Annual Review of Earth and Planetary Sciences, Vol. 38, 2010, pp. 467-491.
38. T. Wilson and R. Bell, "Earth Structure and Geodynamics at the Poles," Understanding Earth's Polar Challenges: International Polar Year 2007-2008, 2011, pp. 273-292.
39. M. Nettles, T. C. Wallace and S. L. Beck, "The March 25, 1998 Antarctic Plate Earthquake," Geophysical Research Letters, Vol. 26, No. 14, 1999, pp. 2097-2100.

Index